STRUCTURAL
STEEL DESIGN

Transfertruss, 150 Federal Street, Boston, Mass. (Courtesy Owen Steel Company, Inc.)

STRUCTURAL STEEL DESIGN: LRFD METHOD

JACK C. McCORMAC
Clemson University

1817

HARPER & ROW, PUBLISHERS, New York
Grand Rapids, Philadelphia, St. Louis, San Francisco,
London, Singapore, Sydney, Tokyo

Sponsoring Editor: James Cook
Project Editor: Ellen MacElree
Cover Design: Wanda Lubelska Design
Text Art: Burmar
Production Manager: Jeanie Berke
Production Assistant: Beth Maglione
Compositor: Science Press
Printer and Binder: R. R. Donnelley & Sons Company

STRUCTURAL STEEL DESIGN: LRFD METHOD

McCormac, Jack C.
    Structural steel design: LRFD method / Jack C. McCormac.
        p.   cm.
    Includes index.
    ISBN 0-06-044346-4
    1. Building, Iron and steel.   2. Steel, Structural.   I. Title.
TA684.M25   1989
624.1'821—dc19   89-1685
                                                                   CIP

89   90   91   92   9   8   7   6   5   4   3   2   1

Contents

Preface   xiii

CHAPTER 1   INTRODUCTION TO STRUCTURAL STEEL
            DESIGN   1

1-1   Advantages of Steel as a Structural Material   1
1-2   Disadvantages of Steel as a Structural Material   3
1-3   Early Uses of Iron and Steel   4
1-4   Steel Sections   5
1-5   Stress-Strain Relationships in Structural Steel   9
1-6   Modern Structural Steels   13
1-7   Uses of High-Strength Steels   17
1-8   Furnishing of Structural Steel   19
1-9   The Structural Designer   19
1-10 Objectives of the Structural Designer   21
1-11 Design of Steel Members   21
1-12 Calculation Accuracy   24

CHAPTER 2   SPECIFICATIONS, LOADS, AND METHODS OF
            DESIGN   25

2-1   Specifications and Building Codes   25
2-2   Loads   26
2-3   Dead Loads   27
2-4   Live Loads   27
2-5   Selection of Design Loads   35
2-6   Elastic and Plastic Design Methods Defined   35
2-7   Load and Resistance Factor Design   36
2-8   Load Factors   37
2-9   Resistance Factors   39
2-10 Discussion of Sizes of Load and Resistance Factors   40
2-11 Reliability and the LRFD Specification   41
2-12 Advantages of LRFD   43

CHAPTER 3   ANALYSIS OF TENSION MEMBERS   45

3-1   Introduction   45
3-2   Design Strength of Tension Members   48
3-3   Net Areas   49
3-4   Effect of Staggered Holes   52
3-5   Effective Net Areas   56
3-6   Connecting Elements for Tension Members   59
3-7   Block Shear   60
      Problems   65

CHAPTER 4   DESIGN OF TENSION MEMBERS   70

4-1   Selection of Sections   70
4-2   Rods and Bars   79
4-3   Eyebar Tension Members   83
4-4   Design for Fatigue Loads   83
         Problems   86

CHAPTER 5   INTRODUCTION TO AXIALLY LOADED
               COMPRESSION MEMBERS   89

5-1   General   89
5-2   Residual Stresses   91
5-3   Sections Used for Columns   92
5-4   Development of Column Formulas   95
5-5   Derivation of the Euler Formula   97
5-6   End Restraint and Effective Lengths of Columns   99
5-7   Stiffened and Unstiffened Elements   102
5-8   Long, Short, and Intermediate Columns   104
5-9   Column Formulas   104
5-10   Maximum Slenderness Ratios   106
5-11   Example Problems   106
         Problems   110

CHAPTER 6   DESIGN OF AXIALLY LOADED COMPRESSION
               MEMBERS   113

6-1   Introduction   113
6-2   LRFD Design Tables   115
6-3   Built-up Members with Components in Contact with
         Each Other   120
6-4   Connection Requirements for Built-up Columns Whose
         Components Are in Contact with Each Other   121
6-5   Flexural-Torsional Buckling of Compression
         Members   125
6-6   Lacing and Tie Plates   130
         Problems   134

CHAPTER 7   DESIGN OF AXIALLY LOADED COMPRESSION
               MEMBERS CONTINUED   137

7-1   Further Discussion of Effective Lengths   137
7-2   Stiffness Reduction Factors   142
7-3   Columns Leaning on Each Other for In-Plane
         Design   145
7-4   Base Plates for Concentrically Loaded Columns   148
         Problems   154

CHAPTER 8　　　INTRODUCTION TO BEAMS　157

8-1　　Types of Beams　157
8-2　　Sections Used as Beams　157
8-3　　Bending Stresses　158
8-4　　Plastic Hinges　159
8-5　　Elastic Design　160
8-6　　The Plastic Modulus　160
8-7　　Theory of Plastic Analysis　163
8-8　　The Collapse Mechanism　164
8-9　　The Virtual-Work Method　165
8-10　　Location of Plastic Hinge for Uniform Loadings　168
8-11　　Continuous Beams　170
　　　　Problems　172

CHAPTER 9　　　DESIGN OF BEAMS FOR MOMENTS　178

9-1　　Introduction　178
9-2　　Plastic Buckling, Zone 1　181
9-3　　Design of Beams, Zone 1　182
9-4　　Lateral Support of Beams　188
9-5　　Introduction to Inelastic Buckling, Zone 2　189
9-6　　Moment Capacities, Zone 2　192
9-7　　Elastic Buckling, Zone 3　194
9-8　　Noncompact Sections　197
　　　　Problems　198

CHAPTER 10　　　DESIGN OF BEAMS CONTINUED　203

10-1　　Design of Continuous Beams　203
10-2　　Shear　205
10-3　　Deflections　209
10-4　　Webs and Flanges with Concentrated Loads　213
10-5　　Unsymmetrical Bending　218
10-6　　Design of Purlins　220
10-7　　The Shear Center　223
10-8　　Beam-Bearing Plates　228
　　　　Problems　231

CHAPTER 11　　　BENDING AND AXIAL FORCE　236

11-1　　Occurrence　236
11-2　　Members Subject to Bending and Axial Tension　237
11-3　　First-Order and Second-Order Moments　239
11-4　　Magnification Factors　240
11-5　　Modification or C_m Factors　241
11-6　　Interaction Equations for Axial Compression Loads and Bending　243
11-7　　Design of Beam-Columns　248
　　　　Problems　253

CHAPTER 12   BOLTED CONNECTIONS   255

12-1   **Introduction   255**
12-2   **Types of Bolts   255**
12-3   **History of High-Strength Bolts   256**
12-4   **Advantages of High-Strength Bolts   257**
12-5   **Snug-Tight and Fully Tensioned Bolts   258**
12-6   **Methods for Fully Tensioning High-Strength Bolts   259**
12-7   **Slip-Resistant Connections and Bearing-Type Connections   261**
12-8   **Mixed Joints   262**
12-9   **Sizes of Holes for Bolts and Rivets   262**
12-10   **Load Transfer and Types of Joints   263**
12-11   **Failure of Bolted Joints   266**
12-12   **Spacing and Edge Distances of Bolts   267**
12-13   **Bearing-Type Connections—Loads Passing Through Center of Gravity of Connections   270**
12-14   **Slip-Critical Connections—Loads Passing Through Center of Gravity of Connections   276**
            Problems   280

CHAPTER 13   BOLTED CONNECTIONS CONTINUED AND HISTORICAL NOTES ON RIVETS   284

13-1   **Bolts Subjected to Eccentric Shear   284**
13-2   **Bolts Subjected to Shear and Tension   296**
13-3   **Tension Loads on Bolted Joints   299**
13-4   **Prying Action   301**
13-5   **Historical Notes on Rivets   305**
13-6   **Types of Rivets   307**
13-7   **Strength of Riveted Connections—Rivets in Shear   308**
            Problems   310

CHAPTER 14   WELDED CONNECTIONS   318

14-1   **General   318**
14-2   **Advantages of Welding   319**
14-3   **Types of Welding   320**
14-4   **Welding Inspection   322**
14-5   **Classification of Welds   325**
14-6   **Welding Symbols   327**
14-7   **Groove Welds   328**
14-8   **Fillet Welds   331**
14-9   **Strength of Welds   332**
14-10   **LRFD Requirements   333**
14-11   **Design of Simple Fillet Welds   337**
14-12   **Design of Fillet Welds for Truss Members   341**

14-13   **Shear and Torsion   345**
14-14   **Shear and Bending   351**
         Problems   352

CHAPTER 15   BUILDING CONNECTIONS   360

15-1   **Selection of Type of Fastener   360**
15-2   **Types of Beam Connections   361**
15-3   **Standard Bolted Beam Connections   366**
15-4   **LRFD Manual Standard Connection Tables   369**
15-5   **Designs of Standard Bolted Framed Connections   369**
15-6   **Designs of Standard Welded Framed Connections   373**
15-7   **Single-Plate Framing Connections   375**
15-8   **Designs of Welded Seated Beam Connections   376**
15-9   **Stiffened Seated Beam Connections   378**
15-10   **Design of Moment-Resisting Connections   378**
15-11   **Column Web Stiffeners   382**
         Problems   385

CHAPTER 16   DESIGN OF STEEL BUILDINGS   388

16-1   **Introduction   388**
16-2   **Types of Steel Frames Used for Buildings   388**
16-3   **Common Types of Floor Construction   393**
16-4   **Concrete Slabs on Open-Web Steel Joists   393**
16-5   **One-Way and Two-Way Reinforced-Concrete
         Slabs   395**
16-6   **Composite Floors   397**
16-7   **Concrete-Pan Floors   398**
16-8   **Steel-Decking Floors   400**
16-9   **Flat Slabs   400**
16-10   **Precast Concrete Floors   401**
16-11   **Types of Roof Construction   403**
16-12   **Exterior Walls and Interior Partitions   404**
16-13   **Fireproofing of Structural Steel   405**

CHAPTER 17   COMPOSITE BEAMS   407

17-1   **Composite Construction   407**
17-2   **Advantages of Composite Construction   408**
17-3   **Discussion of Shoring   410**
17-4   **Effective Flange Widths   411**
17-5   **Shear Transfer   411**
17-6   **Partially Composite Beams   414**
17-7   **Strength of Shear Connectors   414**
17-8   **Number, Spacing, and Cover Requirements for Shear
         Connectors   416**

17-9   **Moment Capacity of Composite Sections**   **417**
17-10   **Design of Composite Sections**   **422**
17-11   **Continuous Composite Sections**   **430**
17-12   **Design of Concrete Encased Sections**   **431**
         Problems   436

CHAPTER 18   COMPOSITE COLUMNS   440

18-1   **Introduction**   **440**
18-2   **Advantages of Composite Columns**   **440**
18-3   **Disadvantages of Composite Columns**   **442**
18-4   **Lateral Bracing**   **443**
18-5   **Specifications for Composite Columns**   **443**
18-6   **Axial Design Strengths of Composite Columns**   **445**
18-7   **LRFD Tables**   **448**
18-8   **Flexural Design Strengths of Composite Columns**   **449**
18-9   **Axial Load and Bending Equation**   **449**
18-10   **Design of Composite Columns Subject to Axial Load and Bending**   **450**
18-11   **Load Transfer at Footings and Other Connections**   **453**
         Problems   454

CHAPTER 19   BUILT-UP BEAMS, BUILT-UP WIDE-FLANGE SECTIONS, AND PLATE GIRDERS   457

19-1   **Cover-Plated Beams**   **457**
19-2   **Built-up Wide-Flange Sections**   **459**
19-3   **Introduction to Plate Girders**   **463**
19-4   **Plate Girder Proportions**   **465**
19-5   **Proportions of Webs**   **466**
19-6   **Design of Plate Girders with Slender Webs But with Full Lateral Bracing for Their Compact Compression Flanges**   **469**
19-7   **Design of Plate Girders with Noncompact Flanges and Without Full Lateral Bracing for Compression Flanges**   **472**
19-8   **Design of Stiffeners**   **476**
19-9   **Flexure-Shear Interaction**   **479**
         Problems   483

CHAPTER 20   MULTISTORY BUILDINGS   485

20-1   **Introduction**   **485**
20-2   **Column Splices**   **487**
20-3   **Discussion of Lateral Forces**   **489**
20-4   **Types of Lateral Bracing**   **491**
20-5   **Analysis of Buildings with Diagonal Wind Bracing for Lateral Forces**   **495**

20-6   **Moment-Resisting Joints   497**
20-7   **Analysis of Buildings with Moment-Resisting Joints for
          Lateral Loads   498**
20-8   **Analysis of Buildings for Gravity Loads   503**
20-9   **Design of Members   506**

Index   509

Preface

This book presents an elementary discussion of structural steel design using the load and resistance factor method (LRFD). My primary objective is to introduce the fundamental theories necessary for the design of simple steel structures in such a way that the reader will become interested in the subject and wish to learn more.

For quite a few decades structural steel designers have proportioned steel members and their connections using an allowable stress design method. Today, however, the profession is moving rapidly toward a limit states design philosophy called load and resistance factor design. The basic idea of limit states design is that the combined effect of the various types of loads must not be permitted to exceed the resistance of the structure. The estimated ultimate loads must not be greater than the load-carrying ability of the structure nor must the service or working loads cause the structure to deflect or vibrate excessively.

The designs of tension members, columns, beams, beam columns, and connections are presented in the first 15 chapters of the text. A general discussion of common types of steel buildings is presented in Chapter 16 while Chapters 17 and 18 present the design of composite beams and columns. Chapter 19 is concerned with built-up beam sections and plate girders while Chapter 20 provides an introductory discussion of tall steel buildings.

It is hoped that the reader will find the subject of structural steel design as interesting as do those persons presently in the profession.

The author gratefully acknowledges the aid he has received from several sources. He is indebted to L. B. Burgett, Andrew K. Courtney, R. E. Elling, Douglas A. Foutch, Peter B. Keating, R. J. Schmidt, Donald W. White, and A. H. Zureick, who have, by their suggestions and criticisms, directly contributed to the preparation of this manuscript; and to his own professors J. B. Wilbur and B. B. Williams who patiently instructed him in their design classes. Several organizations have kindly submitted photographs, acknowledgements for which are given with the pictures.

Jack C. McCormac

Chapter 1

Introduction to Structural Steel Design

1-1 ADVANTAGES OF STEEL AS A STRUCTURAL MATERIAL

A person traveling in the United States might quite understandably decide that steel was the perfect structural material. He or she would see an endless number of steel bridges, buildings, towers, and other structures—comprising, in fact, a list too lengthy to enumerate. After seeing these numerous steel structures this traveler might be quite surprised to learn that steel was not economically made in the United States until late in the nineteenth century and the first wide-flange beams were not rolled until 1908.

The assumption of the perfection of this metal, perhaps the most versatile of structural materials, would appear to be even more reasonable when its great strength, light weight, ease of fabrication, and many other desirable properties are considered. These and other advantages of structural steel are discussed in detail in the following paragraphs.

High Strength

The high strength of steel per unit of weight means that structure weights will be small. This fact is of great importance for long-span bridges, tall buildings, and structures having poor foundation conditions.

Uniformity

The properties of steel do not change appreciably with time as do those of a rein-forced-concrete structure.

Elasticity

Steel behaves closer to design assumptions than most materials because it follows Hooke's law up to fairly high stresses. The moments of inertia of a steel structure can be definitely calculated while the values obtained for a reinforced-concrete structure are rather indefinite.

Permanence

Steel frames that are properly maintained will last indefinitely. Research on some of the newer steels indicates that under certain conditions no painting maintenance whatsoever will be required.

Ductility

The property of a material by which it can withstand extensive deformation without failure under high tensile stresses is said to be its *ductility*. When a *mild* or *low-carbon* steel member is being tested in tension, a considerable reduction in cross section and a large amount of elongation will occur at the point of failure before the actual fracture occurs. A material that does not have this property is probably hard and brittle and might break if subjected to a sudden shock.

In structural members under normal loads, high stress concentrations develop at various points. The ductile nature of the usual structural steels enables them to yield locally at those points, thus preventing premature failures. A further advantage of ductile structures is that when overloaded their large deflections give visible evidence of impending failure (sometimes jokingly referred to as "running time").

Fracture Toughness

Structural steels are tough; that is, they have both strength and ductility. A steel member loaded until it has large deformations will still be able to withstand large forces. This is a very important characteristic because it means that steel members can be subjected to large deformations during fabrication and erection without fracture—thus allowing them to be bent, hammered, sheared, and have holes punched in them without visible damage. The ability of a material to absorb energy in large amounts is called *toughness*.

Additions to Existing Structures

Steel structures are quite well suited to having additions made to them. New bays or even entire new wings can be added to existing steel frame buildings, and steel bridges may often be widened.

Miscellaneous

Several other important advantages of structural steel are: (a) ability to be fastened together by several simple connection devices including welds, bolts, and rivets, (b) adaptation to prefabrication, (c) speed of erection, (d) ability to be rolled into a wide variety of sizes and shapes as described in Section 1.4, (e) fatigue strength, (f) possible reuse after a structure is disassembled, and (g) scrap value even though not reusable in its existing form.

1-2 DISADVANTAGES OF STEEL AS A STRUCTURAL MATERIAL

In general steel has the following disadvantages.

Maintenance Costs

Most steels are susceptible to corrosion when freely exposed to air and water and must therefore be periodically painted. The use of weathering steels, in suitable design applications, tends to eliminate this cost.

Fireproofing Costs

Although structural members are incombustible, their strength is tremendously reduced at temperatures commonly reached in fires when the other materials in a building burn. Many disastrous fires have occurred in empty buildings where the only fuel for the fires were the buildings themselves. Furthermore steel is an excellent heat conductor such that nonfireproofed steel members may transmit enough heat from a burning section or compartment of a building to ignite materials with which they are in contact in adjoining sections of the building. As a result of these facts the steel frame of a building may have to be protected by materials with certain insulating characteristics, or the building may have to include a sprinkler system if it is to meet the building code requirements of the locality in question.

Susceptibility to Buckling

The longer and slenderer the compression members, the greater the danger of buckling. As previously indicated, steel has a high strength per unit of weight but when used for steel columns is sometimes not very economical because considerable material has to be used merely to stiffen the columns against buckling.

Fatigue

Another undesirable property of steel is that its strength may be reduced if it is subjected to a large number of stress reversals or even to a large number of variations of tensile stress. (We have fatigue problems only when tension is involved.) The present practice is to reduce the estimated strengths of such members if it is anticipated that they will have more than a prescribed number of cycles of stress variation.

1-3 EARLY USES OF IRON AND STEEL

Although the first metal used by human beings was probably some type of copper alloy such as bronze (made with copper and tin and perhaps some other additives), the most important metal developments throughout history have occurred in the manufacture and use of iron and its famous alloy named steel. Today, iron and steel comprise almost 95 percent of all the tonnage of metal produced in the world.[1]

Despite diligent efforts for many decades archaeologists have been unable to discover when iron was first used. They did find an iron dagger and an iron bracelet in the Great Pyramid in Egypt which they claim had been there undisturbed for at least 5000 years. The use of iron has had a great influence on the course of civilization since the earliest times and may very well continue to do so in the centuries ahead. Since the beginning of the iron age in about 1000 B.C. the progress of civilization in peace and war has been heavily dependent on what people have been able to make with iron. On many occasions its use has decidedly affected the outcome of military engagements. For instance in 490 B.C. in Greece at the Battle of Marathon the greatly outnumbered Athenians killed 6400 Persians and only lost 192 of their own men. Each of the victors wore 57 pounds of iron armor in the battle. (This was the battle where the runner Pheidippides ran the approximately 25 miles to Athens and died while shouting news of the victory.) This battle supposedly saved Greek civilization for many years.

According to the classic theory concerning the first production of iron in the world there was once a great forest fire on Mount Ida in Ancient Troy (now Turkey) near the Aegean Sea. The land surface supposedly had a rich content of iron and the heat of the fire is said to have produced a rather crude form of iron which could be hammered into various shapes. Many historians believe however that human beings first learned to use iron that fell to the earth in the form of meteorites. Frequently the iron in meteorites is combined with nickel with the result that a harder metal is produced. Perhaps early human beings were able to hammer and chip this material into the shape of crude tools and weapons.

Steel is defined as a combination of iron with a small amount of carbon, usually less than 1 percent. It also contains small percentages of some other elements. Although some steel has been made for at least 2000 or 3000 years there was really no economical production method available until the middle of the nineteenth century.

The first steel was surely obtained when the other elements necessary for producing it were accidentally present when iron was heated. As the years went by steel was probably made by heating iron in contact with charcoal. The surface of the iron absorbed some carbon from the charcoal which was then hammered into the hot iron. Repeating this process several times resulted in a case-hardened exterior of steel. In this way the famous swords of Toledo and Damascus were produced.

[1]American Iron and Steel Institute, *The Making of Steel* (Washington, D.C., not dated), p. 6.

The first large volume process for producing steel was named after Sir Henry Bessemer of England. He received an English patent for his process in 1855 but his efforts to obtain a U.S. patent for the process in 1856 were unsuccessful as it was proved that William Kelly of Eddyville, Kentucky, had made steel by the same process seven years before Bessemer applied for his English patent. Although Kelly was given the patent, the name Bessemer was used for the process.[2]

Kelly and Bessemer learned that a blast of air through molten iron burned out most of the impurities in the metal. Unfortunately at the same time the blow eliminated some desirable elements such as carbon and manganese. It was later learned that these needed elements could be restored by adding spiegeleisen, which is an alloy of iron, carbon, and manganese. It was further learned that the addition of limestone in the converter resulted in the removal of the phosphorus and most of the sulfur.

The Bessemer converter was commonly used in the United States until after the turn of the century, but since that time it has been replaced with better methods such as the open-hearth process and the basic oxygen process.

As a result of the Bessemer process structural carbon steel could be produced in quantity by 1870 and by 1890 steel had become the principal structural metal used in the United States.

The first use of metal for a sizable structure occurred in England in Shropshire (about 140 miles northwest of London) in 1779 when cast iron was used for the construction of the 100-ft Coalbrookdale Arch Bridge over the River Severn. It is said that this bridge (which still stands) was a turning point in engineering history because it changed the course of the Industrial Revolution by introducing iron as a structural material. This iron was supposedly four times as strong as stone and thirty times as strong as wood.[3]

A number of other cast-iron bridges were constructed in the following decades; but soon after 1840 the more malleable wrought iron began to replace cast iron. The development of the Bessemer process and subsequent advances such as the open-hearth process permitted the manufacture of steel at competitive prices which encouraged the beginning of the almost unbelievable developments of the last 100 years with structural steel.

1-4 STEEL SECTIONS

The first structural shapes made in the United States were angle irons rolled in 1819. Steel I beams were first rolled in the United States in 1884 and the first skeleton frame structure (the Home Insurance Company Building of Chicago) was erected that same year. Credit for inventing the "skyscraper" is usually given to an engineer William LeBaron Jenny, who planned this building appar-

[2]American Iron and Steel Institute, *Steel '76* (Washington, D.C., 1976), pp. 5–11.

[3]M. H. Sawyer, "World's First Iron Bridge," *Civil Engineering* (New York: ASCE, December 1979), pp. 46–49.

Erection of tower for the Holy Name Church, Edensburg, Pa. (Courtesy of the Lincoln Electric Company.)

ently during a bricklayers' strike. Up until this time tall buildings in the United States were constructed with several-foot-thick load-bearing brick walls.

For the exterior walls of this 10-story building Jenny used cast-iron columns encased in brick. The beams for the lower six floors were made from wrought iron while structural steel beams were used for the upper floors. The first building completely framed with steel was the second Rand-McNally building completed in Chicago in 1890.

An important feature of the 985-ft wrought-iron Eiffel tower constructed in 1889 was the use of mechanically operated passenger elevators. The availability of these machines along with Jenny's skeleton frame idea led to the construction of thousands of high-rise buildings throughout the world in the next 100 years.

During these early years the various mills rolled their own individual shapes and published catalogs which provided the dimensions, weight, and other properties of these shapes. In 1896 the Association of American Steel Manufacturers (now the American Iron and Steel Institute, AISI) made the first efforts to standardize shapes. Today nearly all structural shapes are standardized, though their exact dimensions may vary just a little from mill to mill.[4]

Structural steel can be economically rolled into a wide variety of shapes and sizes without appreciably changing its physical properties. Usually the most desirable members are those which have large moments of inertia in proportion to their areas. The **I, T,** and **[** shapes, so commonly used, fall into this class.

Steel sections are usually designated by the shapes of their cross sections. As examples, there are angles, tees, zees, and plates. It is necessary, however, to make a definite distinction between American standard beams (called S beams) and wide-flange beams (called W beams) as they are both I-shaped. The inner surface of the flange of a W section is either parallel to the outer surface or nearly so with a maximum slope of 1 to 20 on the inner surface, depending on the manufacturer.

The S beams, which were the first beam sections rolled in America, have a slope on their inside flange surfaces of 1 to 6. It might be noted that the constant, or nearly constant, thickness of W beam flanges as compared to the tapered S beam flanges may facilitate connections. Wide-flange beams comprise nearly 50 percent of the tonnage of structural steel shapes rolled today. The W and S sections are shown in Fig. 1-1 together with several other familiar steel sections. The uses of these various shapes will be discussed in detail in the chapters to follow.

Constant reference is made throughout this book to the *Manual of Steel Construction Load & Resistance Factor Design* published by the American

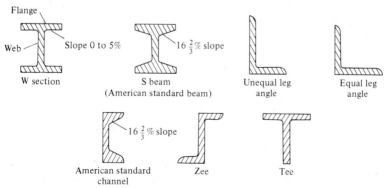

Figure 1-1　Rolled-steel shapes.

[4]W. McGuire, *Steel Structures* (Englewood Cliffs, N.J.: Prentice-Hall, 1968), pp. 19–21.

Institute of Steel Construction (AISC). This manual, which provides detailed information for structural steel shapes, is referred to hereafter as the LRFD Manual or just the Manual or the Handbook. Reference is made here to the 1st edition of the handbook which is based on the September 1, 1986, "Load and Resistance Factor Design Specification for Structural Steel Buildings." Structural shapes are abbreviated by a certain system described in the handbook for use in drawings, specifications, and designs. This system is standardized so that all steel mills can use the same identification for purposes of ordering, billing, etc. In addition so much work is handled today with computers and other automated equipment that it is necessary to have a letter and number system which can be printed out with a standard keyboard (as opposed to the old system where certain symbols were used for angles, channels, etc.). Examples of this abbreviation system are as follows:

1. A W27 × 114 is a W section approximately 27 in. deep weighing 114 lb/ft.
2. An S12 × 35 is an S section 12 in. deep weighing 35 lb/ft.
3. An HP12 × 74 is a bearing pile section which is approximately 12 in. deep weighing 74 lb/ft. Bearing piles are made with the regular W rolls but with thicker webs to provide better resistance to the impact of pile driving.
4. An M8 × 6.5 is a miscellaneous section 8 in. deep weighing 6.5 lb/ft. It is one of a group of doubly-symmetrical H-shaped members which cannot by dimensions be classified as a W, S, or HP section.
5. A C10 × 30 is a channel 10 in. deep weighing 30 lb/ft.
6. An MC18 × 58 is a miscellaneous channel which cannot be classified as a C shape by dimensions.
7. An L6 × 6 × ½ is an equal leg angle, each leg being 6 in. long and ½ in. thick.
8. A WT18 × 140 is a tee obtained by splitting a W36 × 280. This type of section is known as a structural tee.

The student should refer to the LRFD Manual for information concerning other rolled shapes, such as the distinction between bars and plates, pipe and structural tubing, and other items. Additional sections will be mentioned here as it becomes necessary.

In Part 1 of the LRFD Manual the dimensions and properties of W, S, C, and other shapes are given. The dimensions of the members are given in decimals (for the use of designers) and in fractions to the nearest sixteenth of an inch (for the use of craftsmen and steel detailers or draftsmen). Also provided for the use of designers are such items as moments of inertia, section moduli, radii of gyration, and other cross-sectional properties discussed later in this text.

There are variations present in any manufacturing process and the steel industry is certainly no exception. As a result the cross-sectional dimensions of steel members may vary somewhat from the values specified in the LRFD Manual. Maximum tolerances for the rolling of steel shapes are specifically set up by

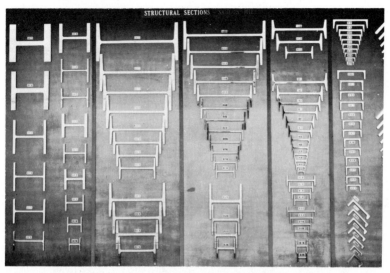

Structural shapes rolled at Bethlehem Steel Company's Bethlehem, Pa., plant. (Courtesy of Bethlehem Steel Corporation.)

the American Society for Testing and Materials (ASTM) A6 Specification and are quoted in Part 1 of the Manual. As a result calculations can be made on the basis of the properties given in the Manual regardless of the manufacturer.

Through the years there have been changes in the sizes of steel sections. For instance, there may be insufficient demand to continue rolling a certain shape; an existing shape may be dropped because a similar sized but more efficient shape has been developed; and so forth. Occasionally the designer may need to know the properties of one of the discontinued shapes which are no longer listed in the latest edition of the Handbook or in other tables normally available to him or her. As an example, it may be desired to add another floor to an existing building which was constructed with shapes no longer rolled.In 1953 the AISC published a book entitled *Iron and Steel Beams* 1873 *to* 1952 which gives a complete listing of iron and steel beams and their properties rolled in the United States during that period. Since that book was published there have been many additional shape changes. As a result the wise structural designer will carefully preserve old editions of the manual so as to have them available when needed.

1-5 STRESS-STRAIN RELATIONSHIPS IN STRUCTURAL STEEL

To understand the behavior of steel structures it is absolutely essential for the designer to be familiar with the properties of steel. Stress-strain diagrams present a valuable part of the information necessary to understand how steel will behave in a given situation. Satisfactory steel design methods cannot be developed unless complete information is available concerning the stress-strain relationships of the material being used.

Steel erection for the Chase Manhattan Bank Building, New York City. (Courtesy of Bethlehem Steel Corporation.)

If a piece of mild structural steel is subjected to a tensile force it will begin to elongate. If the tensile force is increased at a constant rate the amount of elongation will increase constantly within certain limits. In other words, elongation will double when the stress goes from 6000 to 12,000 psi (pounds per square inch). When the tensile stress reaches a value roughly equal to one-half of the ultimate strength of the steel the elongation will begin to increase at a greater rate without a corresponding increase in the stress.

The largest stress for which Hooke's law applies or the highest point on the straight-line portion of the stress-strain diagram is the *proportional limit*. The largest stress which a material can withstand without being permanently deformed is called the *elastic limit*. This value is seldom actually measured and for most engineering materials including structural steel is synonymous with the proportional limit. For this reason the term *proportional elastic limit* is sometimes used.

The stress at which there is a decided increase in the elongation or strain without a corresponding increase in stress is said to be the *yield point*. It is the first point on the stress-strain diagram where a tangent to the curve is horizontal. The yield point is probably the most important property of steel to the designer as so many design procedures are based on this value. Beyond the yield point there is a range in which a considerable increase in strain occurs without increase in stress. The strain that occurs before the yield point is referred to as the *elastic*

Newport Bridge between Jamestown and Newport, R.I. (Courtesy of Bethlehem Steel Corporation.)

strain; the strain that occurs after the yield point, with no increase in stress, is referred to as the *plastic strain.* These latter strains are usually from 10 to 15 times the elastic strains.

Yielding of steel without stress increase may be thought to be a severe disadvantage when in actuality it is a very useful characteristic. It has often performed the wonderful service of preventing failure due to omissions or mistakes on the designer's part. Should the stress at one point in a ductile steel structure reach the yield point, that part of the structure will yield locally without stress increase, thus preventing premature failure. This ductility allows the stresses in a steel structure to be readjusted. Another way of describing this phenomenon is to say that very high stresses caused by fabrication, erection, or loading will tend to equalize themselves. It might also be said that a steel structure has a reserve of plastic strain that enables it to resist overloads and sudden shocks. If it did not have this ability, it might suddenly fracture, like glass or other vitreous substances.

Following the plastic strain there is a range where additional stress is necessary to produce additional strain and this is called *strain-hardening.* This portion of the diagram is not too important to today's designer. A familiar stress-strain diagram for mild or low-carbon structural steel is shown in Fig. 1-2. Only the initial part of the curve is shown here because of the great deformation which occurs before failure. At failure in the mild steels the total strains are from 150 to 200 times the elastic strains. The curve will actually continue up to its maximum stress value and then "tail off" before failure. A sharp reduction in the cross section of the member takes place (called "necking") followed by failure.

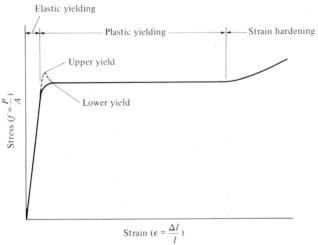

Figure 1-2 Typical stress-strain diagram for a mild or low-carbon structural steel.

The stress-strain curve of Fig. 1-2 is typical of the usual ductile structural steel and is assumed to be the same for members in tension or compression. (The compression members must be stocky because slender compression members subjected to compression loads tend to bend laterally and their properties are greatly affected by the bending moments so produced.) The shape of the diagram varies with the speed of loading, the type of steel, and the temperature. One such variation is shown in the figure by the dotted line which is marked *upper yield*. This shape stress-strain curve is the result when a mild steel has the load applied rapidly while the *lower yield* is the case for slow loading.

A very important property of a structure which has not been stressed beyond its yield point is that it will return to its original length when the loads are removed. Should it be stressed beyond this point it will return only part of the way to its original position. This knowledge leads to the possibility of testing an existing structure by loading and unloading. If after the loads are removed the structure will not resume its original dimensions, it has been stressed beyond its yield point.

Steel is an alloy consisting almost entirely of iron (usually over 98 percent). It also contains small quantities of carbon, silicon, manganese, sulfur, phosphorus, and other elements. Carbon is the material that has the greatest effect on the properties of steel. The hardness and strength increase as the carbon percentage is increased but unfortunately the resulting steel is more brittle and its weldability is adversely affected. A smaller amount of carbon will make the steel softer and more ductile but also weaker. The addition of such elements as chromium, silicon, and nickel produces steels with considerably higher strengths. These steels, however are appreciably more expensive and are often not as easy to fabricate.

A typical stress-strain diagram for a brittle steel is shown in Fig. 1-3. Unfortunately, low ductility or brittleness is a property usually associated with high strengths in steels (although not entirely confined to high-strength steels). As it is desirable to have both high strength and ductility, the designer may have

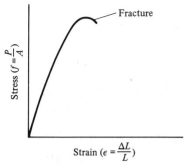

Figure 1-3 Typical stress-strain diagram for a brittle steel.

to decide between the two extremes or compromise between them. A brittle steel may fail suddenly without warning when overstressed and during erection could possibly fail due to the shock of erection procedures.

1-6 MODERN STRUCTURAL STEELS

The properties of steel can be greatly changed by varying the quantities of carbon present and by adding other elements such as silicon, nickel, manganese, and copper. A steel that has a significant amount of the latter elements is referred to as an alloy steel. Although these elements do have a great effect on the properties of steel the actual quantities of carbon or other alloying elements are quite small. For instance, the carbon content of steel is almost always less than 0.5 percent by weight and is normally from 0.2 to 0.3 percent.

   The chemistry of steel is extremely important in its effect on the properties of the steel such as weldability, corrosion resistance, resistance to brittle fracture, and so on. The carbon present in steel increases its hardness and strength but at the same time reduces its ductility as do the impurities phosphorus and sulfur. The ASTM specifies the exact maximum percentages of carbon, manganese, silicon, etc., which are permissible for a number of structural steels. Although the physical and mechanical properties of steel sections are primarily determined by their chemical composition they are also influenced to a certain degree by the rolling process and by their stress history and heat treatment.

   Perhaps 50 percent of the steel used in engineering structures in the United States is a structural carbon steel designated as A36 by the ASTM, but many other steels are available and demand for them is rising rapidly. A higher-strength steel designated as A572 by the ASTM and described later in this section is today being used almost as much as A36.

   In recent decades the engineering and architecture professions have been continually requesting stronger steels, steels with more corrosion resistance, steels with better welding properties, and various other requirements. Research by the steel industry during this period has supplied several groups of new steels which satisfy many of the demands and today there are quite a few structural steels designated by the ASTM and included in the LRFD Specification.

One half of a 170-ft clear span roof truss for the Athletic and Convention Center, Lehigh University, Bethlehem, Pa. (Courtesy of Bethlehem Steel Corporation.)

Structural steels are generally grouped into several major ASTM classifications: the *all-purpose carbon steel* (A36), the *structural carbon steel* (A529), the *high-strength low-alloy structural steels* (A441 and A572), the *atmospheric corrosion-resistant high-strength low-alloy structural steels* (A242 and A588) and *quenched and tempered alloy steel plate* (A514).

In the paragraphs that follow a few general remarks are made about these steel classifications and then the seven ASTM steels mentioned here are listed in Table 1-1 together with a few general remarks concerning their uses and properties. (As you look at this table notice the phenomenon that the thinner steel is rolled the stronger it becomes. Thicker members tend to be more brittle and their slower cooling rate causes a coarser microstructure of the steel.)

Carbon Steels

These steels have as their principal strengthening agents carefully controlled quantities of carbon and manganese. In detail the carbon steels are those which

TABLE 1-1 PROPERTIES OF STRUCTURAL STEELS

ASTM designation	Type of steel	Forms	Recommended uses	Minimum yield stress F_y, ksi[a]	Specified minimum tensile strength F_u, ksi[b]
A36	Carbon	Shapes, bars, and plates	Bolted, welded, or riveted buildings and bridges and other structural uses	36 but 32 if thickness >8 in.	58–80
A529	Carbon	Shapes, plates to ½ in.	Similar to A36	42	60–85
A441	High-strength low-alloy	Shapes, plates, and bars to 8 in.	Similar to A36	40–50	60–70
A572	High-strength low-alloy	Shapes, plates, and bars to 6 in.	Bolted, welded, and riveted construction. Not for welded bridges for F_y grades 55 and above	42–65	60–80
A242	Atmospheric corrosion-resistant high-strength low-alloy	Shapes, plates, and bars to 4 in.	Bolted, welded, or riveted construction; welding technique very important	42–50	63–70
A588	Atmospheric corrosion-resistant high-strength low-alloy	Plates and bars	Bolted and riveted construction	42–50	63–70
A514	Quenched and tempered alloy	Plates only to 4 in.	Welded structures with great attention to technique, discouraged if ductility important	90–100	100–130

[a] F_y values vary with thickness and group (see Tables 1 and 2, Part 1, LRFD Manual).
[b] F_u values vary by grade and type.

have the following maximum amounts of elements: 1.7 percent carbon, 1.65 percent manganese, 0.60 percent silicon, and 0.60 percent copper. These steels are divided into four categories depending on carbon percentages as follows.

1. Low carbon steel < 0.15 percent.
2. Mild carbon steel 0.15 to 0.29 percent. (The structural carbon steels fall into this category.)
3. Medium carbon steel 0.30 to 0.59 percent.
4. High carbon steel 0.60 to 1.70 percent.

The very common steel, A36, which has a yield stress of 36,000 psi, is suitable for bolted, welded, or riveted bridges and buildings. It is used for the majority of the design problems in this text.

High-Strength Low-Alloy Steels

There are a large number of high-strength low-alloy steels and they are included under several ASTM numbers. In addition to carbon and manganese these steels obtain their higher strengths and other properties by the addition of one or more alloying agents such as columbium, vanadium, chromium, silicon, copper, nickel, and others. Included are steels with yield stresses as low as 40,000 psi and as high as 70,000 psi. These steels generally have much greater atmospheric corrosion resistance than the carbon steels.

The term low-alloy is used to arbitrarily describe steels for which the total of all the alloying elements does not exceed 5 percent of the total composition of the steel.

Atmospheric Corrosion-Resistant High-Strength Low-Alloy Structural Steels

When steels are alloyed with small percentages of copper they become more corrosion-resistant. When exposed to the atmosphere the surfaces of these steels oxidize and form a very tightly adherent film (sometimes referred to as a "tightly bound patina") which prevents further oxidation and thus eliminates the need for further painting. After this process takes place within 18 months to 3 years (depending on the type of exposure—rural, industrial, direct or indirect sunlight, etc.) the steel reaches a deep reddish to brown to black color.

Supposedly the first steel of this type was developed in 1933 by the U.S. Steel Corporation to provide resistance to the severe corrosive conditions of railroad coal cars.

You can see the many uses which can be made of such a steel particularly for structures with exposed members that are difficult to paint—bridges, electrical transmission towers, and others. This steel is not considered to be satisfactory for use in locations where it is frequently subject to salt water sprays or fogs, or continually submerged in water (fresh or salt) or the ground or in locations where there are severe corrosive industrial fumes. It is also not satisfactory in very dry areas of the world as in some western parts of the United States. For the "patina" to form the steel must be subjected to a wetting and drying cycle. Otherwise it will continue to look like unpainted steel.

Quenched and Tempered Alloy Steels

These steels have alloying agents in excess of those used in the carbon steels and they are heat-treated by quenching and tempering to obtain strong and tough steels with yield strengths in the 80,000 to 110,000 psi range. Quenching consists of rapid cooling of the steel with water or oil from temperatures of at least 1650°F down to about 300 or 400°F. In tempering the steel is reheated to at least 1150°F and allowed to cool.

The quenched and tempered steels do not exhibit well-defined yield points as do the carbon steels and the high-strength low-alloy steels. As a result their yield strengths are usually defined as the stress at 0.2 percent offset strain. (In other words a line is drawn parallel to the straight-line portion of the stress-strain diagram from a strain of 0.002 until it intersects the stress-strain curve. The stress at that point is the assumed yield point for 0.2 percent offset strain.) The quenched and tempered steels in Tables 2-1 are listed under the ASTM designation A514 which has yield stresses of 90,000 or 100,000 psi depending on thickness.

In Section A3.1 of Part 6 of the LRFD Manual seven other ASTM grades of steel are listed (A53, A500, A501, A570, A606, A607, and A618). These grades cover pipe, cold- and hot-formed tubing, sheet, and strip.

A set of actual stress-strain curves for the three major types of steels described here (carbon, high-strength low-alloy, and quenched and tempered) are shown in Fig. 1-4. The reader will be able to clearly see that the first two of these steels have well-defined yield points while the quenched and tempered ones do not. As a result their yield strengths are usually defined as indicated in the figure as the stress at 0.2 percent offset strain.

1-7 USES OF HIGH-STRENGTH STEELS

There are indeed other groups of high-strength steels such as the ultra-high-strength steels which have yield strengths from 160,000 to 300,000 psi. These steels have not been included in the LRFD Manual because they have not been assigned ASTM numbers.

It is said that there are today more than 200 steels on the market that provide yield stresses in excess of 36,000 psi. The steel industry is now experimenting with steels with yield stresses varying from 200,000 to 300,000 psi and this may only be the beginning. Many people in the steel industry feel that steels with 500,000 psi yield strengths will be made available within a few years. The theoretical binding force between iron atoms has been estimated to be in excess of 4,000,000 psi.[5]

[5]L. S. Beedle et al., *Structural Steel Design* (New York: Ronald Press, 1964), p. 44.

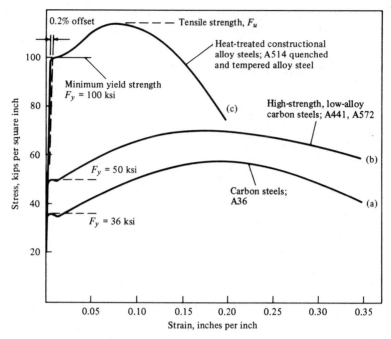

Figure 1-4   Stress-strain curves. (From C.G. Salmon and J.E. Johnson, *Steel Structures,* 2nd ed., Copyright © 1980, Harper & Row, Publishers, Inc.)

Although the prices of steels increase with increasing yield points the per-centage of price increase does not keep up with the percentage of yield-point increase. The result is that the use of the stronger steels will quite frequently be economical for tension members, beams, and columns. Perhaps the greatest economy can be realized with tension members (particularly those without bolt or rivet holes). They may provide a great deal of saving for beams if deflections are not important or if deflections can be controlled by methods described in later chapters. In addition considerable economy can frequently be achieved with high-strength steels for short and medium-length stocky columns. Another application which can provide considerable saving is in hybrid construction. In this type of construction two or more steels of different strengths are used, the weaker steels being used where stresses are smaller and the stronger steels where stresses are higher.

Among the other factors that might lead to the use of high-strength steels are the following.

1. Superior corrosion resistance.
2. Possible savings in shipping, erection, and foundation costs caused by weight saving.
3. Use of shallower beams permitting smaller floor depths.
4. Possible savings in fireproofing because smaller members can be used.

The first thought of most engineers in choosing a type of steel is the direct cost of the members. Such a comparison can be made quite easily, but the economy of which strength grade to use cannot be obtained unless consideration is given to weights, sizes, deflections, maintenance, and fabrication. To make an accurate general comparison of the steels is probably impossible—rather it is necessary to have a specific job to consider.

1-8 FURNISHING OF STRUCTURAL STEEL

The furnishing of structural steel consists of the rolling of the steel shapes, the fabrication of the shapes for the particular job (including cutting to the proper dimensions and punching of holes necessary for field connections), and their erection. Very rarely will a company perform all three of these functions, and the average company performs only one or two of them. For instance, many companies fabricate structural steel and erect it while others may only be steel fabricators or steel erectors. There are approximately 400 to 500 companies in the United States which make up the fabricating industry for structural steel. Most of them do fabrication and erection.

Steel fabricators normally carry very little steel in stock because of the high interest and storage charges. When they get a job they may order the shapes to certain lengths directly from the rolling mill or they may obtain them from service centers. Service centers, which are an increasingly important factor in the supply of structural steel, buy and stock large quantities of structural steel which they buy at the best prices they can find anywhere in the world.

The design of structural steel is usually made by an engineer in collaboration with an architectural firm. The designer makes design drawings that show member sizes, controlling dimensions, and any unusual connections. The company that is to fabricate the steel makes the detailed drawings subject to the engineer's approval. These drawings provide all the information necessary to fabricate the members correctly. They show the dimensions for each member, the locations and sizes of holes, the positions and sizes of connections, and the like.

The erection of steel buildings is more a matter of assembly than nearly any other line of construction work. Each of the members is marked in the shop with letters and numbers to distinguish them from the other members to be used. The erection is performed in accordance with a set of erection plans. These plans are not detailed drawings but are simple line diagrams showing the position of the various members in the building. A piece mark is placed on the left end of each member as it appears on the detail drawing. Directions (north, south, east, or west) are usually painted on column faces. These marks enable steel workers to quickly find the position of each member in the field. (Those persons performing steel erection are often called *ironworkers,*which is a name held over from the days before structural steel.)

1-9 THE STRUCTURAL DESIGNER

Structural designers can take great pride in their part in the development of our country. Cities, farmlands, and industrial areas of the United States are filled

Steel erection for Transamerica Pyramid, San Francisco, Calif. (Courtesy of Kaiser Steel Corporation.)

with the amazing structures designed by members of their profession. But even this remarkable array of structures will be merely child's play compared to the structural endeavors of the next few generations. These structures of the future should provide many opportunities in the structural field for young engineers.

The structural designer arranges and proportions structures and their parts so that they will satisfactorily support the loads to which they may feasibly be subjected. It might be said that he or she is involved with the following: the general layout of structures; studies of the possible structural forms that can be used; consideration of loading conditions; analysis of stresses, deflections, etc.; design of parts; and the preparation of design drawings. More precisely, the word design pertains to the proportioning of the various parts of a structure after the forces have been calculated, and it is this process which will be emphasized throughout the text using structural steel as the material.

1-10 OBJECTIVES OF THE STRUCTURAL DESIGNER

The structural designer must learn to arrange and proportion the parts of structures so that they can be practically erected and will have sufficient strength and reasonable economy. These items are discussed briefly below.

Safety

Not only must the frame of a structure safely support the loads to which it is subjected, but it must support them in such a manner that deflections and vibrations are not so great as to frighten the occupants or cause unsightly cracks.

Cost

The designer needs to keep in mind the items causing lower cost without sacrifice of strength. These items which are discussed in more detail throughout the text include the use of standard-size members, simple connections and details, and the use of members and materials that will not require an unreasonable amount of maintenance through the years.

Practicality

Another objective is the design of structures that can be fabricated and erected without great problems arising. Designers need to understand fabrication methods and should try to fit their work to the fabrication facilities available.

Designers should learn everything possible about the detailing, the fabrication, and the field erection of steel. The more the designer knows about the problems, the tolerances, and the clearances in shop and field the more probable it is that reasonable, practical, and economical designs will be produced. This knowledge should include information concerning the transportation of the materials to the job site (such as the largest pieces that can practically be transported by rail or truck), labor conditions, and the equipment available for erection. Perhaps the designer should ask the question, "Could I get this thing together if I were sent out to do it?"

Finally he or she needs to proportion the parts of the structure so that they will not unduly interfere with the mechanical features of the structure (pipes, ducts, etc.) or the architectural effects.

1-11 DESIGN OF STEEL MEMBERS

The design of a steel member involves much more than a calculation of the properties required to support the loads and the selection of the lightest section providing these properties. Although at first glance this procedure would seem to give the most economical designs, many other factors need to be considered. Among these are:

1. The designer needs to select steel sections of sizes which are usually rolled. Steel beams and bars and plates of unusual sizes will be difficult to obtain during boom periods and will be expensive during any period. A little study on the designer's part will enable him or her to avoid these expensive shapes. Steel fabricators are constantly supplied with information from the steel companies and the steel warehousemen as to the sizes of sections available. This information includes the section sizes as well as the standard lengths available. (Most structural shapes can be obtained in lengths from 60 to 75 ft depending on the producer, while it is possible under certain conditions to obtain some shapes up to 120 ft in length.)

2. A blind assumption that the lightest section is the cheapest one may be in considerable error. A building frame designed by the "lightest-section" procedure will consist of a large number of different shapes and sizes of members. Trying to connect these many-sized members and fit them in the building will be quite complicated and the pound price of the steel will in all probability be rather high. A more reasonable approach would be to smooth out the sizes by selecting many members of the same sizes although some of them may be slightly overdesigned.

3. The beams usually selected for the floors in buildings will normally be the deeper sections because these sections for the same weights have the largest moments of inertia and the greatest resisting moments. As building heights increase, however, it may be economical to modify this practice. As an illustration of this fact the erection of a 20-story building, for which each floor has a minimum clearance, is considered. It is assumed that the depths of the floor beams may be reduced by 6 in. without an unreasonable increase in beam weights. The beams will cost more but the building height will be reduced by 20×6 in. $= 120$ in. or 10 ft, with resulting savings in walls, elevator shafts, column heights, plumbing, wiring, and footings.

4. For larger sections, particularly the built-up ones, the designer needs to have information pertaining to transportation problems. The desired information includes the greatest lengths and depths that can be shipped by truck or rail, clearances available under bridges and power lines leading to the project, and allowable loads on bridges. It may be possible to fabricate a steel roof truss in one piece, but is it possible to transport it to the job site and erect it in one piece?

5. Sections should be selected which are reasonably easy to erect and which have no conditions present that will make them difficult to maintain. As an example, it is necessary to have access to all exposed surfaces of steel bridge members so that they may be periodically painted (unless one of the special corrosion-resistant steels is being used).

6. Buildings are often filled with an amazing conglomeration of pipes, ducts, conduits, and other items. Every effort should be made to select steel members that will fit in with the requirements made by these items.

7. The members of a steel structure are often exposed to the public, partic-

Steel erection for Transamerica Pyramid, San Francisco, Calif. (Courtesy of Kaiser Steel Corporation.)

ularly in the case of steel bridges and auditoriums. Appearance may often be the major factor in selecting the type of structure to be used as where a bridge is desired which will fit in and actually contribute to the appearance of an area. Exposed members may be surprisingly graceful when a simple arrangement and perhaps curved members are used, but other arrangements may create a terrible eyesore. The student has certainly seen illustrations of each case. It is very interesting to know that beautiful structures in steel are usually quite reasonable in cost.

1-12 CALCULATION ACCURACY

A most important point which many students with their superb pocket calculators and microcomputers have difficulty in understanding is that structural design is not an exact science for which answers can confidently be calculated to eight places. Among the reasons for this fact are: the methods of analysis are based on partly true assumptions, the strengths of materials used vary appreciably, and maximum loadings can only be approximated. With respect to this last sentence, how many of the users of this book could estimate within 10 percent the maximum load in pounds per square foot that will ever occur on the building floor which they are now occupying? Calculations to more than two or three significant figures are obviously of little value and may actually be harmful in that they mislead the student by giving him or her a fictitious sense of precision.

Chapter 2

Specifications, Loads, and Methods of Design

2-1 SPECIFICATIONS AND BUILDING CODES

The design of most structures is controlled by specifications. Even if not so controlled, the designer will probably refer to them as a guide. No matter how many structures a person has designed it is impossible for him or her to have encountered every situation, and by referring to specifications he or she is making use of the best available material on the subject. Engineering specifications which are developed by various organizations present the best opinion of those organizations as to what represents good practice.

Municipal and state goverments concerned with the safety of the public have established building codes with which they control the construction of various structures within their jurisdiction. These codes, which are actually laws or ordinances, specify design loads, design stresses, construction types, material quality, and other factors. They vary considerably from city to city, a fact that causes some confusion among architects and engineers.

Several organizations publish recommended practices for regional or national use. Their specifications are not legally enforceable unless they are embodied in the local building code or made a part of a particular contract. Among these organizations are the AISC and AASHTO (American Association of State Highway and Transportation Officials). Nearly all municipal and state building codes have adopted the AISC Specification and nearly all state highway and transportation departments have adopted the AASHTO Specifications.

Some people feel that specifications prevent engineers from thinking for themselves—and there may be some basis for the criticism. They say that the ancient engineers who built the great pyramids, the Parthenon, and the great Roman bridges were controlled by few specifications, which is certainly true. On

Hackensack River Bridge, New Jersey Turnpike. (Courtesy of USX Corporation.)

the other hand, it should be said that only a few score of these great projects were built over many centuries and they were apparently built without regard to cost of material, labor, or human life. They were probably built by intuition and by certain rules of thumb developed by observing the minimum size or strength of members which would just fail under given conditions. Their probably numerous failures are not recorded in history; only their successes endured.

Today, however, there are hundreds of projects being constructed at any one time in the United States which rival in importance and magnitude the famous structures of the past. It appears that if all engineers in our country were allowed to design projects such as these without restrictions there would be many disastrous failures. *The important thing to remember about specifications, therefore, is that they are not written for the purpose of restricting engineers but for the purpose of protecting the public.*

No matter how many specifications are written it is impossible for them to cover every possible situation. As a result, no matter which code or specification is or is not being used, the ultimate responsibility for the design of a safe structure lies with the structural designer.

2-2 LOADS

Perhaps the most important and most difficult task faced by the structural designer is the accurate estimation of the loads which may be applied to a structure during its life. No loads that may reasonably be expected to occur may be overlooked. After loads are estimated the next problem is to decide the worst possible combinations of these loads which might occur at one time. For instance, would a highway bridge completely covered with ice and snow be simultaneously subjected to fast-moving lines of heavily loaded trailer trucks in every lane and to a 90-mile lateral wind, or is some lesser combination of these loads more feasible?

The next two sections of this chapter provide a brief introduction to the types of loads with which the structural designer needs to be familiar. The purpose of these sections is not to discuss loads in great detail but rather to give the reader a "feel" for the subject. As will be seen, loads are classed as being dead loads or live loads.

2-3 DEAD LOADS

Dead loads are loads of constant magnitude that remain in one position. They consist of the structural frame's own weight and other loads that are permanently attached to the frame. For a steel-frame building some dead loads are the frame, walls, floors, roof, plumbing, and fixtures.

To design a structure it is necessary for the weights or dead loads of the various parts to be estimated for use in the analysis. The exact sizes and weights of the parts are not known until the structural analysis is made and the members of the structure selected. The weights, as determined from the actual design, must be compared with the estimated weights. If large discrepancies are present, it will be necessary to repeat the analysis and design using better estimated weights.

Reasonable estimates of structure weights may be obtained by referring to similar types of structures or to various formulas and tables available in several locations. The weights of many materials are given in Part 7 of the LRFD Manual. Even more detailed information on dead loads is provided in Tables A1 and A2 of the ANSI Code.[1] An experienced designer can estimate very closely the weights of most materials and will spend little time repeating designs because of poor estimates.

2-4 LIVE LOADS

Live loads are loads that may change in position and magnitude. Simply stated all loads that are not dead loads are live loads. Live loads that move under their own power are said to be moving loads, such as trucks, people, and cranes, whereas those loads that may be moved are movable loads, such as furniture, warehouse materials, and snow. Other live loads include those caused by construction operations, wind, rain, earthquakes, blasts, soils, and temperature

---

[1]*American National Standard Minimum Design Loads for Buildings and Other Structures,* ANSI A58.1-1982 (New York: American National Standards Institute, Inc.).

changes. A brief description of some of these loads follows:

1. *Floor loads.* The minimum gravity live loads to be used for building floors are usually clearly specified by the applicable building code. Unfortunately, however, the values given in these various codes vary from city to city and the designer must be sure that his or her designs meet the requirements of that locality. A few of the typical values for floor loadings are listed in Table 2-1. These values were adopted from the ANSI Code.[2] In the absence of a governing code this seems to be an excellent one to follow.

2. *Snow and ice.* In the colder states snow and ice loads are often quite important. One inch of snow is equivalent to approximately 0.5 psf (pounds per square foot) but it may be higher at lower elevations where snow is denser. For roof designs, snow loads of from 10 to 40 psf are used, the magnitude depending primarily on the slope of the roof and to a lesser degree on the character of the roof surface. The larger values are used for flat roofs and the smaller ones for sloped roofs. Snow tends to slide off sloped roofs, particularly those with metal or slate surfaces. A load of approximately 10 psf might be used for 45° (degree) slopes and a 40-psf load for flat roofs. Studies of snowfall records in areas with severe winters may indicate the occurrence of snow loads much greater than 40 psf with values as high as 100 psf in northern Maine.

Snow is a variable load which may cover an entire roof or only part of it. There may be drifts against walls or buildup in valleys or between parapets. Snow may slide off one roof onto a lower one. The wind may blow it off one side of a

TABLE 2-1   TYPICAL MINIMUM UNIFORM LIVE LOADS FOR DESIGN OF BUILDINGS

Type of building	LL (psf)
Apartment houses	
Apartments	40
Public rooms	100
Dining rooms and restaurants	100
Garages (passenger cars only)	50
Gymnasiums, main floors, and balconies	100
Office buildings	
Lobbies	100
Offices	50
Schools	
Classrooms	40
Corridors first floor	100
Corridors above 1st floor	80
Storage warehouses	
Light	125
Heavy	250
Stores (retail)	
First floor	100
Other floors	75

[2]*American National Standard Minimum Design Loads for Buildings and Other Structures,* ANSI A58.1-1982 (New York: American National Standards Institute, Inc.).

sloping roof or the snow may crust over and remain in position even during very heavy winds.

Bridges are generally not designed for snow loads, since the weight of the snow is usually not significant in comparison with truck and train loads. In any case it is doubtful that a full load of snow and maximum traffic would be present at the same time. Bridges and towers are sometimes covered with layers of ice from 1 to 2 in. (inches) thick. The weight of the ice runs up to about 10 psf. Another factor to be considered is the increased surface area of the ice-coated members as it pertains to wind loads.

3. *Rain.* Though snow loads are a more severe problem than rain loads for the usual roof the situation may be reversed for flat roofs, particularly those in warmer climates. If water on a flat roof accumulates faster than it runs off, the result is called *ponding* because the increased load causes the roof to deflect into a dish shape that can hold more water, which causes greater deflections, and so on. This process continues until equilibrium is reached or until collapse occurs. Ponding is a serious matter as illustrated by the large number of flat-roof failures that occur during rain storms every year in the United States.

Ponding will occur on almost any flat roof to a certain degree even though roof drains are present. Roof drains may very well be used but they may be inadequate during severe storms or they may become partially or completely clogged. The best method of preventing ponding is to have an appreciable slope of the roof

Cold-storage warehouse, Grand Junction, Colo. (Courtesy of the American Institute of Steel Construction, Inc.)

(¼ in./ft or more) together with good drainage facilities. In addition to the usual ponding another problem may occur for very large flat roofs (with perhaps an acre or more of surface area). During heavy rainstorms strong winds frequently occur. If there is a great deal of water on the roof a strong wind may very well push a large quantity of it toward one end. The result can be a dangerous water depth as regards the load in psf on that end. For such situations *scuppers* are sometimes used. These are large holes or tubes in walls or parapets that enable water above a certain depth to quickly drain off the roof.

    4. *Traffic loads for bridges.* Bridges are subjected to series of concentrated loads of varying magnitude caused by groups of truck or train wheels.

    5. *Impact loads.* Impact loads are caused by the vibration of moving or movable loads. It is obvious that a crate dropped on the floor of a warehouse or a truck bouncing on uneven pavement of a bridge causes greater forces than would occur if the loads were applied gently and gradually. Impact loads are equal to the difference between the magnitude of the loads actually caused and the magnitude of the loads had they been dead loads.

    The LRFD Specification (A4.2) requires that structures that are to support live loads that tend to cause impact are to have their assumed nominal loads for design increased by the following percentages if not otherwise specified:

For supports of elevators and elevator machinery	100%
For supports of light machinery, shaft or motor driven, not less than	20%
For supports of reciprocating machinery or power driven units, not less than	50%
For hangers supporting floors and balconies	33%
For cab-operated traveling crane support girders and their connections	25%
For pendant-operated traveling crane support girders and their connections	10%

    6. *Lateral loads.* Lateral loads are of two main types: wind and earthquake. A survey of engineering literature for the past 150 years reveals many references to structural failures caused by wind. Perhaps the most infamous of these have been bridge failures such as those of the Tay Bridge in Scotland in 1879 (which caused the deaths of 75 persons) and the Tacoma Narrows Bridge (Tacoma, Washington) in 1940. But there have also been some disastrous building failures due to wind during the same period such as that of the Union Carbide Building in Toronto in 1958. It is important to realize that a large percentage of building failures due to wind have occurred during their erection.[3]

    a. *Wind loads.* A great deal of research has been conducted in recent years on the subject of wind loads. Nevertheless a great deal more work needs to be done as the estimation of these forces can by no means be classified as an exact science. The magnitudes of wind loads vary with geographical locations, heights above ground, types of terrain surrounding the buildings including other nearby structures, and other factors.

    Wind pressures are usually assumed to be uniformly applied to the windward surfaces of buildings and are assumed to be capable of coming from any

[3]Wind Forces on Structures, Task Committee on Wind Forces. Committee on Loads and Stresses, Structural Division, ASCE, Final Report, *Transactions ASCE* 126, Part II (1961): 1124–1125.

direction. These assumptions are not very accurate because wind pressures are not very even over large areas, the pressures near the corners of buildings being probably greater than elsewhere due to wind rushing around the corners, and so on. From a practical standpoint, therefore, all of the possible variations cannot be considered in design although today's specifications are becoming more and more detailed in their requirements.

When the designer working with large stationary buildings makes poor wind estimates the results are probably not too serious, but this is not the case when tall slender buildings (or long flexible bridges) are being considered. For many years the average designer ignored wind forces for buildings whose heights were not at least twice their least lateral dimensions. For such cases as these it was felt that the floors and walls provided sufficient lateral stiffnesses to eliminate the need for definite wind bracing systems. A better practice for designers to follow, however, is to consider all the possible loading conditions that a particular structure may have to resist. If one or more of these conditions (as perhaps wind loading) seem of little significance they may then be neglected. Should buildings have their walls and floors constructed with modern lightweight materials and/ or should they be subjected to unusually high wind loads (as in coastal or mountaineous areas) they will probably have to be designed for wind loads even if the height/least lateral dimension ratios are less than 2.

Building codes do not usually provide for the estimated forces during tornadoes. The average designer considers the forces created directly in the paths of tornadoes to be so violent that it is not economically feasible to design buildings to resist them. This feeling is undergoing some change, however, as it has been found that the wind resistance of structures (even for small buildings including houses) can be greatly increased at very reasonable costs by using better prac-

Access bridge, Renton, Wash. (Courtesy of the Bethlehem Steel Corporation.)

tices of tying the parts of structures together from the roofs through the walls to the footings, and by making appropriate connections of window frames to walls and perhaps other parts of the structure.[4,5]

Wind forces act as pressures on vertical windward surfaces, pressures or suction on sloping windward surfaces (depending on the slope) and suction on flat surfaces and on leeward vertical and sloping surfaces (due to the creation of negative pressures or vacuums). The student may have noticed this definite suction effect where shingles or other roof coverings have been lifted from the leeward roof surfaces of buildings. Suction or uplift can easily be demonstrated by holding a piece of paper at two of its corners and blowing above it. For some common structures uplift may be as large as 20 to 30 psf or even more.

During the passing of a tornado or hurricane, a sharp reduction in atmospheric pressure occurs. This decrease in pressure is not reflected inside airtight buildings, and the inside pressures, being greater than the external pressures, cause outward forces against the roofs and walls. Nearly everyone has heard stories of the walls of a building "exploding" outward during a storm.

The average building code in the United States makes no reference to wind velocities in the area or to shapes of the building or to other factors. It just probably requires the use of some specified wind pressure in design such as 20 psf on projected area in elevation up to 300 ft with an increase of 2.5 psf for each additional 100-ft increase in height. The values given in these codes are thought to be rather inaccurate for modern structural design.

For a period of several years the ASCE Task Committee on Wind Forces made a detailed study of existing information concerning wind forces. A splendid report of this information entitled *Wind Forces on Structures*[6] was presented by the committee in 1961. As stated in the report, its purpose was to provide a compact source of information which could be practically used by the design profession.

In the report much information is presented concerning wind-pressure coefficients for various types of structures, information concerning maximum wind velocities for particular geographical areas, and much additional data which can be of great value in realistically estimating wind forces.

The wind pressure on a building can be estimated with the expression to follow in which p is the pressure in pounds per square foot acting on vertical surfaces, C_s is a shape coefficient, and V is the basic wind velocity in miles per hour estimated from weather bureau records for that area of the country.

$$p = 0.002558 \ C_s V^2$$

The coefficient C_s depends on the shape of the structure, primarily the roof. For box-type structures C_s is 1.3 of which 0.8 is for the pressure on the windward

---

[4]P.R. Sparks, "Wind Induced Instability in Low-Rise Buildings," *Proceedings of the 5th U.S. National Conference on Wind Engineering,* Lubbock, Texas, November 6–18, 1985.

[5]P.R. Sparks, "The Risk of Progressive Collapse of Single-Story Buildings in Severe Storms," *Proceedings of the ASCE Structures Congress,* Orlando, Florida, August 17–20, 1987.

[6]Ibid., pp. 1124–1198.

side and 0.5 is for suction on the leeward side. For such a building the total pressure on the two surfaces equals 20 psf for a wind velocity of 77.8 mph.

As described in the foregoing paragraphs, the accurate determination of the most critical wind loads on a building or bridge is an extremely involved problem; however, sufficient information is available today to permit satisfactory estimates on a reasonably simple basis.

b. *Earthquake loads.* Many areas of the world fall in earthquake territory and in these areas it is necessary to consider seismic forces in design for tall or short buildings. During an earthquake there is an acceleration of the ground surface. This acceleration can be broken down into vertical and horizontal components. Usually the vertical component of the acceleration is assumed to be negligible, but the horizontal component can be severe.

Most buildings can be designed with little extra expense to withstand the forces caused during an earthquake of fairly severe intensity. On the other hand, earthquakes during recent years have clearly shown that the average building that is not designed for earthquake forces can be destroyed by earthquakes that are not particularly severe. Various formulas are used to change earthquake acclerations into static horizontal forces that are dependent on the mass of the structure. The forces are expressed as a percentage of the gravity load of the structure and its contents and depend on the location of the structure on an earthquake probability map (see Fig. 2-1), the type of framing system, and other items. The designer may as an alternative perform a dynamic analysis of the structure. An excellent reference on the subject of seismic forces is a publication by the Structural Engineers Association of California entitled *Recommended Lateral Force Requirements and Commentary.*[7]

It should be noted that many persons look upon the seismic loads to be used in designs as being merely percentage increases of the wind loads. This thought is not correct, however, as seismic loads are different in their action and as they are not proportional to the exposed area but to the building weight above the level in question.

The effect of the horizontal acceleration increases with the distance above the ground because of the "whipping effect" of the earthquake, and design loads should be increased accordingly. Obviously, towers, water tanks, and penthouses on building roofs occupy precarious positions during an earthquake.

Figure 2-1, printed with the permission of The International Conference of Building Officials,[8] shows the seismic risk zones in various locations in the United States. These zones were established on the basis of past earthquakes.

7. *Longitudinal loads.* Longitudinal loads are another type of load that needs to be considered in designing some structures. Stopping a train on a railroad bridge or a truck on a highway bridge causes longitudinal forces to be applied. It is not difficult to imagine the tremendous longitudinal force developed when the driver of a 40-ton-trailer truck traveling 60 mph suddenly has to apply

---

[7]*Recommended Lateral Force Requirements and Commentary* [San Francisco, Calif.: Structural Engineers Association of California (SEAOC), 1987].

[8]*Uniform Building Code* (Whittier, Calif.: International Conference of Building Officials, 1979).

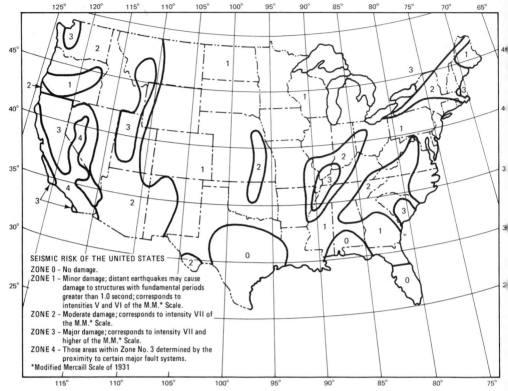

SEISMIC RISK OF THE UNITED STATES
ZONE 0 – No damage.
ZONE 1 – Minor damage; distant earthquakes may cause
        damage to structures with fundamental periods
        greater than 1.0 second; corresponds to
        intensities V and VI of the M.M.* Scale.
ZONE 2 – Moderate damage; corresponds to intensity VII of
        the M.M.* Scale.
ZONE 3 – Major damage; corresponds to intensity VII and
        higher of the M.M.* Scale.
ZONE 4 – Those areas within Zone No. 3 determined by the
        proximity to certain major fault systems.
*Modified Mercaill Scale of 1931

Figure 2-1   Seismic risk zones in the 48 contiguous states. (Reproduced from the 1982
Uniform Building Code.)

Hungry Horse Dam and Reservoir, Rocky Mountains, in northwestern Montana. (Courtesy of the Montana Travel Promotion Division.)

the brakes while crossing a highway bridge. There are other longitudinal load situations, such as ships running into docks and the movement of traveling cranes that are supported by building frames.

8. *Other live loads.* Among the other types of live loads with which the structural designer will have to contend are *soil pressures* (such as the exertion of lateral earth pressures on walls or upward pressures on foundations); *hydrostatic pressures* (as water pressure on dams, inertia forces of large bodies of water during earthquakes, and uplift pressures on tanks and basement structures); *blast loads* (caused by explosions, sonic booms, and military weapons); *thermal forces* (due to changes in temperature resulting in structural deformations and resulting structural forces); and *centrifugal forces* (as those caused on curved bridges by trucks and trains or similar effects on roller coasters, etc.).

2-5 SELECTION OF DESIGN LOADS

To assist the designer in estimating the magnitudes of live loads with which he or she should proportion structures, various records have been assembled through the years in the form of building codes and specifications. These publications provide conservative estimates of live load magnitudes for various situations. One of the most widely used design load specifications for buildings is that published by the previously mentioned American National Standards Institute. Some other commonly used specifications are:

1. For railroad bridges, American Railway Engineering Association (AREA).[9]
2. For highway bridges, American Association of State Highway and Transportation Officials.[10]
3. For buildings, the Uniform Building Code (UBC).

These specifications will on many occasions clearly prescribe the loads for which structures are to be designed. Despite the availability of this information the designer's ingenuity and knowledge of the situation are often needed to predict what loads a particular structure will have to support in years to come. Over the past several decades insufficient estimates of future traffic loads by bridge designers have resulted in a great amount of replacement with wider and stronger structures.

2-6 ELASTIC AND PLASTIC DESIGN METHODS DEFINED

Almost all existing steel structures were designed by the *elastic design* methods. The designer estimates the *working* or *service* loads, that is, the loads that the

[9]*Specifications for Steel Railway Bridges* (Chicago, Ill.: American Railway Enginering Association (AREA), 1980).

[10]*Standard Specifications for Highway Bridges* 13th ed. [Washington, D.C.: American Association of State Highway and Transportation Officials (AASHTO), 1983].

structure may feasibly have to support, and proportions the members on the basis of certain allowable stresses. These allowable stresses are usually some fraction of the specified minimum yield point of the steel. Although the term "elastic design" is very commonly used to describe this method the terms *allowable-stress design* or *working-stress design* are definitely more appropriate. Many of the provisions of the specifications for this method are actually based on plastic or ultimate-strength behavior and not on elastic behavior.

The ductility of steel has been shown to give it a reserve strength and the realization of this fact is the theory behind *plastic design*. In this method the working loads are estimated and multiplied by certain load or overcapacity or safety factors and the members designed on the basis of collapse strengths. Another name for this method is *collapse design*.

The design profession has long been aware that the major portion of the stress-strain curve lies beyond the steel's elastic limit. Furthermore, tests through the years have made it clear that steels can resist stresses appreciably beyond their yield points, and that in cases of overload, statically indeterminate structures have the happy facility of spreading the load out due to the steel's ductility. On the basis of this information many plastic-design proposals have been made in recent decades. It is undoubtedly true that for certain types of structures, plastic design results in a more economical use of steel than does elastic design.

2-7 LOAD AND RESISTANCE FACTOR DESIGN

The structural steel design profession is moving rapidly in the direction of load and resistance factor design (LRFD). This method includes many of the features of the design procedures commonly associated with ultimate design, plastic design, limit design, or collapse design.

Load and resistance factor design is based on a *limit states* philosophy. The term limit state is used to describe a condition at which a structure or some part of that structure ceases to perform its intended function. There are actually two categories of limit states, strength and serviceability.

Strength limit states are based on the safety or load-carrying capacity of structures and include plastic strengths, buckling, fracture, fatigue, overturning, and so on.

Serviceability limit states refer to the performance of structures under normal service loads and are concerned with the uses and/or occupancy of structures including such items as excessive deflections, slipping, vibrations, and cracking.

Not only must the structure be capable of supporting the design or ultimate loads but it must also be able to support the service or working loads in such a manner as to meet the requirements of the users or occupants of the structure. For instance, a high-rise building must be designed so that lateral deflections or drift are not excessive during ordinary storms; that is, the occupants must not become uncomfortable (frightened or seasick). As to the strength limit state, the frame will be designed so that it will safely resist the ultimate load occurring during that 50-year storm, although there may be some minor damage to the building and the occupants may suffer some discomfort.

The LRFD Specification concentrates on very specific requirements relating to the strength limit state and allows the designer some freedom of judgment in the serviceability area. This doesn't mean that the serviceability limit state is not significant, but by far the most important consideration (as in all structural specifications) is the life and property of the public. As a result public safety is not left to the judgment of the individual designer.

In LRFD the working or service loads (Q_i) are multiplied by certain load or safety factors (λ_i) almost always larger than 1.0 and the resulting "factored loads" used for designing the structure. The magnitudes of load factors vary depending on the type and combination of the loads and are discussed in detail in Section 2-8.

The structure is proportioned to have a design ultimate strength sufficient to support the factored loads. This strength is considered to equal the nominal or theoretical strength of the member (R_n) multiplied by a resistance or overcapacity factor (ϕ), which is normally less than 1.0, with which the designer attempts to account for the uncertainties in material strengths, dimensions, and workmanship. Furthermore these factors have been adjusted somewhat to ensure more uniform reliability in design as described in Section 2-11. Values of ϕ factors for different situations are discussed in detail in Section 2-9.

For a particular member the preceding information can be summarized as follows: (Sum of products of load effects and load factors) ≤ (resistance factor)(nominal resistance)

$$\Sigma \lambda_i Q_i \le \phi R_n$$

The left-hand side of this expression refers to the load effects on the structure while the right-hand side refers to the resistance or capacity of the member.

In Canada a probability-based limit states design method has been in use for several years for both hot-rolled and cold-formed steel structures, while much progress has been made in Europe toward the goal of formulating similar procedures for various national codes. Efforts are being made to prepare similar procedures for timber and reinforced-concrete structures in both Europe and America.[11]

At Washington University in St. Louis, MO, a research project was conducted on LRFD from 1969 through 1976 under the direction of T.V. Galambos and M.K. Ravindra. At the conclusion of the project, a document was published entitled "Proposed Criteria for Load and Resistance Factor Design of Steel Building Structures."[12]

2-8 LOAD FACTORS

The purpose of load factors is to increase the loads to account for the uncertainties involved in estimating the magnitudes of dead or live loads. For instance:

---

[11]M.K. Ravindra and T.V. Galambos, "Load and Resistance Factor Design for Steel," *ASCE Journal of Structural Division* 104, no. ST9 (September 1978), pp. 1337–1353.

[12]Research Report 45, C.E. Dept., Washington University, St. Louis, Mo., May 1976.

"how close in percent could you the reader estimate the worst wind or snow loads that will ever be applied to the building which you are now occupying?"

The value of the load factor used for dead loads is smaller than the one used for live loads because designers can estimate so much more accurately the magnitudes of dead loads than they can the magnitudes of live loads. In this regard the student will notice that loads which remain in place for long time periods will be less variable in magnitude while those which are applied for brief periods such as wind loads will have larger variations. It is hoped that the LRFD design procedure will make the designer more conscious of load variations than he or she is with the allowable stress design method.

The LRFD Specification presents load factors and load combinations that were selected to be used with the recommended minimum loads given in the American National Standards Institute (ANSI) Specification A58.1-1982.[13]

The usual load combinations to be considered with the LRFD are given in their specification A4.1 by formulas A4-1 and A4-2. In these expressions the abbreviations used are D for dead loads, L for live loads, L_r for roof live loads, S for snow loads, and R for initial rainwater or ice, not including a ponding contribution. The letter U represents the ultimate load.

$$U = 1.4D \qquad\qquad\qquad\qquad \text{(LRFD Formula A4-1)}$$

$$U = 1.2D + 1.6L + 0.5(L_r \text{ or } S \text{ or } R) \qquad \text{(LRFD Formula A4-2)}$$

Impact loads will only be included in the second of these combinations. Should wind (W) or earthquake (E) forces be involved, the following combinations need to be considered:

$$U = 1.2D + 1.6(\, L_r \text{ or } S \text{ or } R) + (0.5L \text{ or } 0.8W) \quad \text{(LRFD Formula A4-3)}$$

$$U = 1.2D + 1.3W + 0.5L + 0.5(L_r \text{ or } S \text{ or } R) \qquad \text{(LRFD Formula A4-4)}$$

$$U = 1.2D + 1.5E + (0.5L \text{ or } 0.2S) \qquad\qquad \text{(LRFD Formula A4-5)}$$

It is necessary to consider impact loading only in combination A4-5 of this group. There is a change in the value of the load factor for L for combinations A4-3, A4-4, and A4-5 for garages, public assembly areas, and all areas where the live load exceeds 100 psf. For such cases 1.0 is to be used.

To account for possibilities of uplift, another load combination is given in the LRFD Specification. This condition is included to cover cases where tension forces develop owing to overturning moments. It will govern only for tall buildings where high lateral loads are present. In this combination the dead loads are reduced by 10 percent to take into account situations where they may have been overestimated.

$$U = 0.9D - (1.3W \text{ or } 1.5E) \qquad\qquad \text{(LRFD Formula A4-6)}$$

The magnitudes of the loads (D, L, L_r, etc.) should be obtained from the governing building code or from ANSI 58.1-1982 (*Minimum Design Loads for*

---

[13]*American National Standard Minimum Design Loads for Buildings and Other Structures* (New York: American National Standards Institute, Inc.), ANSI A58.1-1982.

Buildings and Other Structures). Wherever applicable the live loads used for design should be the reduced values specified for large floor areas, multistory buildings, and so on.

A brief illustration of the calculation of the critical loading for a building column using the LRFD load combinations is presented in Example 2-1. The largest value obtained is to be used for the design.

EXAMPLE 2-1

The various axial loads for a building column have been computed according to the applicable building code with the following results: dead load = 200 k, load from roof = 50 k (live load or snow or rainwater), live load from floors (has been reduced as applicable for large floor area and multistory columns) = 250 k, wind = 80 k, and earthquake = 60 k. Determine the critical design load using the six LRFD load combinations.

Solution

A4-1 $U = (1.4)(200) = 280$ k

A4-2 $U = (1.2)(200) + (1.6)(250) + (0.5)(50) = 665$ k ←

A4-3(a) $U = (1.2)(200) + (1.6)(50) + (0.5)(250) = 445$ k

A4-3(b) $U = (1.2)(200) + (1.6)(50) + (0.8)(80) = 384$ k

A4-4 $U = (1.2)(200) + (1.3)(80) + (0.5)(250) + (0.5)(50) = 494$ k

A4-5 $U = (1.2)(200) + (1.5)(60) + (0.5)(250) = 455$ k

A4-6(a) $U = (0.9)(200) - (1.3)(80) = 76$ k

A4-6(b) $U = (0.9)(200) - (1.5)(60) = 90$ k

The critical factored load combination or design strength required for this column is 665 k as determined by LRFD Formula A4-2. It will be noted that the results of Formula A4-6 do not indicate an uplift problem.

2-9 RESISTANCE FACTORS

To accurately estimate the ultimate strength of a structure it is considered necessary to take into account the uncertainties in material strengths, dimensions, and workmanship. With a resistance factor the designer attempts to show that the strength of a member cannot be computed exactly because of imperfections in analysis theory (remember the imperfect assumptions you made to analyze trusses and other members), variations in material properties, and imperfect dimensions of members.

This is done by multiplying the theoretical ultimate strength (called the *nominal strength* here) of each member by a resistance or overcapacity factor ϕ, which is almost always less than 1.0. These values are 0.85 for columns, 0.75 or 0.90 for tension members, 0.90 for bending or shear in beams, and so on.

Typical resistance factors from the LRFD Specification are shown in Table 2-2. Some terms used in this table (such as groove and fillet welds, slip-resistant bolts, and web crippling) will be defined in later chapters of this text. The magnitudes of the resistance factors given in the LRFD Specification are based on research recommendations from investigators at Washington University in St. Louis.

2-10 DISCUSSION OF SIZES OF LOAD AND RESISTANCE FACTORS

Students may feel that it is foolish and uneconomical to design structures with such large load factors and such small resistance factors. As the years go by, however, they will learn that these factors are subject to so many uncertainties that they may spend sleepless nights wondering if those they have used are sufficient (and they may join other designers in calling them "factors of ignorance"). Some of the uncertainties affecting these factors are:

1. Material strengths may initially vary appreciably from their assumed values and they will vary more with time due to creep, corrosion, and fatigue.

2. The methods of analysis are often subject to appreciable errors.

3. The so-called beggaries of nature or acts of God (hurricanes, earthquakes, etc.) cause conditions difficult to predict.

4. The stresses produced during fabrication and erection are often severe. Workmen in shop and field seem to treat steel shapes with reckless abandon. They drop them. They ram them. They force the members into position to line up the bolt holes. In fact, the stresses during fabrication and erection may exceed those which occur after the structure is completed. The floors for the rooms of apartment houses and office buildings are probably designed for service live loads varying from 40 to 80 psf (pounds per square foot). During the erection of such buildings the contractor may have 10 ft of bricks or concrete blocks or other construction materials or equipment piled up on some of the floors, causing loads of

TABLE 2-2   TYPICAL RESISTANCE FACTORS

Resistance or ϕ factors	Situations
1.00	Bearing on the projected areas of pins, web yielding under concentrated loads, slip-resistant bolt shear values
0.90	Beams in bending and shear, fillet welds with stress parallel to weld axis, groove welds base metal
0.85	Columns, web crippling, edge distance and bearing capacity at holes
0.80	Shear on effective area of full-penetration groove welds, tension normal to effective area of partial-penetration groove welds
0.75	Bolts in tension, plug, or slot welds, fracture in the net section of tension members
0.65	Bearing on bolts (other than A307)
0.60	Bearing on A307 bolts, bearing on concrete foundations

several hundred pounds per square foot. This discussion is not intended to criticize the practice (not that it is a good one), but rather to make the student aware of the things that happen during construction. (It is probable that the majority of steel structures are overloaded somewhere during construction but hardly any of them fail. On many of these occasions the ductility of the steel has surely saved the day.)

5. There are technological changes that affect the magnitude of live loads. The constantly increasing traffic loads applied to bridges through the years is one illustration. The wind also seems to blow harder as the years go by, or at least building codes keep raising the minimum design wind pressures as more is learned about the subject.

6. Although the dead loads of a structure can usually be estimated quite closely, the estimate of the live loads is more inaccurate. This is particularly true in estimating the worst possible combination of live loads occurring at any one time.

7. Other uncertainties are the presence of residual stresses and stress concentrations, variations in dimensions of member cross sections, etc.

2-11 RELIABILITY AND THE LRFD SPECIFICATION

The word *reliability* as used in this textbook refers to the estimated percentage of times that the strength of a structure will equal or exceed the maximum loading applied to that structure during its estimated life (say 50 years).

In this section the author briefly describes how:

1.   LRFD investigators developed a procedure for estimating the reliability of given designs.
2.   They set what to them were desirable reliability percentages for different situations.
3.   They were able to adjust resistance or ϕ factors so steel designers are able to obtain the reliability percentages established in 2 above.

Before proceeding with this discussion a few comments are presented concerning the word *failure* as it is used in this discussion of reliability. Let us assume that a designer states that his or her designs are 99.7 percent reliable (and this is the approximate value obtained with most LRFD designs). This means that if this person were to design 1000 different structures 3 of them would probably be overloaded at some time during their estimated 50-year lives and would fail. The reader probably and quite reasonably thinks this to be an unacceptably high rate of failure.

To the author 99.7 percent reliability doesn't mean that 3 of 1000 structures are going to fall flat on the ground. To him it means that those structures at some time will be loaded into the plastic range and perhaps the strain-hardening range. As a result deformations may be quite large during the overloading and some slight damage may occur. It is not anticipated that any of these structures will completely collapse. (The reader who is unfamiliar with statistics might say he or she wants 100 percent reliability in design, but this is statistically an impossible goal, as we will see in the paragraphs to follow.)

For this discussion it is assumed that we make a study of the reliability of a large number of steel structures designed at various times and with different past editions of the AISC Specification. To do this we will compute the resistance or strength, R, of each structure as well as the maximum loading, Q, expected during the life of the structure. The structure will be deemed to be safe if $R \geq Q$.

The actual values of R and Q are random variables, and it is therefore impossible to say with 100 percent certainty that R is always equal to or larger than Q for a particular structure. No matter how carefully a structure is designed and constructed there will always be some little chance that Q will be greater than R or that the strength limit state will be exceeded. The goal of the preparers of the LRFD Specification was to keep this chance to a very small and consistent percentage.

Thus the magnitudes of both resistances and loads are uncertain. If we were to plot a curve of R/Q values for a large number of structures the result would be a typical bell-shaped probability curve with mean values R_m and Q_m and a standard deviation. If at any location $R < Q$ we will have exceeded the strength limit state.

For convenience such a curve is plotted logarithmically and shown in Fig. 2-2. It is to be remembered that the logarithm of 1.0 is 0 and thus if $\ln R/Q < 0$ the strength limit state has been exceeded. Such a situation is represented by the shaded area on the diagram. The smaller the shaded area the more reliable is the structure. Another way of expressing this fact is to say that the larger the number of standard deviations from the mean values to the shaded area the greater the reliability. In the figure the number of standard deviations is represented by β and is called the *reliability index*.

Even though the probable values of R and Q are not known very well, a formula has been developed with which values of β can be reasonably calcu-

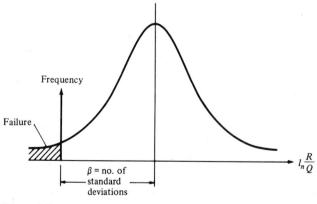

Figure 2-2

lated.[14,15,16] The formula follows:

$$\beta = \frac{\ln (R_m/Q_m)}{\sqrt{V_R^2 + V_Q^2}}$$

In this expression R_m and Q_m are respectively the mean resistance and load effects while V_R and Q_R are respectively the corresponding coefficients of variation.

As a result of the preceding work it is now possible to design a particular element in accordance with a certain edition of the AISC Specification and with the appropriate statistical information compute the value of β for the design. This process is called *calibration*.

The results of our study of the designs of these steel structures will show that the percentage of structures for which the design strengths equal or exceed the worst anticipated loading will vary as we examine designs made in accordance with the requirements of different editions of the AISC Specification. Furthermore our calculations will show that this reliability will vary for the designs of different types of members (such as columns and beams) made with the same edition of the AISC Specification.

Based upon the calculations of reliability described here the investigators decided to use consistent β values in this new specification. These were the values they selected:

1. $\beta = 3.00$ for members subject to gravity loading.
2. $\beta = 4.50$ for connections. (This value reflects the usual practice that connections should be stronger than the members they connect.)
3. $\beta = 2.5$ for members subject to gravity plus wind loadings. (This value reflects an old practice that safety factors do not have to be as large for cases where lateral loads, which are of shorter duration, are involved.)
4. $\beta = 1.75$ for members subject to gravity loads plus earthquake loads.

Then the values of ϕ for the various parts of the specification were adjusted so the β values shown here would be obtained in design. In effect this causes most LRFD designs to be almost identical with those obtained by the allowable stress method when the live to dead load ratio is 3.

2-12 ADVANTAGES OF LRFD

The average person looking at this material will ask "Will LRFD save money compared with allowable stress design (ASD)?" The answer is that it probably will, particularly if the live loads are small compared with the dead loads.

[14]C.A. Cornell, "A Probability-Based Structural Code," *Journal, American Concrete Institute,* vol. 66, no. 12 (December 1969).

[15]B. Ellingwood, T.V. Galambos, J.G. MacGregor, and C.A. Cornell, "Development of a Probability-Based Load Criterion for American National Standard A58," National Bureau of Standards Special Publication 577 (June 1980).

[16]M.K. Ravindra and T.V. Galambos, "Load and Resistance Factor Design for Steel," *Journal of the Structural Division,* vol. 104, no ST9, (September 1978), pp. 1337–1353.

It should be noted, however, that the AISC has introduced LRFD not for the specific purpose of obtaining immediate economic advantages but because it helps provide a more uniform reliability for all steel structures whatever the loads, and it is written in a form that facilitates the introduction of the advances in knowledge that will occur through the years in structural steel design.

Despite this stated objective of the AISC, the author would like to continue for a little while with the topic of possible steel weight savings with LRFD. In ASD the same safety factor was used for dead loads and live loads whereas in LRFD a much smaller safety or load factor is used for dead loads (because they generally can be determined so much more accurately than can live loads). As a result of this fact a weight of steel comparison for ASD and LRFD designs will necessarily depend on the ratio of the live loads to dead loads.

For the usual building the live load to dead load ratio varies from approximately 0.25 to as high as 4.0 or even a little higher for some very light structures. Low-rise steel buildings and preengineered buildings generally will fall in the upper range of these ratios. In allowable stress design we use the same safety factors for dead and live loads regardless of their ratio to each other. Thus with ASD heavier members will result and the factor of safety will increase as the live load to dead load ratio decreases.

It can be shown that for the lower range of L to D values, that is, less than 3, there can be steel weight savings with LRFD perhaps as much as one-sixth for tension members and columns and perhaps as much as one-tenth for beams. On the other hand if we have very high L and D ratios there is almost no increase in steel weight with LRFD compared with ASD.[17]

[17]J.A. Edinger, "Introduction to the Proposed AISC Load and Resistance Factor Design Specification," *Engineering Journal,* AISC, 21 no. 1 (first quarter, 1984), pp. 62–65.

Chapter 3

Analysis of Tension Members

3-1 INTRODUCTION

Tension members are found in bridge and roof trusses, towers, bracing systems, and in situations where they are used as tie rods. The selection of a section to be used as a tension member is one of the simplest problems encountered in design. As there is no danger of buckling the designer needs only to compute the factored force to be carried by the member and divide that force by a design stress to determine the effective cross-sectional area required. Then it is necessary to select a steel section that provides the required area. Though these introductory calculations for tension members are quite simple they do serve the important tasks of getting students started with design ideas and getting their "feet wet" with the massive LRFD Manual.

   One of the simplest forms of tension members is the circular rod, but there is some difficulty in connecting it to many structures. The rod has been used frequently in the past but has only occasional uses today in bracing systems, light trusses, and in timber construction. One important reason why rods are not popular with designers is that they have been used improperly so often in the past that they have a bad name; but if they are designed and installed correctly they are satisfactory for many situations.

   The average size rod has very little stiffness and may quite easily sag under its own weight injuring the appearance of the structure. The threaded rods formerly used in bridges often worked loose and rattled. Another disadvantage of rods is the difficulty of fabricating them with the exact lengths required and the consequent difficulties of installation.

   When rods are used in wind bracing it is a good practice to produce initial tension in them, as this will tighten up the structure and reduce rattling and swaying. To obtain initial tension the members may be detailed shorter than their required lengths, a method that gives the steel fabricator very little trouble. A common rule of thumb used is to detail the rods about $\frac{1}{16}$ in. short for each 20 ft. of length. (Approximate stress $f = \epsilon E = [\frac{1}{16}/(12)(20)] (29 \times 10^6) = 7550$ psi.) Another very satisfactory method involves tightening the rods with some sort of sleeve net or turnbuckle. Part 5 of the LRFD Manual provides detailed information for these devices.

The preceding discussion on rods should illustrate why rolled shapes such as angles have supplanted rods for most applications. In the early days of steel structures, tension members consisted of rods, bars, and perhaps cables. Today, although the use of cables is increasing for suspended-roof structures, tension members usually consist of single angles, double angles, tees, channels, W sections, or sections built up from plates or rolled shapes. These members look better than the old ones, are stiffer and easier to connect. Another type of tension section often used is the welded tension plate or flat bar which is very satisfactory for use in transmission towers, signs, foot bridges, and similar structures.

The tension members of steel roof trusses may consist of single angles as small as $2\frac{1}{2} \times 2 \times \frac{1}{4}$ for minor members. A more satisfactory member is made from two angles placed back to back with sufficient space between them to permit the insertion of plates for connection purposes. Where steel sections are used back-to-back in this manner, they should be connected every 4 or 5 ft to prevent rattling, particularly in bridge trusses. Single angles and double angles are probably the most common types of tension members in use. Structural tees make very satisfactory chord members for welded trusses because web members can conveniently be connected to them.

For bridges and large roof trusses tension members may consist of channels, W or S shapes, or even sections built up from some combination of angles, channels, and plates. Single channels are frequently used as they have little eccentricity and are conveniently connected. Although, for the same weight, W sections are stiffer than S sections, they may have a connection disadvantage in their varying depths. For instance, the W12×79, W12×72, and W12×65 all have slightly different depths (12.38 in., 12.25 in., and 12.12 in., respectively), while the S sections of a certain nominal size all have the same depths. For instance, the S12×50, the S12×40.8, and the S12×35 all have 12.00-in. depths.

Although single structural shapes are a little more economical than built-up sections, the latter are occasionally used when the designer is unable to obtain sufficient area or rigidity from single shapes. Where built-up sections are used it is important to remember that field connections will have to be made and paint applied; therefore, sufficient space must be available to accomplish these things.

Members consisting of more than one section need to be tied together. Tie plates (also called tie bars) located at various intervals or perforated cover plates serve to hold the various pieces in their correct positions. These plates serve to correct any unequal distribution of loads between the various parts. They also keep the slenderness ratios (to be discussed) of the individual parts within limitations and they may permit easier handling of the built-up members. Long individual members such as angles may be inconvenient to handle due to flexibility,

Erection of Federal Reserve bank building, Minneapolis, Minn. (Courtesy of Inryco, Inc.).

but when four angles are laced together into one member as shown in Fig. 3-1, the member has considerable stiffness. None of the intermittent tie plates may be considered to increase the effective areas of the sections. As they do not theoretically carry portions of the force in the main sections, their sizes are usually governed by specifications and perhaps by some judgment on the designer's part. Perforated cover plates (see Fig. 6-10) are an exception to this rule as part of their areas can be considered as being effective in resisting axial load.

A few of the various types of tension members in general use are illustrated

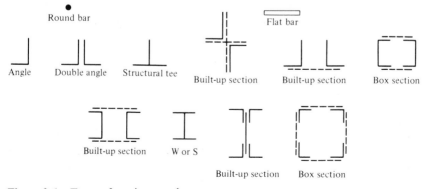

Figure 3-1 Types of tension members.

Federal Reserve bank building, Minneapolis, Minn. (Courtesy of Inryco, Inc.)

in Fig. 3-1. In this figure the dotted lines represent the intermittent tie plates or bars used to connect the shapes.

Steel cables are made with special steel alloy wire ropes which are cold-drawn to the desired diameter. The resulting wires with strengths of about 200,000 to 250,000 psi can be economically used for suspension bridges, cable supported roofs, ski lifts, and other similar applications.

Normally to select a cable tension member the designer uses a manufacturer's catalog. From the catalog the yield point of the steel and the cable size required for the design force are determined. It is also possible to select clevises or other devices to use for connectors at the cable ends.

3-2 DESIGN STRENGTH OF TENSION MEMBERS

A ductile steel member without holes subject to a tensile load can resist without fracture a load larger than its gross cross-sectional area times its yield stress because of strain hardening. However, a tension member loaded until strain hardening is reached will lengthen a great deal before fracture, a fact that will in all probability take away the member's usefulness and may even cause failure of the structural system of which the member is a part.

If, on the other hand, we have a tension member with bolt holes it can possibly fail by fracture at the net section through the holes. This failure load may very well be smaller than the load required to yield the gross section away from the holes. It is to be realized that the portion of the member where we have a reduced cross-sectional area due to the presence of holes normally is very short compared with the total length of the member. Though the strain-hardening situation is quickly reached at the net section portion of the member, yielding there may not really be a limit state of significance because the overall change in length of the member due to yielding in this small part of the member length may be negligible.

As a result of the preceding information the LRFD Specification (D1) states that the design strength of a tension member, $\phi_t P_n$, is to be the smaller of the values obtained by substituting into the following two expressions:

For the limit state of yielding in the gross section (which is intended to prevent excessive elongation of the member)

$$P_u = \phi_t F_y A_g \qquad \text{with } \phi_t = 0.9$$

For fracture in the net section where bolt or rivet holes are present

$$P_u = \phi_t F_u A_e \qquad \text{with } \phi_t = 0.75$$

In the preceding expression F_u is the specified minimum tensile stress and A_e is the effective net area that can be assumed to resist tension at the section through the holes. This area may be somewhat smaller than the actual net area, A_n, because of stress concentrations and other factors that are discussed in section 3.5 of this chapter.

The design strengths presented here are not applicable to threaded steel rods or to members with pin holes (as in eyebars). These situations are discussed in Section 4-2.

It is not likely that stress fluctuations will be a problem in the average building frame because the changes in load in such structures usually occur only occasionally and produce relatively minor stress variations. Full design wind or earthquake loads occur so infrequently that they are not considered in fatigue design. Should there, however, be frequent variations or even reversals in stress the matter of fatigue must be considered. This subject is presented in Section 4-4.

3-3 NET AREAS

The presence of a hole obviously increases the unit stress in a tension member even if the hole is occupied by a bolt or rivet. (When high-strength bolts are used there may be some disagreement with this statement under certain conditions.) There is less area of steel to which the load can be distributed and there will be some concentration of stress along the edges of the hole.

Tension is assumed to be uniformly distributed over the net section of a tension member, although photoelastic studies show there is a decided increase in stress intensity around the edges of holes, sometimes equaling several times the stresses if the holes were not present. For ductile materials, however, a uniform stress distribution assumption is reasonable when the material is loaded beyond its yield point. Should the fibers around the holes be stressed to their yield point they will yield without further stress increase, with the result that there is a redistribution or balancing of stresses. At ultimate load it is reasonable to assume a uniform stress distribution. The importance of ductility on the strength of bolted or riveted tension members has been clearly demonstrated in tests. Tension members (with bolt or rivet holes) made from ductile steels have proved to be as much as one-fifth to one-sixth stronger than similar members made from brittle steels with the same ultimate strengths.

This discussion is applicable only for tension members subjected to relatively static loading. Should tension members be designed for structures subjected to fatigue-type loadings considerable effort should be made to minimize the items causing stress concentrations such as points of sudden change of cross section and sharp corners. In addition, as described in Section 4-4, the members may have to be enlarged.

The term "net cross-sectional area" or simply "net area" refers to the gross cross-sectional area of a member minus any holes, notches, or other indentations. In considering the area of such items as these it is important to realize that it is usually necessary to subtract an area a little larger than the actual hole. For instance, in fabricating structural steel which is to be connected with bolts the holes are usually punched $\frac{1}{16}$ in. larger than the diameter of the bolt. Furthermore, the punching of the hole is assumed to damage or even destroy $\frac{1}{16}$ in. (1.6 mm) more of the surrounding metal; therefore, the area of the holes subtracted is $\frac{1}{8}$ in. (3 mm) larger than the diameter of the bolt. The area of the holes subtracted is rectangular and equals the diameter of the hole times the thickness of the metal.

For steel much thicker than bolt diameters it is difficult to punch out the holes to the full sizes required without excessive deformation of the surrounding material. These holes may be subpunched (with diameters $\frac{3}{16}$ in. undersized) and then reamed out to full size after the pieces are assembled. Very little material is damaged by this quite expensive process as the holes are even and smooth and it is considered unnecessary to subtract the $\frac{1}{16}$ in. for damage to the sides. Sometimes when very thick pieces are being connected the holes may be drilled to the diameter of the bolts or rivets plus $\frac{1}{32}$ in. This is a very expensive process and should be avoided if possible.

It may be necessary to have an even greater latitude in meeting dimensional tolerances during erection and for high-strength bolts larger than $\frac{5}{8}$ in. in diameter, holes larger than the standard ones may be used without reducing the performance of the connections. These oversized holes can be short-slotted or long-slotted as described in Chapter 12.

Example 3-1 illustrates the calculations necessary for determining the net area of a plate type of tension member.

EXAMPLE 3-1

Determine the net area of the $\frac{3}{8} \times 8$-in. plate shown in Fig. 3-2. The plate is connected at its end with two lines of $\frac{3}{4}$-in. bolts.

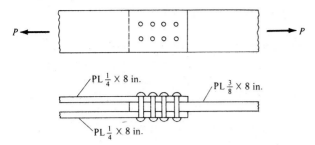

Figure 3-2

Solution

$$\text{Net area} = A_n = (\tfrac{3}{8})(8) - (2)(\tfrac{3}{4} + \tfrac{1}{8})(\tfrac{3}{8}) = 2.34 \text{ in.}^2 \ (1510 \text{ mm}^2)$$

The connections of tension members should be arranged so that no eccentricity is present. (An exception to this rule is permitted by the AISC for certain bolted and welded connections as described in Chapter 14.) If this arrangement is possible the stress is assumed to be spread uniformly across the net section of a member. Should the connections have eccentricities, moments will be produced that will cause additional stresses in the vicinity of the connection. Unfortunately it is often quite difficult to arrange connections without eccentricity. Although specifications cover some situations, the designer may have to give consideration to eccentricities in some cases by making special estimates.

The lines of action of truss members meeting at a joint are assumed to coincide. Should they not coincide, eccentricity is present and secondary stresses are the result. The centers of gravity of truss members are assumed to coincide with the lines of action of their respective forces. No problem is present in a symmetrical member as its center of gravity is at its center line, but for unsymmetrical members the problem is a little more difficult. For these members the center line is not the center of gravity, but the usual practice is to arrange the members at a joint so their gage lines coincide. If a member has more than one gage line, the one closest to the actual center of gravity of the member is used in detailing. Figure 3-3 shows a truss joint in which the gage lines of all the members pass through the same point.

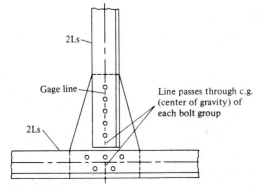

Figure 3-3

3-4 EFFECT OF STAGGERED HOLES

Should there be more than one row of bolt or rivet holes in a member it is often desirable to stagger them in order to provide as large a net area as possible at any one section to resist the load. In the preceding paragraphs tensile members have been assumed to fail transversely as along line AB in either Fig. 3-4(a) or 3-4(b). Figure 3-4(c) shows a member in which a failure other than a transverse one is possible. The holes are staggered, and a failure along section $ABCD$ is possible unless the holes are a large distance apart.

To determine the critical net area in Fig. 3-4(c), it might seem logical to compute the area of a section transverse to the member (as AE) less the area of one hole and then the area along section $ABCD$ less two holes. The smallest value obtained along these sections would be the critical value, but this method is at fault. Along the diagonal line from B to C there is a combination of direct stress and shear and a somewhat smaller area should be used. The strength of the member along section $ABCD$ is obviously somewhere between the strength obtained by using a net area computed by subtracting one hole from the transverse cross-sectional area and the value obtained by subtracting two holes from section $ABCD$.

Tests on joints show that little is gained by using complicated theoretical formulas to consider the staggered-hole situation, and the problem is usually handled with an empirical equation. The LRFD Specification (B2) and other specifications use a very simple method for computing the net width of a tension member along a zigzag section.[1] The method is to take the gross width of the member regardless of the line along which failure might occur, subtract the diameter of the holes along the zigzag section being considered, and add for each inclined line the quantity given by the expression $s^2/4g$.

In this expression s is the longitudinal spacing (or pitch) of any two holes and g is the transverse spacing (or gage) of the same holes. The values of s and g are shown in Fig. 3-4(c). There may be several paths any one of which may be critical at a particular joint. Each possibility should be considered and the one giving the least value should be used. The smallest net width obtained is multiplied by the plate thickness to give the net area, A_n. Example 3-2 illustrates the method of computing the critical net area of a section which has three lines of bolts. (For angles, the gage for holes in opposite legs is considered to be the sum of the gages from the back of the angle minus the thickness of the angle.)

Holes for bolts and rivets are normally punched in steel angles at certain

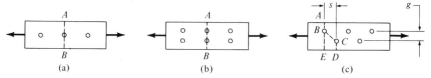

Figue 3-4

[1]V. H. Cochrane, "Rules for Riveted Hole Deductions in Tension Members," *Engineering News-Record* (New York, November 16, 1922), pp. 847–848.

standard locations. These locations or gages are dependent on the angle leg widths and on the number of lines of holes. The LRFD Manual in Part 5 in its table entitled "Usual Gages for Angles" provides these dimensions. It is unwise for the designer to require different gages from those given in the table unless unusual situations are present because of the appreciably higher fabrication costs that will result.

EXAMPLE 3-2

Determine the critical net area of the $\frac{1}{2}$-in.-thick plate shown in Fig. 3-5 using the LRFD Specification (Section B2). The holes are punched for $\frac{3}{4}$-in. bolts.

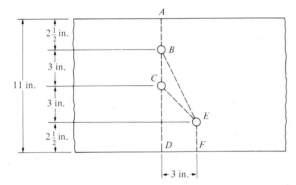

Figure 3-5

Solution

The critical section could possibly be *ABCD*, *ABCEF*, or *ABEF*. Hole diameters to be subtracted are $\frac{3}{4} + \frac{1}{8} = \frac{7}{8}$ in. The net widths for each case are as follows:

$$ABCD = 11 - (2)\left(\frac{7}{8}\right) = 9.25 \text{ in.}$$

$$ABCEF = 11 - (3)\left(\frac{7}{8}\right) + \frac{(3)^2}{(4)(3)} = 9.125 \text{ in.} \quad \text{(controls)}$$

$$ABEF = 11 - (2)\left(\frac{7}{8}\right) + \frac{(3)^2}{(4)(6)} = 9.625 \text{ in.}$$

The reader should note that it is a waste of time to check route *ABEF* for this plate. Two holes need to be subtracted for routes *ABCD* and *ABEF*. As *ABCD* is a shorter route it obviously controls over *ABEF*.)

$$A_n = (9.125)(\tfrac{1}{2}) = 4.56 \text{ in.}^2 \qquad \qquad Ans.$$

The problem of determining the minimum pitch of staggered bolts such that no more than a certain number of holes need be subtracted to determine the net section is handled in Example 3-3. The LRFD Handbook (Part 5) has a chart entitled "Net Section of Tension Members," which can be used to determine the values of $s^2/4g$. This chart can also be used to handle the type of problem solved in Example 3-3.

EXAMPLE 3-3

For the two lines of bolt holes shown in Fig. 3-6 determine the pitch which will give a net area *DEFG* equal to the one along *ABC*. The problem may also be stated as follows: determine the pitch that will give a net area equal to the gross area less one bolt hole. The holes are punched for $\frac{3}{4}$-in. bolts.

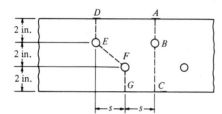

Figure 3-6

Solution

$$ABC = 6 - (1)\left(\frac{7}{8}\right) = 5.125 \text{ in.}$$

$$DEFG = 6 - (2)\left(\frac{7}{8}\right) + \frac{s^2}{(4)(2)} = 4.25 + \frac{s^2}{8}$$

$$ABC = DEFG$$

$$5.125 = 4.25 + \frac{s^2}{8}$$

$$s = 2.65 \text{ in.} \qquad\qquad Ans.$$

The $s^2/4g$ rule is merely an approximation or simplification of the complex stress variations which occur in members with staggered arrangements of bolts or rivets. Steel specifications can only provide minimum standards and designers will have to logically apply such information to complicated situations which the specifications could not cover in their attempts at brevity and simplicity. The next few paragraphs present a discussion and numerical examples of the $s^2/4g$ rule applied to situations not specifically addressed in the LRFD Specification.

The LRFD Specification does not include a method to be used for determining the net widths of sections other than plates and angles. For channels, W sections, S sections, and others the web and flange thicknesses are not the same. As a result it is necessary to work with net areas rather than net widths. If the holes are placed in straight lines across such a member the net area can simply be obtained by subtracting the cross-sectional areas of the holes from the gross area of the member. If the holes are staggered it is necessary to multiply the $s^2/4g$ values by the applicable thickness to change it to an area. Such a procedure is illustrated for a W section in Example 3-4 where bolts pass through the web only.

EXAMPLE 3-4

Determine the net area of the W12×16 (A_g = 4.71 in.²) shown in Fig. 3-7 assuming the holes are for 1-in. bolts.

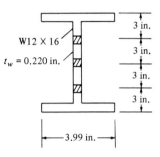
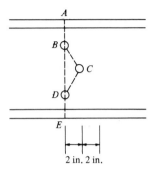

Figure 3-7

Solution

Net areas:

$$ABDE = 4.71 - (2)(1\tfrac{1}{8})(0.220) = 4.21 \text{ in.}^2$$

$$ABCDE = 4.71 - (3)(1\tfrac{1}{8})(0.220) + (2)\frac{(2)^2}{(4)(3)}(0.220) = 4.11 \text{ in.}^2$$

If the zigzag line goes from a web hole to a flange hole the thickness changes at the junction of the flange and web. In Example 3-5 the author has computed the net area of a channel that has bolt holes staggered in its flanges and web. The channel is assumed to be flattened out into a single plate as shown in parts (b) and (c) of Fig. 3-8. The net area along route

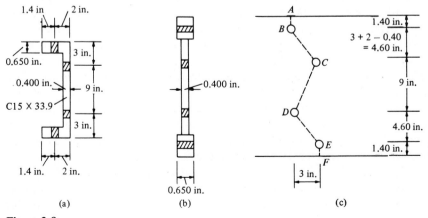

Figure 3-8

ABCDEF is determined by taking the area of the channel minus the area of the holes along the route in the flanges and web plus the $s^2/4g$ values for each zigzag line times the appropriate thickness. For line *CD*, $s^2/4g$ has been multiplied by the thickness of the web. *For lines BC and DE (which run from holes in the web to holes in the flange) an approximate procedure has been used in which the $s^2/4g$ values have been multiplied by the average of the web and flange thicknesses.*

EXAMPLE 3-5

Determine the net area along route *ABCDEF* for the C15×33.9 (A_g = 9.96 in.^{2}) shown in Fig. 3-8. Holes are for $\frac{3}{4}$-in. bolts.

Solution

$$\text{Approximate net } A \text{ along } ABCDEF = 9.96 - (2)\left(\frac{7}{8}\right)(0.650)$$

$$- (2)\left(\frac{7}{8}\right)(0.400)$$

$$+ \frac{(3)^2}{(4)(9)}(0.400)$$

$$+ (2)\frac{(3)^2}{(4)(4.60)}\left(\frac{0.650 + 0.400}{2}\right)$$

$$= 8.736 \text{ in.}^2$$

3-5 EFFECTIVE NET AREAS

When a member other than a flat plate or bar is loaded in axial tension until failure occurs across its net section its actual tensile failure stress will probably be less than the coupon tensile strength of the steel *unless all of the various elements which make up the section are connected so stress is transferred uniformly across the section.* The reason for the reduced strength of the member is the concentration of shear stress, called *shear lag,* in the vicinity of the connection. In such a situation the flow of tensile stress between the full member cross section and the smaller connected cross section is not 100 percent effective. As a result the LRFD Specification (B3) states that the effective net area, A_e, of such a member is to be determined by multiplying its net area (if bolted or riveted) or its gross area (if welded) by a reduction factor U. The use of a factor such as U accounts for the nonuniform stress distribution in a simple manner. An explanation of the way in which U factors are determined follows.

The angle shown in Fig. 3-9(a) is connected at its ends to only one leg. You can easily see that its area effective in resisting tension can be appreciably increased by shortening the width of the unconnected leg and lengthening the width of the connected leg as shown in Fig. 3-9(b).

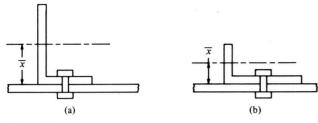

Figure 3-9

Investigators have found that a convenient measure of the effectiveness of a member such as an angle connected by one leg is the distance $\bar{x}$ measured from the plane of the connection to the centroid of the area of the whole section.[2,3] The smaller the value of $\bar{x}$ the larger is the effective area of the member.

The values of U in LRFD Specification B3 were computed from the empirical expression $1-\bar{x}/L$ where L is the length of the connection. The specification in effect reduces the length L of a connection with shear lag to a shorter effective length L'. The value of U then equals L'/L or $1-\bar{x}/L$. Several values of $\bar{x}$ are shown in Fig. 3-10. The U values given in the specification are the approxi-

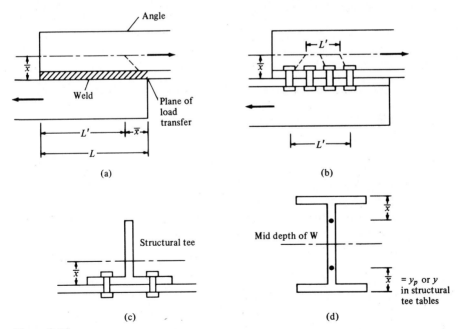

Figure 3-10

[2] E. H. Gaylord, Jr., and C. N. Gaylord, *Design of Steel Structures,* 2d ed. (New York: McGraw-Hill Book Company, 1972), pp. 119–123.

[3] W. H. Munse and E. Chesson, Jr., "Riveted and Bolted Joints: Net Section Design," *Journal of the Structural Division,* ASCE, 89, ST1 (February 1963).

mate lower bound values of the results obtained when the $1-\bar{x}/L$ expression is computed for different shear lag situations.

If it is desired to calculate U for a W section connected by its flanges only, we will assume first that the section is split into two structural tees. Then the value of $\bar{x}$ used in the $1-\bar{x}/L$ expression will be the distance from the outside edge of the flange to the c.g. of the structural tee as shown in part (d) of Fig. 3-10.

Detailed LRFD requirements for tension members for which all of the parts are not connected follow.

General If the force is transmitted directly to each of the cross-sectional elements of a member by connectors the effective net area, A_e, is equal to its net area, A_n.

Bolted or Riveted Members If the load is transmitted by bolts or rivets through some but not all of the elements of the member the value of A_e is to be determined with the expression to follow:

$$A_e = UA_n$$

The following values of U are to be used unless larger values can be justified on the basis of tests or recognized theory. It will be noted from the values shown that as the number of fasteners in a line is increased the shear lag decreases.

a. W, M, or S shapes with flange widths not less than two-thirds the depth, and structural tees cut from these shapes, provided the connection is to the flanges. Bolted or riveted connections shall have no fewer than three fasteners per line in the direction of stress $U = 0.90$
b. W, M, or S shapes not meeting the conditions of subparagraph a, structural tees cut from these shapes and all other shapes, including built-up cross sections. Bolted or riveted connections shall have no fewer than three fasteners per line in the direction of stress $U = 0.85$
c. All members with bolted or riveted connections having only two fasteners per line in the direction of stress $U = 0.75$

Welded members

1. If the load is transmitted by welds through some but not all of the elements of a tension member the effective net area is to be determined by multiplying the reduction coefficient, U, times the *gross area* of the member.

$$A_e = UA_g$$

The values of U to be used are the same as the ones for bolted or riveted members except that the provision about members having only two fasteners per line has no significance. In addition there are some special requirements for cases where loads are transmitted by transverse or longitudinal welds. These situations are described in items 2 and 3 to follow.

2. If a tensile load is transmitted by transverse welds to some but not all of the elements of W, M, or S shapes and structural tees cut from those shapes the effective net area, A_e, is to be set equal to the area of the directly connected parts.

3. Tests have shown that when flat plates or bars connected by longitudinal fillet welds (a term to be described in Chapter 14) are used as tension members they may fail prematurely by shear lag at the corners if the welds are too far apart. Therefore, the LRFD Specification states that when such situations are encountered the length of the welds may not be less than the width of the plates or bars and the effective net area is to equal UA_g. For such situations the following values of U are to be used:

When $L > 2w$	$U = 1.0$
When $2w > L > 1.5w$	$U = 0.87$
When $1.5w > L > w$	$U = 0.75$

where L = weld length, in.
w = plate width (distance between welds), in.

Example 3-6 illustrates the calculations necessary for determining the effective net area of a bolted W section connected only to its flanges. In addition the design strength of the member is computed.

EXAMPLE 3-6

Determine the design strength of a W10 × 45 with two lines of $\frac{3}{4}$-in.-diameter bolts in each flange using A36 steel and the LRFD Specification. There are assumed to be at least three bolts in each line, and the bolts are not staggered with respect to each other.

Solution

Using a W10×45 (A_g = 13.3 in.^{2}, d = 10.10 in., b_f = 8.020 in., t_f = 0.620 in.)

(a) $P_u = \phi_t F_y A_g = (0.90)(36)(13.3) = 430.9 \text{ k} \leftarrow$

(b) $A_n = 13.3 - (4)\left(\frac{7}{8}\right)(0.620) = 11.13 \text{ in.}^2$

$U = 0.90 \text{ since } b_f > \frac{2}{3}d$

$A_e = UA_n = (0.90)(11.13) = 10.02 \text{ in.}^2$

$P_u = \phi_t F_u A_e = (0.75)(58)(10.02) = 435.9 \text{ k}$

$$\underline{\underline{\text{Design strength } P_u = 430.9 \text{ k}}}$$

3-6 CONNECTING ELEMENTS FOR TENSION MEMBERS

When splice or gusset plates are used as statically loaded tensile connecting elements their strength shall be determined from the following expression (LRFD Specification J5.2).

For yielding of welded, bolted, or riveted connection elements

$$\phi_t R_n = \phi_t A_g F_y \qquad \text{with } \phi_t = 0.9$$

For fracture of bolted or riveted connection elements

$$\phi_t R_n = \phi_t A_n F_u \qquad \text{with } \phi_t = 0.75 \text{ and } A_n \leq 0.85\, A_g$$

The A_n to be used in the second of these expressions may not exceed 85 percent of A_g. Tests have shown for decades that bolted or riveted tension connection elements rarely have an efficiency greater than 85 percent even if the holes represent a very small percentage of the gross area of the elements. In Example 3.7 the strength of a pair of tensile connecting plates is computed.

EXAMPLE 3-7

The A36 tension member of Example 3-6 is assumed to be connected at its ends with two $\frac{3}{8} \times 12$-in. plates as shown in Fig. 3-11. If two lines of $\frac{3}{4}$-in. bolts are used in each plate determine the maximum tensile force which the plates can transfer.

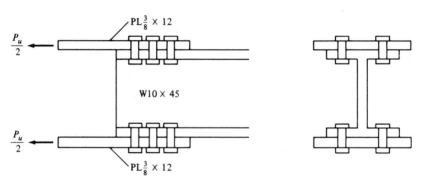

Figure 3-11

Solution

$$P_u = \phi_t F_y A_g = (0.9)(36)(2 \times \tfrac{3}{8} \times 12) = 291.6 \text{ k}$$

$$A_n \text{ of 2 plates} = (\tfrac{3}{8} \times 12 - \tfrac{7}{8} \times 2 \times \tfrac{3}{8})\, 2 = 7.69 \text{ in.}^2$$

$$0.85\, A_g = (0.85)(2 \times \tfrac{3}{8} \times 12) = 7.65 \text{ in.}^2 = A_n$$

$$P_u = \phi_t F_u A_n = (0.75)(58)(7.65) = 332.7 \text{ k}$$

$$\underline{\underline{P_u = 291.6 \text{ k}}}$$

3-7 BLOCK SHEAR

The design strength of a tension member is not always controlled by $\phi_t F_y A_g$ or $\phi_t F_u A_e$ or by the strength of the bolts or welds with which the member is connected. It may instead be controlled by its *block shear* strength as described in this section.

The failure of a member may occur along a path involving tension on one plane and shear on a perpendicular plane as shown in Fig. 3-12, where several possible block shear failures are shown.

It is unlikely that fracture will occur on both planes at the same time. It seems logical to assume that the load will cause the yield strength to be reached on one plane while the other plane is stressed past yielding and on to fracture. Thus it does not seem realistic to add the fracture strengths on both planes together to determine the block shear strength of a particular member.

The member shown in Fig. 3-13 (a) has a large shear area and a small tensile area, and the primary resistance to a block shear failure is shearing and not tensile. The LRFD Specification states that it is logical to assume that when a shear fracture occurs on this large shear-resisting area the small tensile area has yielded.

In part (b) of Fig. 3-13 a member is shown which, so far as block shear goes, has a large tensile area and a small shearing area. The LRFD feels that for this case the primary resisting force against a block shear failure will be tensile and not shearing. Thus a block shear failure cannot occur until the tensile area fractures. At that time it seems logical to assume that the shear area has yielded.

Based on the preceding discussion the LRFD Specification (J5.2c) states that the block shear design strength of a particular member is to be determined by (1) computing the tensile fracture strength on the net section in one direction

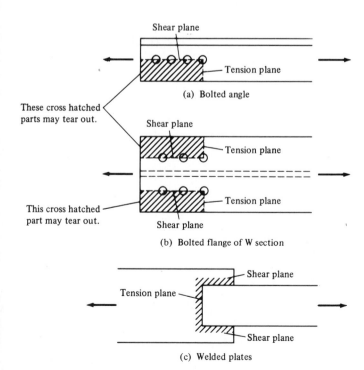

(a) Bolted angle

(b) Bolted flange of W section

(c) Welded plates

Figure 3-12

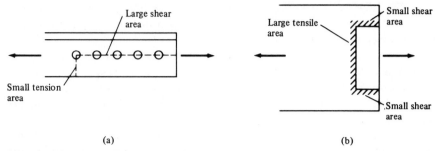

(a)   (b)

Figure 3-13

and adding to that value the shear yield strength on the gross area on the perpendicular segment and (2) computing the shear fracture strength on the gross area subject to tension and adding it to the tensile yield strength on the net area subject to shear on the perpendicular segment. The larger value determined in (1) and (2) is the block shear strength.

Test results show that this procedure gives good results. Furthermore it is consistent with the calculations previously used for tension members where gross areas are used for one limit state of yielding $(\phi_t F_y A_g)$ and net area for the fracture limit state $(\phi_t F_u A_e)$. The LRFD Commentary (J4) states that the block shear strength of a member, P_{bs}, is to be taken as the larger of the values given at the end of this paragraph.

Tension fracture and shear yielding

$P_{bs} = \phi$ (tensile fracture strength + shear yielding)

$= \phi \left(F_u A_{nt} + 0.6 F_y A_{vg}\right)$ (LRFD Formula C-J4-1)

Shear fracture and tension yielding

$P_{bs} = \phi$ (shear fracture strength + tension yielding)

$= \phi \left(F_y A_{tg} + 0.6 F_u A_{ns}\right)$ (LRFD Formula C-J4-2)

In which

$$\phi = 0.75$$

A_{vg} = gross area subjected to shear

A_{tg} = gross area subjected to tension

A_{ns} = net area subjected to shear

A_{nt} = net area subjected to tension

Examples 3-8 and 3-9 illustrate the determination of the block shear strengths for two members. This topic of block shear is continued in the connection chapters of this text.

EXAMPLE 3-8

The A36 tension member shown in Fig. 3-14 is connected with three $\frac{3}{4}$-in. bolts. Determine the block shearing strength of the member and its tensile strength.

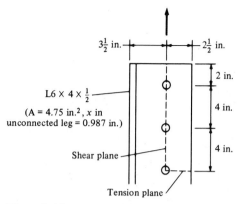

Figure 3-14

Solution

Tension fracture + shear yielding

$$P_{bs} = \phi(F_u A_{nt} + 0.6\,F_y A_{vg})$$
$$= 0.75\,[(58)\,(\tfrac{1}{2})\,(2.50 - \tfrac{1}{2} \times \tfrac{7}{8}) + (0.6)\,(36)\,(10)\,(\tfrac{1}{2})]$$
$$= 125.9\ k$$

Shear fracture and tension yielding

$$P_{bs} = \phi(F_y A_{tg} + 0.6\,F_u A_{ns})\ \text{noting that there are}$$
$$2\tfrac{1}{2}\ \text{holes in the net area shear plane shown in Fig. 3-12}$$
$$= 0.75[(36)\,(2.5)\,(\tfrac{1}{2}) + (0.6)\,(58)\,(10.0 - 2.5 \times \tfrac{7}{8})\,(\tfrac{1}{2})]$$
$$135.7\ k \leftarrow$$

Tensile strength of angle

(a) $P_u = \phi_t F_y A_g = (0.9)(36)(4.75) = 153.9\ k \leftarrow$

(b) $A_n = 4.75 - (1)(\tfrac{7}{8})(\tfrac{1}{2}) = 4.31\ \text{in.}^2$

$$U = 1 - \frac{0.987}{8} = 0.88 \qquad (0.85\ \text{given in LRFD})$$

$$P_u = \phi_t F_u A_e = (0.75)(58)(0.88 \times 4.31) = 165\ k$$

P_u for member = larger of P_{bs} values 135.7 k or smaller of tensile strength values 153.9 k.

$$\underline{\underline{P_u = 135.7\ k}}$$

EXAMPLE 3-9
Determine the block shear design strength of the A36 welded member shown in Fig. 3-15.

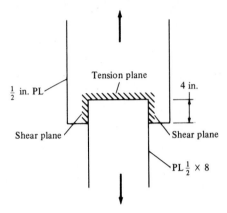

Figure 3-15

Solution

Tension fracture and shear yielding

$$P_{bs} = \phi(F_u A_{nt} + 0.6\,F_y A_{vg})$$
$$= 0.75\,[(58\,(\tfrac{1}{2} \times 8) + (0.6)(36)(\tfrac{1}{2} \times 8)] = 238.8 \text{ k}$$

Shear fracture and tensile yielding

$$P_{bs} = \phi(F_y A_{tg} + 0.6\,F_y A_{ns})$$
$$= 0.75\,[(36)(\tfrac{1}{2} \times 8) + (0.6)(58)(\tfrac{1}{2} \times 8)] = 212.4 \text{ k}$$

Tensile strength of plate

$$P_u = \theta_t F_y A_g = (0.9)(36)(\tfrac{1}{2} \times 8) = 129.6 \text{ k} \longleftarrow$$

<u>Design strength of plate $= 129.6$ k</u>

Sometimes cases are encountered where it is not altogether clear what sections should be considered for block shear calculations. For such situations designers will have to use their own judgment. One such case is shown in Fig. 3-16. In part (a) of the figure it is first assumed that web tear-out will occur along the lines *abcdef*. An alternate tear-out possibility for the same member along

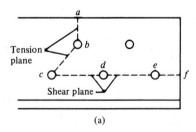

(a)   (b)

Figure 3-16

lines *abdef* is shown in part (b) of the figure. For this connection it is assumed that the load to be resisted is distributed equally between the five bolts. Thus, when web tear out is considered for case (b), we will assume only $\frac{4}{5} P_u$ is carried by the section in question because one of the bolts is outside of the tear-out area. To compute the width of the tension planes *abc* and *abd* for these two cases, it seems logical to make use of the $s^2/4g$ expression presented in Section 3-4.

PROBLEMS (Use standard size bolt holes for all problems.)

3-1 to 3-4. Compute the net area of each of the given members.

3-1. An L6 × 4 × $\frac{3}{4}$ with one line of $\frac{3}{4}$-in. bolts in each leg. (*Ans.* 5.63 in.²)

3-2. A pair of 8 × 6 × $\frac{3}{4}$Ls with two rows of $\frac{7}{8}$-in. bolts in the long legs and one row in the short legs.

3-3. A W27×84 with two holes in each flange and two in the web all for 1-in. bolts (*Ans.* 20.88 in.²)

3-4. The $\frac{3}{4}$ × 12PL shown in the accompanying illustration. The holes are for $\frac{3}{4}$-in. bolts.

Problem 3-4

3-5. For the plate and bolts shown in the accompanying illustration find the minimum pitch *s* for which only two bolts need be subtracted at any one section in calculating the net area. Bolts are $\frac{7}{8}$-in. (*Ans.* 2.83 in.)

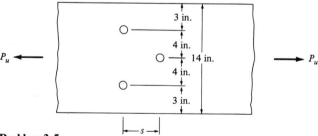

Problem 3-5

3-6. Find the minimum pitch *s* in the plate of Prob. 3-5 so that only two and one-half bolts need be subtracted at any one section.

3-7. An L8 × 6 × 1 is used as a tension member with one gage line of $\frac{7}{8}$-in. bolts in each leg at the usual gage location. What is the minimum amount of stagger necessary so that only one bolt need be subtracted from the gross area of the angle? Compute the net area of this member if the holes are staggered at 2 in. (*Ans.* s = 5.29 in., A_n = 11.14 in.²)

3-8. An L8 × 4 × $\frac{3}{4}$ is shown in the accompanying illustration. Two rows of $\frac{7}{8}$-in. bolts are used in the long leg and one in the short leg. Determine the minimum stagger (or pitch s in the figure) necessary so that only two holes need be subtracted in determining the net area.

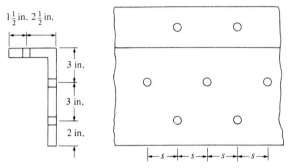

Problem 3-8

3-9. A 6 × 4 × $\frac{1}{2}$ angle has one line of holes in each leg for $\frac{3}{4}$-in. bolts. Determine the minimum pitch so that only one and one-half holes need be deducted to obtain the net area. (Use the usual gage for angles as given in the LRFD Manual (*Ans.* 3.10 in.)

3-10. Determine the effective net area of the C12×30 shown in the accompanying illustration. Assume the holes are for 1-in. bolts.

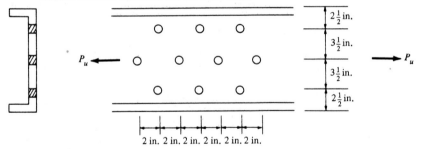

Problem 3-10

3-11. Determine the effective net cross-sectional area of the C12×25 shown in the accompanying illustration. Holes are for $\frac{3}{4}$-in. bolts. (*Ans.* 6.11 in.^{2})

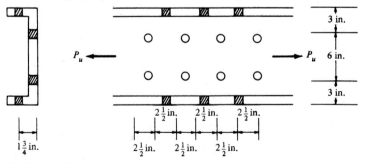

Problem 3-11

3-12. Compute the effective net area of the built-up section shown in the accompanying illustration if the holes are punched for $\frac{7}{8}$-in. bolts. Assume at least three bolts in each line.

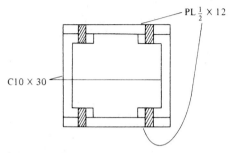

Problem 3-12

3-13. A C10×15.3 is connected through its web with three gage lines of $\frac{7}{8}$-in. bolts. The gage lines are $2\frac{1}{2}$ in. on centers and the bolts are spaced 3 in. on centers along the gage line. If the center row of bolts is staggered with respect to the outer row determine the effective net cross-sectional area of the channel. Assume there are at least three bolts in each line (*Ans.* 3.30 in.^{2})

3-14. Using A36 steel determine the design tensile strength P_u of a W10×60 with two lines of $\frac{7}{8}$-in. bolts in each flange (three bolts in each line). Neglect block shear strength.

3-15. Determine the tensile design strength P_u of a W21×122 (A36 steel) if it has two lines of $\frac{3}{4}$-in. bolts in each flange (at least three bolts in each line). Neglect block shear. (*Ans.* 1163.2 k)

3-16. A single-angle tension member ($7 \times 4 \times \frac{5}{8}$) has two gage lines in its long leg and one in the short leg for $\frac{3}{4}$-in. bolts arranged as shown in the accompanying illustration. Determine the tensile design strength P_u of this member if A36 steel is used and if block shear is neglected.

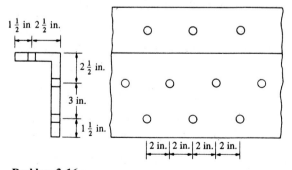

Problem 3-16

3-17. Determine the tensile design strength P_u of the pair of $5 \times 5 \times \frac{7}{8}$ angles made from A242 steel shown in the accompanying illustration. Standard gages are to be used as determined from the LRFD Manual for the $\frac{3}{4}$-in. bolts. Neglect block shear. There are at least 3 bolts in each line. (*Ans.* 622 k.)

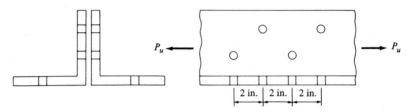

Problem 3-17

3-18. The $7 \times 4 \times \frac{3}{8}$ angle shown is connected with three $\frac{7}{8}$-in. bolts. If the angle consists of A36 steel, determine its block shear strength. Compare the results with the design tensile strength of the member.

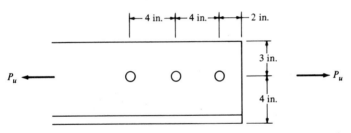

Problem 3-18

3-19. A W10×45 is connected at its ends with the plates shown. Determine the block shear strength of the member if A36 steel is used and if it is connected with six $\frac{3}{4}$-in. bolts in each flange as shown. Compare the result with the tensile design strength of the member. Do not check plate strength. (*Ans.* 430.9 k.)

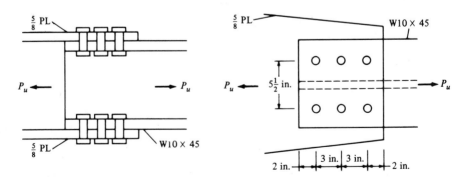

Problem 3-19

3-20. Repeat Prob. 3.14 if A242 steel is used and block shear is included.

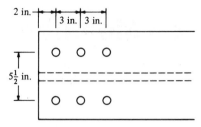

Problem 3-20

3-21. Compute the design strength of the $6 \times 6 \times \frac{1}{2}$ angle shown if it consists of A36 steel. Consider block shear as well as the tensile strength of the angle. Assume vertical leg is unconnected for determining U. (*Ans.* 186.3 k.)

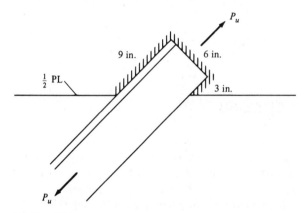

Problem 3-21

3-22. Compute the design strength of the bolted connection shown. Include block shear strength as well as the tensile strength of the angle. The angle is made of A36 steel and the bolts are $\frac{3}{4}$ in .

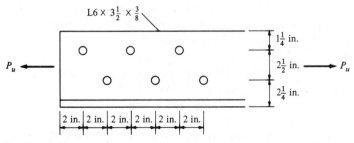

Problem 3-22

Chapter 4

Design of Tension Members

4-1 SELECTION OF SECTIONS

The determination of the design strengths of various tension members was presented in Chapter 3. In this chapter the selection of members to support given tension loads is described. Although the designer has considerable freedom in the selection, the resulting members should have the following properties: (a) compactness, (b) dimensions that fit into the structure with reasonable relation to the dimensions of the other members, and (c) connections to as many parts of the sections as possible to minimize shear lag.

The choice of member type is often affected by the type of connections used for the structure. Some steel sections are not very convenient to bolt together with the required gusset plates, while the same sections may be welded together with little difficulty. Tension members consisting of angles, channels, W or S sections will probably be used when the connections are made with bolts while plates, channels, and structural tees might be used for welded structures.

Various types of sections are selected for tension members in the examples to follow and in each case where bolts are used some allowance is made for holes. Should the connections be made *entirely* by welding, no holes have to be added to the net areas to give the required gross area. *The student should realize, however, that very often welded members may have holes punched in them for temporary bolting during field erection before the permanent field welds are made. These holes need to be considered in design.* It is also to be remembered that in LRFD Formula D1-2 ($P_u = \phi_t F_u A_e$) the value of A_e may be less than A_g depending on the arrangement of welds and on whether all of the parts of the members are connected.

The slenderness ratio of a member is the ratio of its unsupported length to its least radius of gyration. Steel specifications give preferable maximum values of slenderness ratios for both tension and compression members. The purpose of such limitations for tension members is to ensure the use of sections with sufficient stiffness to prevent undesirable lateral deflections or vibrations. Although tension members are not subject to buckling under normal loads stress reversal may occur during shipping and erection and perhaps due to wind or other loads. The specifications recommend that slenderness ratios be kept below certain maximum values in order that some minimum compressive strengths be provided in the members. For tension members other than rods the LRFD Specification (Part 6, Section B7) recommends maximum slenderness ratios of 300. Such members may be subjected to compressive forces caused by wind or earthquake. This specification states that these compressive forces may not exceed 50 percent of the design compressive strengths of the members. (It will later be learned that for slenderness ratios greater than 200 design compressive stresses will be very small, in fact less than 5.33 ksi for A36 steel.)

The *recommended* maximum slenderness ratio of 300 is not applicable to tension rods. Maximum L/r values for rods are left to the designer's judgment. If a maximum value of 300 was specified for them, they would seldom be used because of their extremely small radii of gyration.

Example 4-1 illustrates the design of a bolted tension member with a W section while Example 4-2 illustrates the selection of a bolted single-angle tension member. In both cases the LRFD Specification is used. The design strength P_u is the least of (a) $\phi_t F_y A_g$ or (b) $\phi_t F_u A_e$.

(a) To satisfy the first of these expressions the minimum gross area must be at least equal to the following

$$\text{min } A_g = \frac{P_u}{\phi_t F_y} \tag{1}$$

(b) To satisfy the second expression the minimum value of A_e must be at least

$$\text{min } A_e = \frac{P_u}{\phi_t F_u}$$

And since $A_e = UA_n$ the minimum value of A_n is

$$\text{min } A_n = \frac{\text{min } A_e}{U} = \frac{P_u}{\phi_t F_u U}$$

Then the minimum A_g for the second expression must at least equal the minimum value of A_n plus the estimated hole areas.

$$\text{min } A_g = \frac{P_u}{\phi_t F_u U} + \text{estimated hole areas} \tag{2}$$

The designer can substitute into Equations (1) and (2), taking the larger value of A_g so obtained for an initial size estimate. It is, however, well to notice

Transfer truss, 150 Federal Street, Boston, Mass. (Courtesy Owen Steel Company, Inc.)

that the maximum preferable slenderness ratio L/r is 300. From this value it is easy to compute the least permissible value of r for a particular design, that is, the value of r for which the slenderness ratio will be exactly 300. It is undesirable to consider a section whose least r is less than this value because its slenderness ratio would exceed the preferable maximum value of 300.

$$\min r = \frac{L}{300} \tag{3}$$

For the two examples to follow the author has substituted into the two ordinary gravity load expressions:

$$P_u = 1.4\,D$$

$$P_u = 1.2\,D + 1.6L$$

As the first of these expressions will not control unless the dead load is more than 8 times as large as the live load the first expression is omitted for the remaining problems in this text (unless $D > 8L$).

In Example 4-1 a W section is selected for a given set of tensile loads. For this first application of the tension design formulas the author has narrowed the problem down to one series of W shapes so the reader can concentrate on the application of the formulas and not become lost in considering W8s, W10s, W14s, and so on. Exactly the same procedure can be used for trying these other series as is used here for the W12.

Bridge over Allegheny River at Kittaning, PA. (Courtesy American Bridge Company).

In Example 4-2 a broader example is presented in that the lightest satisfactory angle in the LRFD Manual is selected for a given set of tensile loads.

EXAMPLE 4-1

Select a 30-ft-long W12 section of A36 steel to support a tensile service dead load $P_D = 140$ k and a tensile service live load $P_L = 80$ k. As shown in Fig. 4-1, the member is to have two lines of bolts in each flange for $\frac{7}{8}$-in. bolts (at least three in a line).

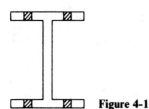

Figure 4-1

Solution

Considering the two ordinary load conditions

$$P_u = 1.4 \, P_D = (1.4)(140) = 196 \text{ k}$$

$$P_u = 1.2 \, P_D + 1.6 \, P_L = (1.2)(140) + (1.6)(80) = 296 \text{ k} \leftarrow$$

Computing the minimum A_g required

(1) $\min A_g = \dfrac{P_u}{\phi_t F_y} = \dfrac{296}{(0.90)(36)} = 9.14 \text{ in.}^2$

(2) $\min A_g = \dfrac{P_u}{\phi_t F_u U} + \text{estimated hole areas}$

Assume $U = 0.90$ and assume flange thickness is about 0.520 in. after looking at W12 sections in LRFD Manual which have areas of about 9.14 in.^{2}

$$\min A_g = \frac{296}{(0.75)(58)(0.90)} + (4)(1.00)(0.520) = 9.64 \text{ in.}^2 \leftarrow$$

(3)   Preferable $\min r = \dfrac{L}{300} = \dfrac{(12)(30)}{300} = 1.2 \text{ in.}$

Try W12 × 35 ($A_g = 10.3$ in.^{2}, $d = 12.50$ in., $b_f = 6.56$ in., $t_f = 0.520$ in., $r_y = 1.54$ in.)

Checking

(1) $P_u = \phi_t F_y A_g = (0.90)(36)(10.3) = 333.7 \text{ k} > 296 \text{ k}$ OK

(2) $P_u = \phi_t F_u U A_n$ with $U = 0.85$ since $\dfrac{b_f}{d} < \dfrac{2}{3}$,

and $A_n = 10.3 - (4)(1.00)(0.520) = 8.22 \text{ in.}^2$

$$P_u = (0.75)(58)(0.85)(8.22) = 303.9 \text{ k} > 296 \text{ k} \qquad \text{OK}$$

(3)　$\dfrac{L}{r} = \dfrac{(12)(30)}{1.54} = 234 < 300$　　　　　　　　　　　　　OK

$$\underline{\underline{\text{Use W12} \times 35}}$$

EXAMPLE 4-2

Design a 9-ft single-angle tension member to support a dead tensile working load of 30 k and a live tensile working load of 40 k. The member is to be
connected to one leg only with $\frac{7}{8}$-in. bolts (at least three in a line). Assume
that only one bolt is to be located at any one cross section. Use A36 steel.

Discussion

There are many different angles listed in the LRFD Manual which will
support the load for the conditions described. As a result it may seem rather
difficult to arrive at the absolutely lightest satisfactory section. It is possible, however, to set up a table with which the various possible sections may
be methodically considered for various angle thicknesses. In the table presented with this solution the author has listed various possible thicknesses
of angles which will provide sufficient areas. He has then found the lightest
angle for each thickness. Finally a glance at the completed table reveals the
lightest angle or the one with the smallest cross-sectional area. A similar
process could be followed for designing tension members from other steel
sections.

Solution

This problem is one which can be conveniently handled with the use of a
table and such a procedure is followed here for a few angle sizes. The limiting values are calculated as follows.

　　　Considering the two ordinary load conditions

$$P_u = (1.4)(30) = 42 \text{ k}$$

$$P_u = (1.2)(30) + (1.6)(40) = 100 \text{ k} \leftarrow$$

(1)　min A_g required $= \dfrac{P_u}{\phi_t F_y} = \dfrac{100}{(0.90)(36)} = 3.09 \text{ in.}^2$

(2)　From Part 6, Section B3 of Manual, $U = 0.85$

　　　min A_n required $= \dfrac{P_u}{\phi_t F_u U} = \dfrac{100}{(0.75)(58)(0.85)} = 2.70 \text{ in.}^2$

(3)　min $r = \dfrac{L}{300} = \dfrac{(12)(9)}{300} = 0.36 \text{ in.}$

Angle t (in.)	Area of one 1-in. bolt hole (in.²)	Gross area required = larger of $P_u/\phi_t F_y$ or $P_u/\phi_t F_u U$ + est. hole area (in.²)	Lightest angles available, their areas (in.²) and least radii of gyration (in.)
$\frac{5}{16}$	0.312	3.09	$6 \times 6 \times \frac{5}{16}\ (A = 3.65, r = 1.20)$
$\frac{3}{8}$	0.375	3.09	$6 \times 3\frac{1}{2} \times \frac{3}{8}\ (A = 3.42, r = 0.767)$
$\frac{7}{16}$	0.438	3.14	$5 \times 3 \times \frac{7}{16}\ (A = 3.31, r = 0.651)$ $4 \times 4 \times \frac{7}{16}\ (A = 3.31, r = 0.785)$
$\frac{1}{2}$	0.500	3.20	$4 \times 3 \times \frac{1}{2}\ (A = 3.25, r = 0.639)\ \leftarrow$ $3\frac{1}{2} \times 3\frac{1}{2} \times \frac{1}{2}\ (A = 3.25, r = 0.683)\ \leftarrow$
$\frac{5}{8}$	0.625	3.33	$4 \times 3 \times \frac{5}{8}\ (A = 3.98, r = 0.637)$

Use L4 $\times$ 3 $\times$ $\frac{1}{2}$ or L3$\frac{1}{2}$$\times$3$\frac{1}{2}$$\times$$\frac{1}{2}$

Example 4-3 illustrates the review of a tension member that is built up from two channels that are separated from each other. Included in the problem is the design of tie plates or tie bars to hold the channels together as shown in part (b) of Fig. 4-2. Part 6, Section D2 of the LRFD Specification states that tie plates (or perforated cover plates) must be used on the open sides of built-up tension members. In this specification empirical rules for the design of ties are presented. These rules are based on many decades of experience with built-up tension members.

In Fig. 4.2 the location of the bolts, that is, the usual gage for these channels, is shown as being $1\frac{3}{4}$ in. from the back of the channels. The LRFD Manual does not provide the usual gages except for angles. For other rolled steel shapes such as Cs, Ws, and Ss it is necessary to refer to a manufacturer's catalog or to old AISC Steel Manuals. The gages necessary to solve problems in this text are furnished by the author. (Though rather standard values of these gages are used by the steel industry for the various rolled shapes, they are not specified in the LRFD Manual in order to give steel fabricators more freedom in placing the holes.)

In Fig. 4-2 the distance between the lines of bolts connecting the tie plates to the channels can be seen to equal 8.50 in. The LRFD Specification (D2) states that the length of tie plates (lengths are always measured parallel to the long direction of the members in this text) may not be less than two-thirds the distance between the lines of connectors. Furthermore their thickness may not be less than one-fiftieth of this distance.

The minimum permissible width of tie plates (not mentioned in the specification) is the width between the lines of connectors plus the necessary edge distance on each side to keep the bolts from splitting the plate. For this example this minimum edge distance is taken as $1\frac{1}{2}$ in. from Part 6, Table J3.7 of the LRFD Specification. (Detailed information concerning edge distances for bolts and riv-

ets is provided in Chapter 12.) The plate dimensions are rounded off to agree
with the plate sizes available from the steel mills as given in the Bars and Plates
section of Part 1 of the LRFD Manual. It is much cheaper to select standard
thicknesses and widths rather than picking odd ones that will require cutting or
other operations.

The LRFD Specification (D2) provides a maximum spacing between tie
plates by stating that the L/r of each individual component of a built-up member
running along by itself between tie plates may not exceed 300. If the designer
substitutes into this expression ($L/r = 300$) the least r of an individual component of the built-up member, the value of L may be computed. This will be the
maximum spacing of the tie plates permitted by the LRFD Specification for this
member.

EXAMPLE 4-3

Two C12 × 30s (see Fig. 4-2) have been selected to support a dead tensile
working load of 120 k and a 240-k live tensile working load. The member is
30 ft long, consists of A36 steel, and has one line of at least three $\frac{7}{8}$-in. bolts

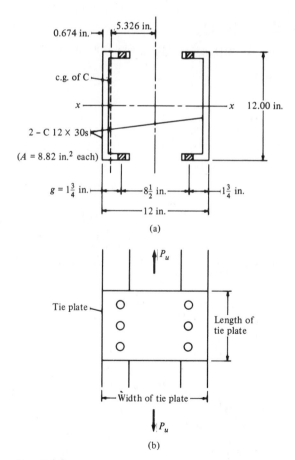

Figure 4-2

in each channel flange. Using the LRFD Specification determine if the member is satisfactory and design the necessary tie plates. Assume centers of bolt holes are $1\frac{3}{4}$ in. from the backs of the channels.

Solution

Using C12 × 30s (A_g = 8.82 in.^{2} each, t_f = 0.501 in., I_x = 162 in.^{4} each, I_y = 5.14 in.^{4} each, y axis 0.674 in. from back of C, r_y = 0.763 in.)

Load to be resisted

$$P_u = (1.2)(120) + (1.6)(240) = 528 \text{ k}$$

Design strengths

$$P_u = \phi_t F_y A_g = (0.90)(36)\,(2 \times 8.82) = 571.5 \text{ k} > 528 \text{ k} \qquad \text{OK}$$

$$A_n = [8.82 - (2.0)(0.501)]\,2 = 15.64 \text{ in.}^2$$

$$P_u = \phi_t F_u A_n U = (0.75)(58)(15.64)(0.85) = 578.3 \text{ k} > 528 \text{ k} \qquad \text{OK}$$

Slenderness ratio

$$I_x = (2)(162) = 324 \text{ in.}^4$$

$$I_y = (2)(5.14) + (2)(8.82)(5.33)^2 = 511 \text{ in.}^4$$

$$r_x = \sqrt{\frac{324}{17.64}} = 4.29 \text{ in.}$$

$$\frac{L}{r} = \frac{(12 \times 30)}{4.29} = 83.9 < 300$$

Design of tie plates (LRFD Part 6, Section D2)

Distance between lines of bolts = $12.00 - (2)(1\frac{3}{4}) = 8.50$ in.

Minimum length of tie plates = $(\frac{2}{3})(8.50) = 5.67$ in. (say 6 in.)

Minimum thickness of tie plates = $(\frac{1}{50})(8.50) = 0.17$ in. (say $\frac{3}{16}$ in.)

Minimum width of tie plates = $8.50 + (2)(1\frac{1}{2}) = 11.5$ in. (say 12 in.)
(Part 6, Table J3.7 LRFD Manual)

Maximum preferable spacing of tie plates
least r of one C = 0.763 in.

Maximum preferable $\dfrac{L}{r}$ = 300

$$\frac{(12)(L)}{0.763} = 300$$

$$L = 19.08 \text{ ft}$$

Use $\frac{3}{16} \times 6 \times 1$ ft 0 in. tie plates 15 ft 0 in. o.c.(on center)

4-2 RODS AND BARS

When rods and bars are used as tension members they may be simply welded at their ends, or they may be threaded and held in place with nuts. The LRFD nominal tensile design stress for threaded rods is given in their Table J3.2 and equals $\phi 0.75\ F_u$ and is to be applied to the gross area of the rod A_D computed with the major thread diameter, that is, the diameter to the outer extremity of the thread. The area required for a particular tensile load can then be calculated from the following expression:

$$A_D = \frac{P_u}{\phi 0.75\ F_u} \qquad \text{with } \phi = 0.75$$

In the table entitled "Threaded Fasteners Screw Threads" the LRFD Manual (Part 5) presents properties of standard threaded rods. Example 4-4 which follows illustrates the selection of a rod using this table. It will be noted that the LRFD Specification (Part 6, Section J1.5) states that the factored load P_u used for connection design may not be less than 10 k except for lacing, sag rods, or girts.

EXAMPLE 4-4

Using A36 steel and the LRFD Specification select a threaded rod of A36 steel to support a tensile working dead load of 10 k and a tensile working live load of 20 k.

Solution

$$P_u = (1.2)(10) + (1.6)(20) = 44 \text{ k} \leftarrow$$

$$A_D = \frac{P_u}{\phi\ 0.75\ F_u} = \frac{44}{(0.75)(0.75)(58)} = 1.35 \text{ in.}^2$$

Use $1\frac{3}{8}$-in.-diameter rod with 6 threads per in. $(A_D = 1.485 \text{ in.}^2)$

Sometimes upset rods are used where the ends are made larger than the regular rod and the threads are placed in the enlarged section so that the area at the root of the thread is larger than that of the regular bar. As a footnote to their Table J3.2 the LRFD Specification states that the nominal tensile capacity of the upset end is equal to $0.75\ F_u$ times the cross-sectional area (A_b) at its major thread diameter. This value must be larger than the nominal body area of the rod before upsetting times F_y. Upsetting permits the designer to use the entire area of the cross section; however, the use of upset bars is probably not economical and should be avoided unless a large order is being made.

A common example of the use of tension rods occurs in steel-frame industrial buildings which have purlins running between their roof trusses to support the roof surface. These types of buildings will also frequently have girts running between the columns along the vertical walls. Sag rods may be required to provide support for the purlins parallel to the roof surface and vertical support for the girts along the walls. For roofs with steeper slopes than 1 vertically to 4 horizontally, sag rods are often considered necessary to provide lateral support for the purlins, particularly where the purlins consist of steel channels. Steel channels are commonly used as purlins but they have very little resistance to lateral bending. Although the resisting moment needed parallel to the roof surface is small an extremely large channel is required to provide such a moment. The use of sag rods for providing lateral support to purlins made from channels is usually economical because of the bending weakness of channels about their y axes. For light roofs (as where trusses support corrugated steel roofs) sag rods will probably be needed at the one-third points if the trusses are more than 20 ft on centers. Sag rods at the midpoints are sufficient if the trusses are less than 20 ft on centers. For heavier roofs such as those made of slate, cement tile, or clay tile, sag rods will probably be needed at closer intervals. The one-third points will probably be necessary if the trusses are spaced at greater intervals than 14 ft and the midpoints will be satisfactory if truss spacings are less than 14 ft. Some designers assume that the load components parallel to the roof surface can be taken by the roof, particularly if it consists of corrugated steel sheets, and that tie rods are unnecessary. This assumption, however, is open to some doubt and definitely should not be followed if the roof is very steep.

It should be noted that out-of-straightness does not affect the strength of tension members very much because the tension loads tend to straighten the members. (The same statement cannot be made for compression members.) For this reason the LRFD specification is a little more liberal in its consideration of tension members including those subject to some compressive forces due to transient loads such as wind or earthquake. As previously discussed in Section 4-1, the LRFD Specification says that such members may be subjected to these types of compressive forces not to exceed 50 percent of the design compressive strength of the members.

New Albany Bridge crossing the Ohio River between Louisville, Ky., and New Albany, Ind. (Courtesy of the Lincoln Electric Company.)

Designers have to use their own judgment in limiting the slenderness values for rods as they will usually be several times the limiting values mentioned for other types of tension members. A common practice of many designers is to use rod diameters no less than 1/500th of their lengths to obtain some rigidity even though design calculations may permit much smaller sizes.

It is usually desirable to limit the minimum size of sag rods to $\frac{5}{8}$ in. because smaller rods than these are often injured during construction. The threads on smaller rods are quite easily injured by overtightening which seems to be a frequent habit of workmen. Sag rods are designed for the purlins of a roof truss in Example 4-5. The rods are assumed to support the simple beam reactions for the components of the gravity loads (roofing, purlins, snow and ice) parallel to the roof surface. Wind forces are assumed to act perpendicular to the roof surfaces and theoretically will not affect the sag rod forces. The maximum force in a sag rod will occur in the top sag rod because it must support the sum of the forces in the lower sag rods. It is theoretically possible to use smaller rods for the lower sag rods but this reduction in size will probably be impractical.

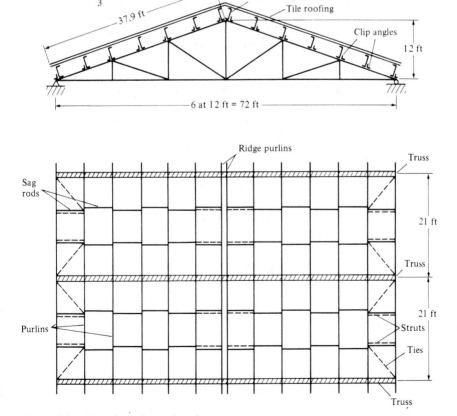

Figure 4-3   Plan of two bays of roof.

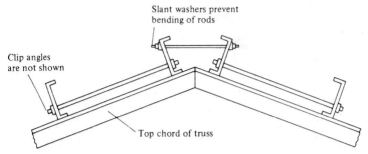

Figure 4-4   Details of sag rod connections.

EXAMPLE 4-5

Design the sag rods for the purlins of the truss shown in Fig. 4-3. Purlins are to be supported at their one-third points between the trusses which are spaced 21 ft on centers. Use A36 steel and assume a minimum size rod of $\frac{5}{8}$ in. is permitted. A clay tile roof weighing 16 psf (0.77 kN/m^2) of roof surface is used and supports a snow load of 20 psf (0.96 kN/m^2) of horizontal projection of roof surface. Details of the purlins and the sag rods and their connections are shown in Figs. 4-3 and 4-4. In these figures the dotted lines represent a common practice of using ties and struts in the end panels in the plane of the roof to give greater resistance to loads located on one side of the roof (a loading situation which might occur when snow is blown off one side of the roof during a severe windstorm).

Solution

Gravity loads in psf of roof surface are as follows.

$$\text{Purlins} = \frac{7 \times 11.5}{37.9} = 2.1 \text{ psf}$$

$$\text{Snow} = 20 \frac{3}{\sqrt{10}} = 19.0 \text{ psf}$$

$$\text{Tile roofing} = 16.0 \text{ psf}$$

$$w_u = (1.2)(2.1 + 16.0) + (0.5)(19.0) = 31.22 \text{ psf}$$

$$w_u = (1.2)(2.1 + 16.0) + (1.6)(19.0) = 52.12 \text{ psf} \leftarrow$$

$$\text{Component of loads} \perp \text{to roof surface} = \left(\frac{1}{\sqrt{10}}\right)(52.12) = 16.48 \text{ psf}$$

$$\text{Load on sag rod} = (37.9)(7)(16.48) = 4372 \text{ lb}$$

$$A_D = \frac{4.372}{(0.75)(0.75)(58)} = 0.134 \text{ in.}^2$$

Use $\frac{5}{8}$-in. rod (minimum practical size),
11 threads per in. ($A_D = 0.307$ in.^{2})

Force in tie rod between ridge parlins

$$T = \frac{\sqrt{10}}{3}(4.372) = 4.608 \text{ k}$$

$$A_D = \frac{4.608}{(0.75)(0.75)(58)} = 0.141 \text{ in.}^2$$

Use $\frac{5}{8}$-in. rod

4-3 EYEBAR TENSION MEMBERS

Until the early years of the twentieth century nearly all bridges in the United States were pin-connected, but today pin-connected bridges are used infrequently because of the advantages of bolted and welded connections. One trouble with the old pin-connected trusses was the wearing of the pins in the holes which caused looseness of the joints.

Pin-connected eyebars are used occasionally today as tension members for long-span bridges and as hangars for some types of bridges and other structures where they are normally subjected to very large dead loads. As a result the eyebars are usually prevented from rattling and wearing under live loads.

Eyebars are generally not made by forging but instead are made by thermally cutting them from plates. The detailed proportions required for the bars are contained in Part 6, Section D3 of the LRFD Specification. These proportions are based on long experience in the steel industry. In this same section of the Manual limit states expressions for the tensile, shearing, and bearing strengths of eyebars are provided.

It has been found that when eyebars and pin-connected members are made from steels with yield stresses greater than 70 ksi there may be a possibility of "dishing." For this reason the LRFD Specification requires stockier member proportions for such situations.

4-4 DESIGN FOR FATIGUE LOADS

It is not likely that fatigue stresses will be a problem in the average building frame because the changes in load in such structures usually occur only occasionally and produce relatively minor stress variations. Should there, however, be frequent variations or even reversals of stress the matter of fatigue must be considered. Fatigue can be a problem in buildings when crane runway girders or heavy vibrating or moving machinery or equipment are supported.

If steel members are subjected to loads that are applied and then removed or changed significantly many thousands of times cracks may occur and they may spread so much as to cause fatigue failures. These cracks tend to occur in places where there are stress concentrations such as where there are holes or rough or damaged edges or poor welds. *Furthermore they are more likely to occur in tension members.*

The LRFD Specification in its Appendix K provides a simple design method for considering repeated loads. With this procedure the number of stress cycles, the expected range of stress (that is, the difference between the maximum and minimum estimated stresses), and the type and location of the member are considered. Based on this information a maximum allowable stress range is given *for service or working loads.*

In detail the maximum calculated stress in a member by the LRFD Specification may not exceed the basic nominal stress in that type of member nor may its maximum range of stress exceed the permissible stress range provided in Appendix K of the Specification.

If it is anticipated that there will be less than 20,000 cycles of loading no consideration needs to be given to fatigue. If the number of loading cycles exceeds 20,000 a permissible stress range is determined as follows.

1. The loading condition is determined from Table A-K4.1 of Appendix K of the LRFD Specification. For instance, if it is anticipated that there will be no less than 100,000 cycles of loading (that is, roughly 10 applications a day for 25 years) and no more than 500,000 cycles, loading condition 2 is to be used as determined from the table.
2. The type and location of material is selected from Fig. A-K4.1 of the Appendix. If a tension member consists of two angles fillet-welded to a plate it should be classified as illustrative example 17 in the figure. (The term fillet weld will be defined in Chapter 14. In such a weld one member is lapped over another and they are welded together.)
3. From Table A-K4.2 the stress category A, B, C, D, E, or F is selected. For a fillet welded tension connection classified as illustrative example 17 the stress category is E.
4. Finally from Table A-K4.3 of the Appendix the allowable stress range for the service loads for stress category E and loading condition 2 is $F_{sr} = 13$ ksi.

Example 4-6 presents the design of a two-angle tension member subjected to fluctuating loads using Appendix K of the LRFD Specification. Stress fluctuations and reversals are an everyday problem in the design of bridge structures. The 1983 AASHTO Specifications in their Article 10.3 provide allowable stress ranges determined in a manner very similar to those of the LRFD Specification.

EXAMPLE 4-6

An 18-ft member is to consist of a pair of equal leg angles with fillet welded end connections. The dead load working tensile force is 30 k while it is estimated that the working live load force may be applied 250,000 times and may vary from a compression of 12 k to a tension of 65 k. Select the angles using A36 steel and the LRFD Specification.

Solution

Referring to Appendix K of the LRFD Specification and selecting the following values:

From Table A-K4.1—loading condition 2
From Fig. A-K4.1 and Table A-K4.2—illustrative example 17 and stress category E
Therefore, from Table A-K4.3—allowable stress range is 13 ksi

Range of P_u

For maximum tension

$$P_u = (1.2)(30) + (1.6)(65) = 140 \text{ k}$$

For compression

$$P_u = (1.4)(30) = 42 \text{ k}$$

$$P_u = (1.2)(30) + (1.6)(-12) = +16.8 \text{ k}$$

Therefore, member remains in tension

Estimated section size

$$A_g = \frac{P_u}{\phi_t F_y} = \frac{140}{(0.90)(36)} = 4.32 \text{ in.}^2$$

Try 2Ls $4 \times 4 \times \frac{5}{16}$ ($A = 4.80$ in.^{2}, $r = 1.24$ in.)

$$\text{Max service load tension } f_t = \frac{30 + 65}{4.80} = 19.79 \text{ ksi}$$

$$\text{Min service load tension } f_t = \frac{30 - 12}{4.80} = 3.75 \text{ ksi}$$

Actual stress range = 16.04 ksi

>13 ksi   NG (means no good)

Try 2Ls $4 \times 4 \times \frac{1}{2}$ ($A = 7.50$ in.^{2}, $r = 1.22$ in.)

$$\text{Max service load } f_t = \frac{30 + 65}{7.50} = 12.67 \text{ ksi}$$

$$\text{Min service load } f_t = \frac{30 - 12}{7.50} = 2.40 \text{ ksi}$$

Actual stress range = 10.27 ksi

<13 ksi   OK

PROBLEMS

4-1. Select the lightest W12 section available to support working tensile loads of P_D = 200 k and P_L = 220 k using A36 steel. The member is to be 25 ft long and is assumed to have two lines of holes for 1-in. bolts in each flange. There will be at least three bolts in each line. (*Ans.* W12×65)

4-2. Repeat Prob. 4-1 using A441 steel.

4-3. Select the lightest W14 available with A36 steel to support a factored tensile load P_u of 300 k. Assume there are two lines of $\frac{7}{8}$-in. bolts in each flange (at least three bolts in each line.) The member is to be 32 ft. long. (*Ans.* W14×34)

4-4. Select the lightest American Standard channel that will safely support the service tensile loads P_D = 75 k and P_L = 100 k. The member is 15 ft long and is assumed to have one line of holes for 1-in. bolts in each flange. Use A36 steel and assume there are at least three holes in each line.

4-5. A welded tension member is to support a design load P_u = 620 k and is to consist of two channels placed 12 in. back to back with the flanges turned in. Select the lightest standard channels available using A36 steel. Assume U = 0.87. The member is to be 30 ft long. (*Ans.* 2 Cs15×33.9)

4-6. A36 steel is to be used in selecting a single angle member to resist service tensile loads of P_D = 80 k and P_L = 90 k. The member is to be 18 ft long and is assumed to be connected with one line of four $\frac{7}{8}$-in. bolts in one leg.

4-7. Repeat Prob. 4-6 using a pair of angles and A36 steel. Assume angle legs are touching and assume one hole for a $\frac{7}{8}$-in. bolt is to be taken out of each angle. Also assume U = 0.85. (*Ans.* 2Ls 4×4×$\frac{1}{2}$ or 2Ls 5×3×$\frac{1}{2}$ long legs back to back)

4-8. Design member L_2L_3 of the truss shown in the accompanying illustration. It is to consist of a pair of angles with a $\frac{3}{8}$-in. gusset plate between the angles at each joint. Use A36 steel and assume one line of three $\frac{3}{4}$-in. bolts in each angle leg. Consider only the angles shown in the double-angle tables of the Manual. For each load P_D = 18 k and P_r = 10 k (roof load).

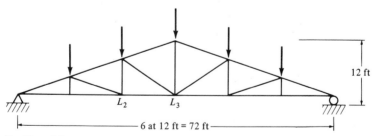

Problem 4-8

4-9. Select a single-angle tension member to resist the service loads P_D = 60 k and P_L = 60 k. The member is to be 15 ft long and is to be connected with one line of four $\frac{7}{8}$-in. bolts. Assume F_y = 40 ksi and F_u = 60 ksi. (*Ans.* L6×6×$\frac{7}{16}$)

4-10. Repeat Prob. 4-5 assuming that one line of $\frac{7}{8}$-in. bolts will be used in each flange with at least three bolts in each line. Also design tie plates. Assume distance or gage from back of channel to center of line of bolts is 2 in. U is to be determined from LRFD specification B3.

4-11. A tension member is to consist of four equal leg angles arranged as shown in the accompanying illustration to support the service loads $P_D = 200$ k and $P_L = 360$ k. The member is assumed to be 30 ft long and is to have one line of three $\frac{7}{8}$-in. bolts in each leg. Design the member with A36 steel including the necessary tie plates. (*Ans.* 4Ls 6×6×$\frac{9}{16}$)

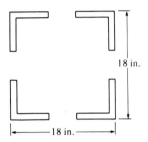

18 in.

18 in.

Problem 4-11

4-12. Select a threaded round rod as a hanger to resist the service loads $P_D = 8$ k and $P_L = 7$ k using A36 steel.

4-13. The horizontal thrust at the base of the three-hinged arch shown in the accompanying illustration is to be resisted by a tie rod of A36 steel. What size threaded round tie rod should be used if the arch supports the working loads shown in the figure? (*Ans.* 2-in. rod)

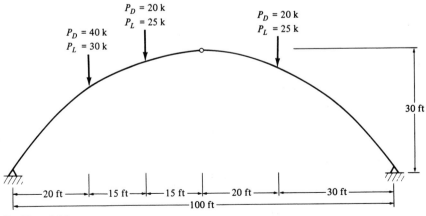

$P_D = 20$ k
$P_L = 25$ k

$P_D = 20$ k
$P_L = 25$ k

$P_D = 40$ k
$P_L = 30$ k

30 ft

20 ft — 15 ft — 15 ft — 20 ft — 30 ft
100 ft

Problem 4-13

4-14. The roof trusses for a particular industrial building are spaced 21 ft on centers, have a roof covering weighing an estimated 6 psf of roof surface, and have purlins spaced as shown in the accompanying illustration and weighing an estimated 3 psf of roof surface. Design sag rods with A36 steel assuming a snow load of 30 psf of horizontal roof surface. The sag rods are to be used at the one-third points.

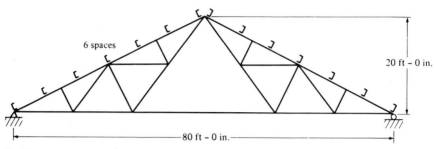

Problem 4-14

Chapter 5

Introduction to Axially Loaded Compression Members

5-1 GENERAL

There are several types of compression members, the column being the best known. Among the other types are the top chords of trusses, bracing members, the compression flanges of rolled beams and built-up beam sections, and members that are subjected simultaneously to bending and compressive loads. Columns are usually thought of as being straight vertical members whose lengths are considerably greater than their thicknesses. Short vertical members subjected to compressive loads are often called struts or simply compression members; however, the terms *column* and *compression member* will be used interchangeably in the pages that follow.

There are two significant differences between tension and compression members. These are

1. Whereas tensile loads tend to hold a member straight compressive loads tend to bend them out of the plane of the loads (a serious situation).
2. The presence of bolt or rivet holes in tension members reduces the areas available for resisting loads; but in compression members the bolts or rivets are assumed to fill the holes (although there may be some very slight initial slippage until the bolts or rivets bear against the adjoining material) and the entire gross areas are available for resisting load.

Tests on all but the shortest columns show that they will fail at P/A stresses well below the elastic limit of the column material because of their tendency to buckle or bend laterally. For this reason their design stresses are reduced in some relation to the danger of buckling. The longer a column becomes for the same cross section the greater becomes its tendency to buckle and the smaller becomes the load it will support. The tendency of a member to buckle is usually measured by its *slenderness ratio,* which has previously been defined as the ratio of the length of the member to its least radius of gyration. The tendency to buckle is also affected by such factors as the types of end connections, eccentricity of load application, imperfection of column material, initial crookedness of columns, residual stresses from manufacture, etc.

The loads supported by a building column are applied by the column section above and by the connections of other members directly to the column. The ideal situation is for the loads to be applied uniformly across the column with the center of gravity of the loads coinciding with the center of gravity of the column. Furthermore, it is desirable for the column to have no flaws, to consist of a homogeneous material, and to be perfectly straight; but these situations are obviously impossible to achieve.

Loads that are exactly centered over a column are referred to as *axial* or *concentric loads.* The dead loads may or may not be concentrically placed over an interior building column and the live loads may never be centered. For an outside column the load situation is probably even more eccentric as the center of gravity of the loads will usually fall well on the inner side of the column. In other words, it is doubtful that a perfect axially loaded column will ever be encountered in practice.

The other desirable situations are also impossible to achieve because of the following: imperfections of cross-sectional dimensions, residual stresses, holes

Guggenheim Laboratories, Princeton University, Princeton, N.J. (Courtesy of Bethlehem Steel Corporation.)

punched for bolts or rivets, erection stresses, and transverse loads. It is difficult to take into account all of these imperfections in a formula.

Slight imperfections in tension members and beams can be safely disregarded as they are of little consequence. On the other hand, slight defects in columns may be of major significance. A column that is slightly bent at the time it is put in place may have serious bending moments. Obviously a column is a more critical member in a structure than is a beam or tension member because minor imperfections in materials and dimensions mean a great deal. This fact can be illustrated by a bridge truss that has some of its members damaged by a truck. The bending of tension members probably will not be serious as the tensile loads will tend to straighten those members; but the bending of any compression members is a serious matter, as compressive loads will tend to magnify the bending in those members.

The preceding discussion should clearly show that column imperfections cause them to bend and the designer must consider stresses due to those moments as well as due to axial loads. Chapters 5 to 7 are limited to a discussion of axially loaded columns while members subjected to a combination of axial loads and bending loads are discussed in Chapter 11.

5-2 RESIDUAL STRESSES

Research at Lehigh University has shown that residual stresses and their distribution are very important factors affecting the strength of axially loaded steel columns. These stresses are of particular importance for columns with slenderness ratios varying from approximately 40 to 120, a range that includes a very large percentage of practical columns. A major cause of residual stress is the uneven cooling of shapes after hot-rolling. For instance, in a W shape the outer tips of the flanges and the middle of the web cool quickly while the areas at the intersection of the flange and web cool more slowly.

The quicker cooling parts of the sections when solidified resist further shortening while those parts that are still hot tend to shorten further as they cool. The net result is that the areas which cooled more quickly have residual compressive stresses while the slower cooling areas have residual tensile stresses. The magnitude of these stresses varies from about 10 to 15 ksi (69 to 103 MPa) although some values greater than 20 ksi (138 MPa) have been found.

When rolled-steel column sections with their residual stresses are tested, their proportional limits are reached at P/A values of only a little more than half of their yield stresses and the stress-strain relationship is nonlinear from there up to the yield stress. Because of the early localized yielding occurring at some points of the column cross sections, buckling strengths are appreciably reduced. Reductions are greatest for columns with slenderness ratios varying from approximately 70 to 90 and may possibly be as high as 25 percent.[1]

As a column load is increased, some parts of the column will quickly reach

---

[1]L.S. Beedle and L. Tall, "Basic Column Strength," *Proc. ASCE* 86 (July 1960), pp. 139–173.

the yield stress and go into the plastic range because of residual compression stresses. The stiffness of the column will be reduced and become a function of the part of the cross section that is still elastic. A column with residual stresses will behave as though it has a reduced cross section. This reduced section or elastic portion of the column will change as the applied stresses change. The buckling calculations for a particular column with residual stresses can be handled by using an effective moment of inertia I_e of the elastic portion of the cross section or by using the tangent modulus. For the usual sections used as columns the two methods give fairly close results.

Residual stresses may also be caused during fabrication when cambering is performed by cold bending and due to cooling after welding. Cambering is the bending of a member in one direction so that it won't look so bad when the service loads bend it in the other direction. For instance, we may bend a beam initially upward so that it will be approximately straight when its normal gravity loads are applied. Welding can produce severe residual stresses in columns which actually can approach the yield point in the vicinity of the weld. Another important fact is that columns may actually be appreciably bent by the welding process decidedly affecting their load carrying ability.

5-3 SECTIONS USED FOR COLUMNS

Theoretically an endless number of shapes can be selected to safely resist a compressive load in a given structure. From a practical viewpoint, however, the number of possible solutions is severely limited by such considerations as sections available, connection problems, and type of structure in which the section is to be used. The paragraphs that follow are intended to give a brief résumé of the sections which have proved to be satisfactory for certain conditions. These sections are shown in Fig. 5-1 and the letters in parentheses in the paragraphs to follow refer to the parts of this figure.

The sections used for compression members are usually similar to those used for tension members with certain exceptions. The exceptions are caused by the fact that the strengths of compression members vary in some inverse relation to the slenderness ratios and stiff members are required. Individual rods, bars, and plates are usually too slender to make satisfactory compression members unless they are very short and lightly loaded.

Single-angle members (a) are satisfactory for bracing and compression members in light trusses. Equal-leg angles may be more economical than unequal-leg angles because their least r values are greater for the same area of steel. The top chord members of bolted roof trusses might consist of a pair of angles back to back (b). There will often be a space between them for the insertion of a gusset or connection plate at the joints necessary for connections to other members. An examination of this section will show that it is probably desirable to use unequal-leg angles with the long legs back to back to give a better balance between the r values about the x and y axes.

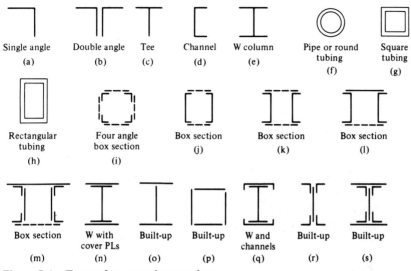

Figure 5-1   Types of compression members.

If roof trusses are welded, gusset plates may be unnecessary and structural tees (c) might be used for the top chord compression members because the web members can be welded directly to the stems of the tees. Single channels (d) are not satisfactory for the average compression member because of their almost negligible r values about their web axes. They can be used if some method of providing extra lateral support in the weak direction is available. The W shapes (e) are the most common shapes used for building columns and for the compression members of highway bridges. Their r values although far from being equal about the two axes are much more nearly balanced than for channels.

Several famous bridges constructed during the nineteenth century (such as the Firth of Forth Bridge in Scotland and Ead's Bridge in St. Louis) made extensive use of tube-shaped members. Their use, however, declined due to connection problems and manufacturing costs, but with the development of economical welded tubing their use is again increasing (although the tube shapes of today are very small compared with the giant ones used in those early steel bridges).

For small and medium loads, pipe sections or round tubing (f) are quite satisfactory. They are often used as columns in long series of windows, as short columns in warehouses, as columns for the roofs of covered walkways, in the basements and garages of residences, etc. Pipe columns have the advantage of being equally rigid in all directions and are usually very economical unless moments are large. The LRFD Handbook furnishes the sizes of these sections and classifies them as being standard, extra strong, and double extra strong.

Square and rectangular tubing (g) and (h) have not been used to a great extent for columns until recently. In fact, for many years only a few steel mills manufactured steel tubing for structural uses. Perhaps the major reason why tubing was not used to a great extent was the difficulty of making connections with rivets or bolts. This problem has been fairly well eliminated, however, by the advent of modern welding. The use of tubing for structural purposes by architects

Bents fabricated from tubing sections in New Jersey. (Courtesy of Bethlehem Steel Corporation.)

and engineers in the years to come will probably be greatly increased for several reasons. These include

1. The most efficient compression member is one that has a constant radius of gyration about its centroid, a property available in round tubing. Square tubing is the next most efficient compression member.
2. Their smooth surfaces permit easier painting.
3. They have excellent torsional resistance.
4. The surfaces of tubing are quite attractive.
5. When exposed the wind resistance of round tubing is only about two-thirds of that of flat surfaces of the same width.

A slight disadvantage that comes into play in certain cases is that the ends of tubing may have to be sealed to protect their inaccessible inside surfaces from corrosion. Although making very attractive exposed members for beams, tubing is at a definite weight disadvantage as compared with W sections which have so much larger resisting moments for the same weights.

Where compression members are designed for very large structures it may be necessary to use built-up secitons. Built-up sections are needed where the members are long and support very heavy loads and/or when there are connection advantages. Generally speaking, a single shape such as a W section is more economical than a built-up section having the same cross-sectional area. With heavy column loads high-strength steels can frequently be used with very economical results if their increased strength permits the use of W sections rather than built-up members.

When built-up sections are used they must be connected on their open sides with some type of lacing (also called lattice bars) to hold the parts together in their proper positions and to assist them in acting together as a unit. The ends of these members are connected with tie plates (also called batten plates or stay plates). Several types of lacing for built-up compression members are shown in Fig. 6-10.

The dotted lines in Fig. 5-1 represent lacing or discontinuous parts and the solid lines represent parts that are continuous for the full length of the members. Four angles are sometimes arranged as shown in (i) to produce large r values. This type of member may often be seen in towers and in crane booms. A pair of channels (j) is sometimes used as a building column or as a web member in a large truss. It will be noted that there is a certain spacing for each pair of channels at which their r values about the x and y axes are equal. Sometimes the channels may be turned out as shown in (k).

A section well suited for the top chords of bridge trusses is a pair of channels with a cover plate on top (l) and with lacing on the bottom. The gusset plates at joints are conveniently connected to the insides of the channels and may also be used as splices. When the largest channels available will not produce a top chord member of sufficient strength a built-up section of the type shown in (m) may be used.

When the rolled shapes do not have sufficient strength to resist the column loads in a building or the loads in a very large bridge truss, their areas may be increased by adding plates to the flange (n). In recent years it has been found that for welded construction a built-up column of the type shown in part (o) is a more satisfactory shape than a W with welded cover plates (n). It seems that in bending (as where a beam frames into the flange of a column) it is difficult to efficiently transfer tensile force through the cover plate to the column without pulling the plate away from the column. For very heavy column loads a welded box section of the type shown in (p) has proved to be quite satisfactory. Some other built-up sections are shown in parts (q) through (s). The built-up sections shown in parts (n) through (q) have an advantage over those shown in parts (i) through (m) in that they do not require the expense of the lattice work necessary in the former. Lateral shearing forces are negligible for the single column shapes and for the nonlatticed built-up sections, *but they are definitely not negligible for the built-up latticed columns.*

Today *composite columns* are being increasingly used. These columns usually consist of steel pipe and structural tubing filled with concrete or of W shapes encased in concrete usually square or rectangular in cross section. These columns are discussed in Chapter 18.

5-4 DEVELOPMENT OF COLUMN FORMULAS

The use of columns goes back before the dawn of history but it was not until 1729 that a paper was published on the subject by Pieter van Musschenbroek, a Dutch mathematician.[2] He presented an empirical column formula for estimating the strength of rectangular columns. A few years later in 1757 Leonhard Euler, a

---

[2]L.S. Beedle et al., *Structural Steel Design* (New York: Ronald Press, 1964), p. 269.

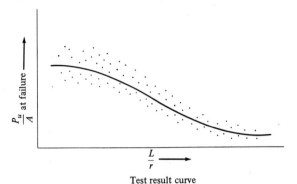

Test result curve

Figure 5-2   Text result curve.

Swiss mathematician, wrote a paper of great value concerning the buckling of columns. He was probably the first person to realize the significance of buckling. The Euler formula, the most famous of all column expressions, which is presented in Section 5-5, marked the real beginning of theoretical and experimental investigation of columns.

Engineering literature is filled with formulas developed for ideal column conditions but these conditions are not encountered in actual practice. Consequently practical column design is based primarily on formulas that have been developed to fit with reasonable accuracy test-result curves. The reasoning behind this procedure is simply the fact that the independent derivation of column expressions does not yield formulas that give results comparing closely with test-result curves for all slenderness ratios.

The testing of columns with various slenderness ratios results in a scattered range of values such as those shown by the broad band of dots in Fig. 5-2. The dots will not fall on a smooth curve even if all of the testing is done in the same laboratory because of the difficulty of exactly centering the loads, lack of perfect uniformity of the materials, varying dimensions of the sections, residual stresses, end restraint variations, etc. The usual practice is to attempt to develop formulas which give results represented by an approximate average of the test results. The student should also realize that laboratory conditions are not field conditions and column tests probably give the limiting values of column strengths.

The magnitudes of the yield stresses of the sections tested are quite important for short columns as their failure stresses are close to those yield stresses. For

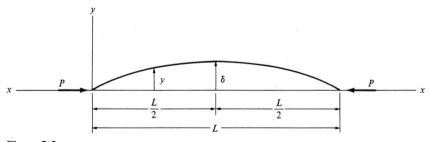

Figure 5-3

columns with intermediate slenderness ratios the yield stresses are of lesser importance in their effect on failure stresses and they are of no significance for long slender columns. For intermediate range columns residual stresses have more effect on the results while the failure stresses for long slender columns are very sensitive to end support conditions. In addition to residual stresses and non-linearity of material another dominant factor in its effect on column strength is member out-of-straightness.

5-5 DERIVATION OF THE EULER FORMULA

The Euler formula is derived in this section for a straight, concentrically loaded, homogeneous, long slender column with rounded ends. It is assumed that this perfect column has been laterally deflected by some means as shown in Fig. 5-3 and that if the concentric load P was removed the column would straighten out completely.

The x and y axes are located as shown in the figure. As the bending moment at any point in the column is $-Py$ the equation of the elastic curve can be written as follows.

$$EI \frac{d^2y}{dx^2} = -Py$$

For convenience in integration both sides of the equation are multiplied by $2\,dy$ and the integration is performed.

$$EI\, 2 \frac{dy}{dx}\, d\, \frac{dy}{dx} = -2Py\,dy$$

$$EI \left(\frac{dy}{dx}\right)^2 = -Py^2 + C_1$$

When $y = \delta$, $dy/dx = 0$, and the value of C_1 will equal $P\delta^2$ and

$$EI \left(\frac{dy}{dx}\right)^2 = -Py^2 + P\delta^2$$

The preceding expression is arranged more conveniently as follows.

$$\left(\frac{dy}{dx}\right)^2 = \frac{P}{EI}(\delta^2 - y^2)$$

$$\frac{dy}{dx} = \sqrt{\frac{P}{EI}} \sqrt{\delta^2 - y^2}$$

$$\frac{dy}{\sqrt{\delta^2 - y^2}} = \sqrt{\frac{P}{EI}}\, dx$$

Integrating this expression, the result is

$$\arc \sin \frac{y}{\delta} = \sqrt{\frac{P}{EI}}\,x + C_2$$

When $x = 0$ and $y = 0$, $C_2 = 0$. The column is bent into the shape of a sine curve expressed by the equation

$$\arc \sin \frac{y}{\delta} = \sqrt{\frac{P}{EI}}\,x$$

When $x = L/2$, $y = \delta$, resulting in

$$\frac{\pi}{2} = \frac{L}{2}\sqrt{\frac{P}{EI}}$$

In this expression P is the *critical buckling load* or the maximum load which the column can support before it becomes unstable. Solving for P

$$P = \frac{\pi^2 EI}{L^2}$$

This expression is the Euler formula but it is usually written in a little different form involving the slenderness ratio. Since $r = \sqrt{I/A}$ and $r^2 = I/A$ and $I = r^2 A$, the Euler formula may be written as

$$\frac{P}{A} = \frac{\pi^2 E}{(L/r)^2} = F_e$$

In the LRFD Manual the Euler stress is called F_e.

The student should carefully note that the buckling load determined from the Euler equation is independent of the strength of the steel used. The equation is only useful if the end support conditions are carefully considered. The results obtained by application of the formula to specific examples compare very well with test results for centrally loaded, long slender columns with rounded ends. Designers, however, do not encounter perfect columns of this type. The columns with which they work do not have rounded ends and are not free to rotate because their ends are bolted, riveted, or welded to other members. These practical columns have different amounts of restraint against rotation varying from slight restraint to almost fixed conditions. For the actual cases encountered in practice where the ends are not free to rotate, different length values can be used in the formula and more realistic values will be obtained.

To successfully use the Euler equation for practical columns the value of L should be the distance between points of inflection in the buckled shape. This distance is referred to as the *effective length* of the column. For a pinned-end column (whose ends can rotate but cannot translate) the points of inflection or zero moment are located at the ends a distance L apart. For columns with different end conditions the effective lengths may be entirely different.

A heavy column showing base plate with combination welded and high-strength bolted connection. (Courtesy of Bethlehem Steel Corporation.)

5-6 END RESTRAINT AND EFFECTIVE LENGTHS OF COLUMNS

End restraint and its effect on the load-carrying capacity of columns is a very important subject indeed. Columns with appreciable end restraint can support considerably more load than those with little end restraint as at hinged ends.

The effective length of a column was defined in the last section as the distance between points of zero moment in the column, that is, the distance between its inflection points. In steel specifications the effective length of a column is referred to as KL where K is the *effective length factor*. K is the number that must be multiplied by the length of the column to obtain its effective length. Its size depends on the rotational restraint supplied at the ends of the column and upon the resistance to lateral movement provided.

The concept of effective lengths is simply a mathematical method of taking a column whatever its end and bracing conditions and replacing it with an equivalent pinned-end braced column. A complex buckling analysis could be made for a frame to determine the critical stress in a particular column. The K factor is determined by finding the pinned-end column with an equivalent length that pro-

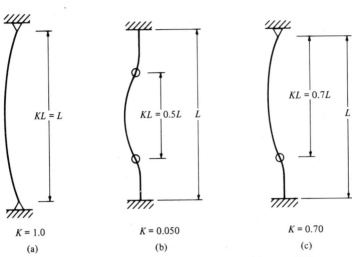

$KL = L$

$KL = 0.5L$ L

$KL = 0.7L$

L

$K = 1.0$

$K = 0.050$

$K = 0.70$

(a)

(b)

(c)

Figure 5-4   Effective lengths for columns in braced frames (sidesway prevented)

vides the same critical stress. The K factor procedure is a method of making simple solutions for complicated frame buckling problems.

Columns with different end conditions have entirely different effective lengths. For this initial discussion it is assumed that no sidesway or joint translation is possible. Sidesway or joint translation means that one or both ends of a column can move laterally with respect to each other. Should a column be connected with frictionless hinges as shown in part (a) of Fig. 5-4 its effective length would be equal to the actual length of the column and K would equal 1.0. If there were such a thing as a perfectly fixed ended column, its points of inflection (or points of zero moment) would occur at its one-fourth points and its effective length would equal $L/2$ as shown in part (b) of Fig. 5-4. As a result its K value would equal 0.50.

Obviously the smaller the effective length of a particular column, the smaller its danger of lateral buckling and the greater its load-carrying capacity. In part (c) of Fig. 5-4 a column is shown with one end fixed and one end pinned. The K for this column is theoretically 0.70.

Actually there are no perfect pin connections nor any perfect fixed ends, and the usual column falls in between the two extremes. This discussion would seem to indicate that column effective lengths always vary from an absolute minimum of $L/2$ to an absolute maximum of L but there are exceptions to this rule. An example is given in Fig. 5-5 (a) where a simple bent is shown. The base of each of the columns is pinned and the other end is free to rotate and move laterally (called sidesway). Examination of this figure will show that the effective length will exceed the actual length of the column as the elastic curve will theoretically take the shape of the curve of a pinned-end column of twice its length and K will theoretically equal 2.0. Notice in part (b) of the figure how much smaller the lateral deflection of column AB would be if it were pinned both top and bottom so as to prevent sidesway.

Structural steel columns serve as parts of frames, and these frames are

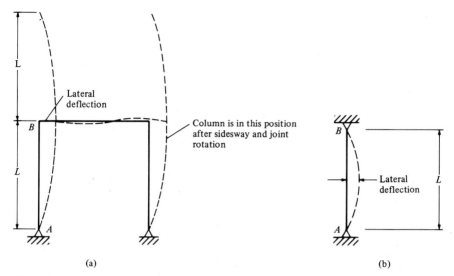

Figure 5-5

sometimes *braced* and sometimes *unbraced*. A braced frame is one for which sideway or joint translation is prevented by means of bracing, shear walls, or lateral support from adjoining structures. An unbraced frame does not have any of these types of bracing supplied and must depend on the stiffness of its own members to prevent lateral buckling. For braced frames K values can never be greater than 1.0 but for unbraced frames the K values will always be greater than 1.0 because of sideway.

Table C-C2.1 of the "Commentary of the AISC LRFD Specification" gives recommended effective length factors when ideal conditions are approximated. This table is reproduced here as Table 5-1 with the permission of the AISC. Two sets of K values are provided in the table, one being the theoretical values and the other being the recommended design values based on the fact that perfectly pinned and fixed conditions are not possible. If the ends of the column of Fig. 5-4 (b) were not quite fixed, the column would be a little freer to bend laterally and its points of inflection would be farther apart. The recommended design K given is 0.65 while the theoretical value is 0.5. As no column ends are perfectly fixed or perfectly hinged, the designer may wish to interpolate between the values given in the table, the interpolation to be based on his or her judgment of the actual restraint conditions.

For many buildings sideway is substantially eliminated by masonry walls, but for buildings built with light curtain walls and large column spacings or for tall buildings built without a positive system of lateral bracing, sideway is appreciable. Such frames are referred to as *unbraced frames*. For such cases the bending stiffness of the structural frames provides most of the lateral support. The effective lengths of columns for such laterally unsupported continuous frames must always be greater than 1.0 because of sideway. This topic is continued in Chapter 7.

TABLE 5-1   COLUMN EFFECTIVE LENGTHS

	(a)	(b)	(c)	(d)	(e)	(f)
Buckled shape of column is shown by dashed line						
Theoretical K value	0.5	0.7	1.0	1.0	2.0	2.0
Recommended design value when ideal conditions are approximated	0.65	0.80	1.2	1.0	2.10	2.0

End condition code	Rotation fixed and translation fixed
	Rotation free and translation fixed
	Rotation fixed and translation free
	Rotation free and translation free

Source: Load and Resistance Factor Design Specification for Structural Steel Buildings September 1, 1986 (Chicago: AISC, 1986), p. 6–151 in the LRFD Manual.

5-7 STIFFENED AND UNSTIFFENED ELEMENTS

Up to this point in the text the author has only considered the overall stability of members and yet it is entirely possible for the thin flanges or webs of a column or beam to buckle locally in compression well before the calculated buckling strength of the whole member is reached. When thin plates are used to carry compressive stresses they are particularly susceptible to buckling about their weak axes due to the small moments of inertia in those directions.

The LRFD Specification in Part 6, section B5 provides limiting values for the width-thickness ratios of the individual parts of compression members and for the parts of beams in their compression regions. The student is quite well aware of the lack of stiffness of thin pieces of cardboard or metal or plastic with free edges. If, however, one of these elements is folded or restrained its stiffness is appreciably increased. For this reason two categories are listed in the LRFD Manual, these being *stiffened elements* and *unstiffened elements*.

An unstiffened element is a projecting piece with one free edge parallel to the direction of the compression force while a stiffened element is supported along the two edges in that direction. These two types of elements are illustrated in Fig. 5-6. In each case the width, b, and the thickness, t, of the elements in question are shown.

Depending on the ranges of different width-thickness ratios for compression elements and depending on whether the elements are stiffened or unstiffened the elements will buckle at different stress situations.

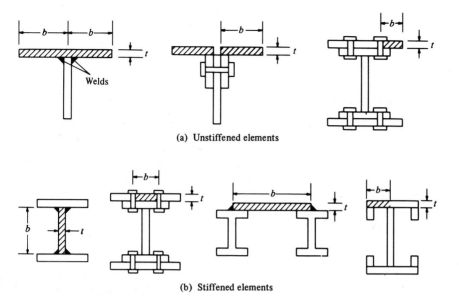

(a) Unstiffened elements

(b) Stiffened elements

Figure 5-6

For establishing width-thickness ratio limits for the elements of compression members the LRFD Specification divides members into three classifications as follows: compact sections, noncompact sections, and slender compression elements. These classifications, which decidedly affect the design compression stresses to be used for columns, are discussed in the paragraphs to follow.

Compact Sections   A compact section is one that has a sufficiently stocky profile so that it is capable of developing a fully plastic stress distribution before buckling. The term plastic means stressed throughout to the yield stress and is discussed at length in Chapter 8. For a compression member to be classified as compact its flanges must be continuously connected to its web or webs and the width-thickness ratios of its compression elements may not be greater than the limiting ratios λ_P given in Table B5.1 of Part 6 of the LRFD Manual.

Noncompact Sections   A noncompact section is one for which the yield stress can be reached in some but not all of its compression elements before buckling occurs. It is not capable of reaching a fully plastic stress distribution. In Table B5.1 the noncompact sections are those which have width-thickness ratios greater than λ_P but not greater than λ_r.

Slender Compression Elements   These elements have width-thickness ratios greater than λ_r and will buckle elastically before the yield stress is reached in any part of the section. For such elements it is necessary to consider elastic buckling strengths. A special design procedure for slender compression elements is provided in Part 6, Appendix B5.3 of the LRFD Manual.

5-8 LONG, SHORT, AND INTERMEDIATE COLUMNS

A column subject to an axial compression load will shorten in the direction of the load. If the load is increased until the column buckles, the shortening will stop and the column will suddenly bend or deform laterally and may at the same time twist in a direction perpendicular to its longitudinal axis.

The strength of a column and the manner in which it fails are greatly dependent on its effective length. A very short stocky steel column may be loaded until the steel yields and perhaps on into the strain hardening range. As a result it can support about the same load in compression that it can in tension.

As the effective length of a column increases its buckling stress will decrease. If the effective length exceeds a certain value the buckling stress will be less than the proportional limit of the steel. Columns in this range are said to fail *elastically.*

As previously shown in Section 5.5 very long steel columns will fail at loads which are proportional to the bending rigidity of the column (EI) and independent of the strength of the steel. For instance, a long column constructed with a 36 ksi yield stress steel will fail at just about the same load as one constructed with a 100 ksi yield stress steel.

Columns are sometimes classed as being long, short or intermediate. A brief discussion of each of these classifications is presented in the paragraphs to follow.

Long Columns   The Euler formula predicts very well the strength of long columns where the axial buckling stress remains below the proportional limit. Such columns will buckle *elastically.*

Short Columns   For very short columns the failure stress will equal the yield stress and no buckling will occur. (For a column to fall into this class it would have to be so short as to have no practical application. Thus no further reference is made to them here.)

Intermediate Columns   For intermediate columns some of the fibers will reach the yield stress and some will not. The members will fail by both yielding and buckling and their behavior is said to be *inelastic.* (For the Euler formula to be applicable for such columns it would have to be modified according to the reduced modulus concept or the tangent modulus concept to account for the presence of residual stresses.)

In Section 5-9 formulas are presented with which the LRFD estimates the strength of columns in these different ranges.

5-9 COLUMN FORMULAS

The LRFD Specification provides one formula (it's the Euler equation) for long columns with inelastic buckling and a parabolic equation for short and intermediate columns. With these equations a critical or buckling stress, F_{cr}, is determined for a compression member. Once this stress is computed for a particular compression member it is multiplied by the cross-sectional area of the member to obtain the member's nominal strength. The design strength of the member can then be determined as follows:

$$P_u = \phi_c P_n = \phi_c F_{cr} A_g \qquad \text{with } \phi_c = 0.85$$

One LRFD formula for F_{cr} is for inelastic buckling and the other is for elastic buckling. In both equations λ_c in easily remembered form is $\sqrt{F_y/F_e}$ where F_e is the Euler stress $\pi^2 E/(KL/r)^2$. Substituting this value for F_e we get the form of λ_c given in the LRFD Manual.

$$\lambda_c = \frac{KL}{r\pi} \sqrt{\frac{F_y}{E}} \qquad \text{(LRFD Formula E2-4)}$$

Both equations for F_{cr} include the estimated effects of residual stresses and initial out-of-straightness of the members. The inelastic formula that follows is an empirical or test result formula.

$$F_{cr} = (0.658^{\lambda_c^2}) \, F_y \text{ for } \lambda_c \le 1.5 \qquad \text{(LRFD Formula E2-2)}$$

The other equation is for elastic or Euler buckling and is the familiar Euler equation multiplied by 0.877 to estimate the effect of out-of-straightness.

$$F_{cr} = \left(\frac{0.877}{\lambda_c^2}\right) F_y \text{ for } \lambda_c > 1.5 \qquad \text{(LRFD Formula E2-3)}$$

These equations are represented graphically in Fig. 5-7.

After looking at these equations the reader might think that their use would be very tedious and time-consuming with a pocket calculator. Such calculations, however, rarely have to be made because the LRFD Manual provides computed $\phi_c F_{cr}$ values for steels with $F_y = 36$ ksi and 50 ksi for KL/r values from 1 to 200 and has shown the results in Tables 3-36 and 3-50 of Part 6 of the LRFD Manual. Furthermore the LRFD has another table (Table 4) from which the user may obtain values for steels with any F_y values.

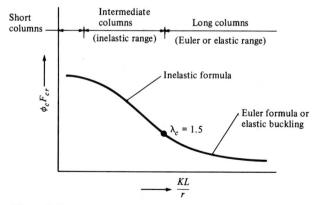

Figure 5-7

5-10 MAXIMUM SLENDERNESS RATIOS

In Part 6, Section B7 the LRFD states that compression members *preferably* should be designed with KL/r ratios not exceeding 200. The reader might note from LRFD Tables 3-36 and 3-50 that design stresses $\phi_c F_{cr}$ for KL/r values of 200 are both 5.33 ksi. Should slenderness ratios larger than these be used the $\phi_c F_{cr}$ values will be very small and will require the user to substitute into the column formulas provided in Section 5-9.

5-11 EXAMPLE PROBLEMS

In this section three simple numerical column problems are presented. In each case the design strength of a column is calculated. In Example 5-1 (a) the author determines the strength of a W section. The value of K is determined as described in Section 5-6, the effective slenderness ratio is computed, and the design stress of the member $\phi_c F_{cr}$ is selected from the appropriate LRFD table and multiplied by the cross-sectional area of the column.

It will be noted that the LRFD in its Part 2 has further simplified the calculations required by computing the column design strength $\phi_c F_{cr} A_g$ for each of the steel shapes normally used as columns for F_y values of 36 and 50 ksi for most of the commonly used effective lengths or KL values given in feet. The use of these tables is illustrated in Example 5-1 (b).

EXAMPLE 5-1

(a)   Using the column design stress values shown in Table 3-36, Part 6 of the LRFD Manual, determine the design strength ($P_u = \phi_c P_n$) of the A36 axially loaded column shown in Fig. 5-8.

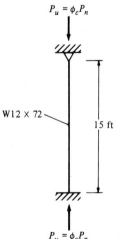

Figure 5-8

(b)   Repeat the problem using the column tables of Part 2 of the Manual.

Solution

(a)   Using a W12×72 ($A = 21.1$ in.^{2}, $r_y = 3.04$ in.)

$K = 0.80$ from Table C-C2.1, Part 6, LRFD Manual

$$\frac{KL}{r} = \frac{(0.80)(12\times15)}{3.04} = 47.37$$

$\phi_c F_{cr} = 27.19$ ksi from Table 3-36, Part 6, LRFD Manual

$$P_u = \phi_c P_n = \phi_c F_{cr} A_g = (27.19)(21.1) = 573.7 \text{ k}$$

(b)   Entering column tables Part 2 of Manual with $K_y L_y$ in feet.

$$K_y L_y = (0.80)(15) = 12 \text{ ft}$$

$$P_u = \phi_c P_n = 574 \text{ k}$$

In Example 5.2 the author illustrates the computations necessary to determine the design strength of a built-up column section. Several special requirements for built-up column sections are described in Chapter 6.

EXAMPLE 5-2

Determine the design strength $\phi_c P_n$ of the axially loaded column shown in Fig. 5-9 if $KL = 19$ ft and A36 steel is used.

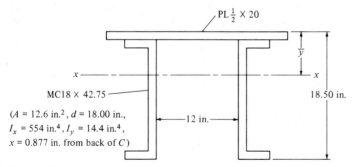

Figure 5-9

Solution

$$A = (20)(\tfrac{1}{2}) + (2)(12.6) = 35.2 \text{ in.}^2$$

$$\bar{y} \text{ from top} = \frac{(10)(0.25) + (2)(12.6)(9.50)}{35.2} = 6.87 \text{ in.}$$

$$I_x = (2)(554) + (25.2)(2.63)^2 + (\tfrac{1}{12})(20)(\tfrac{1}{2})^3 + (10)(6.62)^2 = 1721 \text{ in.}^4$$

$$I_y = (2)(14.4) + (12.6)(6.877)^2(2) + (\tfrac{1}{12})(\tfrac{1}{2})(20)^3 = 1554 \text{ in.}^4$$

$$\text{Least } r = \sqrt{\frac{1554}{35.2}} = 6.64 \text{ in.}$$

$$\frac{KL}{r} = \frac{(12)(19)}{6.64} = 34.34$$

$$\phi_c F_{cr} = 28.76 \text{ ksi}$$

$$\phi_c P_n = P_u = (28.76)(35.2) = \underline{\underline{1012 \text{ k}}}$$

To determine the design compression stress to be used for a particular column it is theoretically necessary to compute both $(KL/r)_x$ and $(KL/r)_y$. The reader will notice, however, that for most of the steel sections used for columns r_y will be much less than r_x. As a result only $(KL/r)_y$ is calculated for most columns and used in the applicable column formulas.

For some columns, particularly the long ones, bracing is supplied perpendicular to the weak axis thus reducing the slenderness or the length free to buckle in that direction. This may be accomplished by framing braces or beams into the sides of a column. For instance, horizontal members called *girts* running parallel to the exterior walls of a building frame may be framed into the sides of columns. The result is stronger columns and ones for which the designer needs to calculate both $(KL/r)_x$ and $(KL/r)_y$. The larger ratio obtained for a particular column indicates the weaker direction and will be used for calculating the design stress $\phi_c F_{cr}$ for that member.

Bracing members must be capable of providing the necessary lateral forces without buckling themselves. The forces to be taken are quite small and are often conservatively estimated to equal 0.02 times the column design loads. These members can be selected as are other compression members. A bracing member must be connected to other members that can transfer the horizontal force by shear to the next restrained level. If this is not done little lateral support will be provided for the original column in question.

If the lateral bracing were to consist of a single bar or rod (⊢─) it would not prevent twisting and torsional buckling of the column (see Chapter 6). As torsional buckling is a difficult problem to handle we should provide lateral bracing that prevents lateral movement and twist.[3]

Steel columns may also be built into substantial masonry walls in such a manner that they are substantially supported in the weaker direction. The designer, however, should be quite careful in assuming complete lateral support parallel to the wall, because a poorly built wall will not provide 100 percent lateral support.

Example 5-3 illustrates the calculations necessary to determine the design strength of a column with two unbraced lengths.

[3]J.A. Yura, "Elements for Teaching Load and Resistance Factor Design" (New York: AISC, August 1987), p. 20.

EXAMPLE 5-3

(a)   Using Table 3-36 of Part 6 of the LRFD Manual determine the design strength $\phi_c P_n$ of the A36 axially loaded W14×90 shown in Fig. 5-10. Because of its considerable length this column is braced perpendicular to its weak or y axis at the points shown in the figure. These connections are assumed to permit rotation of the member but to prevent translation or sidesway.

(b)   Repeat part (a) using the column tables of Part 2 of the Manual.

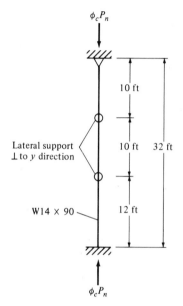

Figure 5-10

Solution

(a) Using a W14×90 ($A = 26.5$ in.^{2}, $r_x = 6.14$ in., $r_y = 3.70$ in.)

Determining effective lengths

$$K_x L_x = (0.80)(32) = 25.6 \text{ ft}$$

$$K_y L_y = (1.0)(10) = 10 \text{ ft} \leftarrow$$

$$K_y L_y = (0.80)(12) = 9.6 \text{ ft}$$

Computing slenderness ratios

$$\left(\frac{KL}{r}\right)_x = \frac{(12)(25.6)}{6.14} = 50.03 \leftarrow$$

$$\left(\frac{KL}{r}\right)_y = \frac{(12)(10)}{3.70} = 32.43$$

$$\phi_c F_{cr} = 26.82 \text{ ksi}$$

$$\phi_c P_n = (26.82)(26.5) = \underline{\underline{710.7 \text{ k}}}$$

(b)   Noting from part (a) solution that there are two different KL values

$$K_x L_x = 25.6 \text{ ft}$$

$$K_y L_y = 10 \text{ ft}$$

Controlling $K_y L_y$ for use in tables is either 10 ft or $\dfrac{K_x L_x}{r_x/r_y}$

$\dfrac{r_x}{r_y}$ for W14×90 from column tables = 1.66

$$\frac{K_x L_x}{r_x/r_y} = \frac{25.6}{1.66} = 15.42 \text{ ft} \leftarrow$$

From column tables with $K_y L_y = 15.42$ ft we find by interpolation

$$\phi_c Pn = \underline{\underline{711 \text{ k}}}$$

PROBLEMS

5-1. Using A36 steel and the LRFD Manual determine the design strength $\phi_c P_n$ for each of the compression members.
(a) A W14×109 with $KL = 9$ ft (*Ans.* 937 k)
(b) A W12×58 with $KL = 17$ ft (*Ans.* 367 k)
(c) An S6×17.25 with $KL = 10$ ft (*Ans.* 34 k)

5-2. Using the LRFD Specification determine the design strength $\phi_c P_n$ for each of the following columns.
(a) A W8×35 with fixed ends, $L = 16$ ft 6 in., A36 steel
(b) A W12×96 with pinned ends, $L = 20$ ft 0 in., A36 steel
(c) A W10×68 with one end fixed and the other pinned, $L = 24$ ft 6 in., $F_y = 50$ ksi
(d) A W14×193 with fixed ends, $L = 22$ ft 0 in., $F_y = 50$ ksi
(e) A Pipe 10 Std. with pinned ends, $L = 20$ ft 0 in., A36 steel
(f) Two 8×8×$\frac{3}{4}$ Ls separated $\frac{3}{8}$ in. (for gusset PLs at ends), pinned ends, $L = 24$ ft 6 in., A36 steel

5-3. A W14×53 with a $\frac{3}{4}$×12-in. cover plate bolted to each flange is to be used for a column with KL = 18 ft. Find its design strength $\phi_c P_n$ if A36 steel is used. (*Ans.* $P_u = 760$ k)

5-4. Using A36 steel determine the design strength $\phi_c P_n$ of the axially loaded compression members shown.

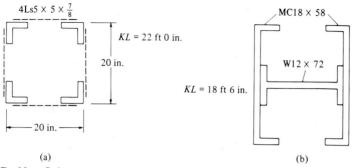

4Ls5 × 5 × $\frac{7}{8}$

KL = 22 ft 0 in.

20 in.

20 in.

(a)

MC18 × 58

W12 × 72

KL = 18 ft 6 in.

(b)

Problem 5-4

5-5. Using a steel with F_y = 50 ksi calculate the design strength $\phi_c P_n$ of the concentrically loaded columns shown.

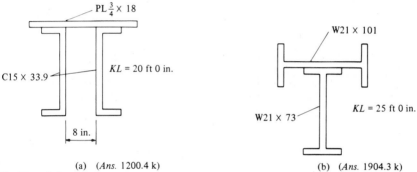

PL$\frac{3}{4}$ × 18

C15 × 33.9

KL = 20 ft 0 in.

8 in.

(a)   (*Ans.* 1200.4 k)

W21 × 101

W21 × 73

KL = 25 ft 0 in.

(b)   (*Ans.* 1904.3 k)

Problem 5-5

5-6. Using A36 steel determine $\phi_c P_n$ for the axially loaded columns shown.

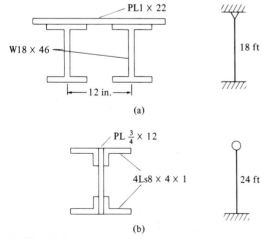

PL1 × 22

W18 × 46

12 in.

(a)

18 ft

PL $\frac{3}{4}$ × 12

4Ls8 × 4 × 1

(b)

24 ft

Problem 5-6

5-7. Determine the design strength $\phi_c P_n$ for each of the axially loaded columns shown. A36 steel.

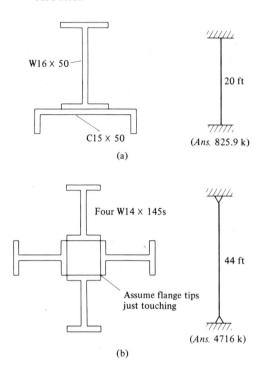

W16 × 50

C15 × 50

(a)

20 ft

(*Ans.* 825.9 k)

Four W14 × 145s

Assume flange tips just touching

(b)

44 ft

(*Ans.* 4716 k)

Problem 5-7

5-8. A 27-ft axially loaded W10×68 column is laterally supported perpendicular to its y axis at middepth. Determine $\phi_c P_n$ for this column if A36 steel is used and if effective length factors = 1.0 are assumed in each case.

Chapter 6

Design of Axially Loaded Compression Members

6-1 INTRODUCTION

In this chapter the designs of several axially loaded columns are presented. Included are the selections of single shapes, W sections with cover plates, and built-up sections constructed with channels. Also included are the designs of sections whose unbraced lengths in the x and y directions are different and the sizing of lacing and tie plates for built-up sections with open sides. Another topic considered is the flexural-torsional buckling of sections.

The design of columns with formulas involves a trial-and-error process. The design stress $\phi_c F_{cr}$ is not known until a column size is selected and vice versa. Once a trial section is assumed the r values for that section can be obtained and substituted into the appropriate column equation to determine its design stress. Examples 6-1, 6-3, and 6-4 illustrate this procedure.

The designer may assume a design stress, divide that stress into the factored column load to give an estimated column area, select a column section, determine its design stress, multiply that stress by the cross-sectional area of the section to obtain the member's design strength to see if the section selected is over- or underdesigned, and if appreciably so, try another size. The student may feel that he or she does not have sufficient background or knowledge to make reasonable initial design stress assumptions. If, however, this student will read the information contained in the next few paragraphs he or she will immediately be able to make excellent estimates.

The effective slenderness ratio (KL/r) for the average column of 10- to 15-ft length will generally fall between about 40 and 60. If for a particular column a KL/r somewhere in this approximate range is assumed and substituted into the appropriate column equation (in the LRFD this usually means looking into the

State Office Tower II, Columbus, Ohio. (Courtesy Owen Steel Company, Inc.)

tables where the design stresses have already been calculated for KL/r values from 0 to 200) the result will usually be a very satisfactory design stress estimate.

In Example 6.1 a column with $KL = 10$ ft is selected using the LRFD formulas. An effective slenderness ratio of 50 is assumed, the design stress for that value is determined from Table 3-36 of Part 6 of the Manual, and the resulting stress is divided into the factored column load to obtain an estimated column area. After a trial section is selected with approximately that area its actual slenderness ratio and design strength are determined. The first estimated size in

Example 6-1 though quite close is a little too small and the next larger section in that series of shapes is tried and found to be satisfactory.

To estimate the effective slenderness ratio for a particular column the designer may estimate a value a little higher than 40 to 60 if the column is appreciably longer than the 10- to 15-ft range and vice versa. A very heavy factored column load, say in the 750- or 1000-k range or higher will require a rather large column for which the radii of gyration will be larger and the designer may estimate a little smaller value of KL/r. For lightly loaded bracing members he or she may estimate high slenderness ratios perhaps over 100.

EXAMPLE 6-1

Using A36 steel select the lightest W14 available for the service loads $P_D = 100$ k and $P_L = 160$ k. $KL = 10$ ft.

Solution

$$P_u = (1.2)(100) + (1.6)(160) = 376 \text{ k}$$

Assume $\dfrac{KL}{r} = 50$

$\phi_c F_{cr}$ from Table 3-36, Part 6 = 26.83 ksi

$$A \text{ required} = \frac{376}{26.83} = 14.01 \text{ in.}^2$$

Try W14×48 ($A = 14.1$ in.^{2}, $r_y = 1.91$ in.)

$$\frac{KL}{r} = \frac{(12)(10)}{1.91} = 62.83$$

$$\phi_c F_{cr} = 24.86 \text{ ksi}$$

$$\phi_c P_n = (24.86)(14.1) = 350 \text{ k} < 376 \text{ k} \qquad\qquad \text{NG}$$

Try W14×53 ($A = 15.6$ in.^{2}, $r_y = 1.92$ in.)

$$\frac{KL}{r} = \frac{(12)(10)}{1.92} = 62.5$$

$$\phi_c F_{cr} = 24.91 \text{ ksi}$$

$$\phi_c P_n = 388.6 \text{ k} > 376 \text{ k} \qquad\qquad \text{OK}$$

$$\text{Use W14×53}$$

6-2 LRFD DESIGN TABLES

For Example 6-2, Part 2 of the LRFD Manual is used to select various column sections from tables without the necessity of using a trial-and-error process. These tables provide axial design strengths ($\phi_c P_n$) for various practical effective lengths of the steel sections commonly used as columns (Ws, Ms, Ss, pipes, tubes, pairs of angles, and structural tees). The values are given with respect to the least radius of gyration for steels with F_y = 36 ksi and 50 ksi (*with the exception of square and rectangular tubes which are available only in 46-ksi steel*).

The resulting tables are very simple to use. The designer takes the *KL* value for the weaker direction in feet and enters the table in question from the left-hand side and moves horizontally across the table. Under each section is listed the design strength $\phi_c P_n$ for that *KL* and the steel yield stress. As an illustration it is assumed that we have a factored design load $P_u = \phi_c P_n$ = 800 k, $K_y L_y$ = 12 ft and we want to select the lightest available W14 section using A36 steel. We enter the tables with *KL* = 12 ft in the left column and read from left to right for A36 steel the numbers 6260, 5700, 5170 k, and so on until several pages later where the consecutive values 823 and 749 k are found. The 749 k value is not sufficient and we go back to the 823 k which falls under the W14×99. A similar procedure can be followed for the other available shapes.

Example 6-2 which follows illustrates the selection of W sections as well as pipes and tubes. It is possible to support a given column load with a standard pipe column; or with an extra strong pipe column (× strong) which has a smaller diameter but thicker walls, thus heavier and more expensive; or with a double extra strong pipe column (×× strong) which has an even smaller diameter and even thicker walls and heavier weight.

EXAMPLE 6-2

Using the LRFD column tables of Part 2 of the Manual,

(a) Select the lightest W section available for the loads, steel, and *KL* of Example 6-1.

(b) Select the lightest standard, extra strong, and double extra strong pipe columns for the situation of part (a) of this problem.

(c) Select the lightest square and rectangular tubes satisfactory for the situation of part (a), except use an F_y of 46 ksi.

Solution

(a)   Enter tables with $K_y L_y$ = 10 ft and $P_u = \phi_c P_n$ = 376 k

W14×53 ($\phi_c P_n$ = 389 k)

W12×53 ($\phi_c P_n$ = 422 k)

W10×49 ($\phi_c P_n$ = 392 k) ←

W8×58 ($\phi_c P_n$ = 441 k)

Use W10×49

(b)   Pipe columns

12 pipe std ($\phi_c P_n$ = 429 k) wt = 49.56 lb/ft

Pipe 10× strong ($\phi_c P_n$ = 465 k) wt = 54.74 lb/ft

Pipe 6×× strong ($\phi_c P_n$ = 399 k) wt = 53.16 lb/ft

Eversharp, Inc., building at Milford, Conn. (Courtesy of Bethlehem Steel Corporation.)

(c)   Square and rectangular tubing ($F_y = 46$ ksi)

$$8 \times 8 \times \tfrac{3}{8} \ (\phi_c P_n = 392 \text{ k}) \text{ wt} = 37.60 \text{ lb/ft}$$

$$8 \times 6 \times \tfrac{1}{2} \ (\phi_c P_n = 404 \text{ k}) \text{ wt} = 42.05 \text{ lb/ft}$$

An axially loaded column is laterally restrained in its weak direction in Fig. 6-1. Example 6-3 illustrates the design of such a column with its different unsupported lengths in the x and y directions. The student can easily solve this problem by trial and error. A trial section can be selected as described in Section 6-1 of this chapter, the slenderness values $(KL/r)_x$ and $(KL/r)_y$ computed, $\phi_c F_{cr}$ determined and multiplied by A_g to obtain $\phi_c P_n$. Then if necessary another size can be tried and so on.

For this discussion it is assumed that K is the same in both directions. Then if we are to have equal strengths about the x and y axes the following relation must hold

$$\frac{L_x}{r_x} = \frac{L_y}{r_y}$$

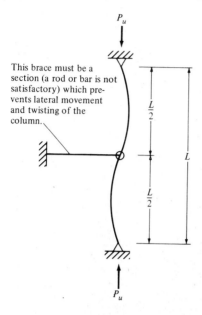

Figure 6-1 A column laterally restrained at middepth in its weaker direction.

For L_y to be equivalent to L_x we would have

$$L_x = L_y \frac{r_x}{r_y}$$

If $L_y(r_x/r_y)$, is less than L_x then L_x controls; if greater than L_x then L_y controls.

Based on the preceding information the LRFD Manual provides a method with which a section can be selected from its tables with little trial and error when the unbraced lengths are different. The designer enters the appropriate table with $K_y L_y$, selects a shape, takes the r_x/r_y value given in the table for that shape, and multiplies it by L_y. If the result is larger than $K_x L_x$, then $K_y L_y$ controls and the shape initially selected is the correct one. If the result of the multiplication is less than $K_x L_x$, then $K_x L_x$ controls and the designer will reenter the tables with a larger $K_y L_y$ equal to $K_x L_x/(r_x/r_y)$ and select the final section.

EXAMPLE 6-3

Select the lightest satisfactory W12 for the following conditions: A36 steel, $P_u = 670$ k, $K_x L_x = 26$ ft, and $K_y L_y = 13$ ft.

(a) By trial and error
(b) Using LRFD tables

Solution

(a) Using trial and error

$$\text{Assume } \frac{KL}{r} = 50$$

$$\phi_c F_{cr} = 26.83 \text{ ksi}$$

$$A \text{ required} = \frac{670}{26.83} = 24.97 \text{ in.}^2$$

Try W12×87 ($A = 25.6$ in.^{2}, $r_x = 5.38$ in., $r_y = 3.07$ in.

$$\left(\frac{KL}{r}\right)_x = \frac{(12)(26)}{5.38} = 57.99 \leftarrow$$

$$\left(\frac{KL}{r}\right)_y = \frac{(12)(13)}{3.07} = 50.81$$

$$\phi_c F_{cr} = 25.63 \text{ ksi}$$

$$\phi_c P_n = (25.63)(25.6) = 656 \text{ k} < 670 \text{ k} \qquad \text{NG}$$

A subsequent check of the next W12 section (96 lb) shows it will work.

$$\underline{\underline{\text{Use W12×96}}}$$

(b) Using LRFD tables

Enter tables with $K_y L_y = 13$ ft

Try W12×87 $\left(\dfrac{r_x}{r_y} = 1.75\right)$

$$(K_y L_y)\left(\frac{r_x}{r_y}\right) = (13)(1.75) = 22.75 < K_x L_x \qquad \text{Therefore, } \underline{K_x L_x \text{ controls,}}$$

Reenter tables with new $K_y L_y = \dfrac{K_x L_x}{r_x/r_y} = \dfrac{26}{1.75} = 14.86$

$$\underline{\underline{\text{Use W12×96}}}$$

6-3 BUILT-UP MEMBERS WITH COMPONENTS IN CONTACT WITH EACH OTHER

Should a column consist of two equal size plates as shown in Fig. 6-2 and should those plates not be connected together, each plate will act as a separate column and each will resist approximately half of the total column load. In other words the total moment of inertia of the column will equal two times the moment of inertia of one plate. The two "columns" will act the same and have equal deformations as shown in part (b) of the figure.

Should the two plates be connected together sufficiently to prevent slippage on each other as shown in Fig. 6-3, they will act as a unit. Their moment of inertia may be computed for the whole built-up section as shown in the figure and will be four times as large as it was for the column of Fig. 6-2 where slipping between the plates was possible. The reader should also notice that the plates of the column of Fig. 6-3 will deform different amounts as the column bends laterally.

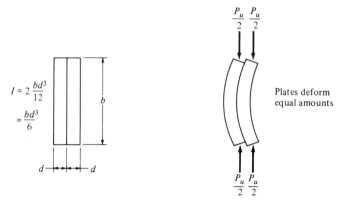

(a) Column cross section   (b) Deformed shape of column

Figure 6-2   Column consisting of two plates not connected to each other.

Should the plates be connected in a few places it would appear that the strength of the resulting column would be somewhere in between the two cases just described.

Reference to Fig. 6-2(b) shows that the greatest displacement between the two plates occurs at their ends and the least displacement occurs at middepth. As a result slip-resistant connectors placed at column ends have the greatest strengthening effect while those placed at middepth have the least effect. (Slip-resistant connections are discussed in Chapter 12.)

Should the plates be fastened together at their ends with slip-resistant connectors those ends will deform together and the column will take the shape shown in Fig. 6-4. As the plates are held together at the ends the column will bend in an S shape as shown in the figure.

If the column were to bend in the S shape shown its K factor would theoretically equal 0.5 and its KL/r value would be the same as the one for the continuously connected column of Fig. 6-3.[1]

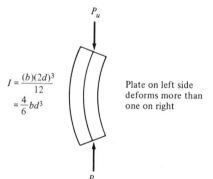

Plate on left side
deforms more than
one on right

Figure 6-3   Column consisting of two plates
fully connected to each other.

[1]J.A. Yura, "Elements for Teaching Load and Resistance Factor Design" (Chicago: AISC, July 1987), pp. 17–19.

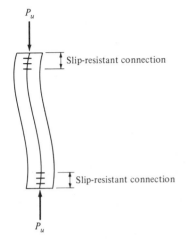

P_u

Slip-resistant connection

Slip-resistant connection

P_u

Figure 6-4

$$\frac{KL}{r} \text{ for the column of Fig. 6-3} = \frac{(1)(L)}{\sqrt{\frac{4}{6}bd^3/2bd}} = 1.732L$$

$$\frac{KL}{r} \text{ for the end-fastened column of Fig. 6-4} = \frac{(0.5)(L)}{\sqrt{\frac{1}{6}bd^3/2bd}} = 1.732L$$

Thus the design stresses are equal for the two cases and the columns would carry the same loads. This is true for the particular case described here but is not applicable for the common case where the parts of Fig. 6-4 begin to separate.

6-4 CONNECTION REQUIREMENTS FOR BUILT-UP COLUMNS WHOSE COMPONENTS ARE IN CONTACT WITH EACH OTHER

Several requirements concerning built-up columns are presented in LRFD Specification E4. When such columns consist of different components which are in contact with each other and which are bearing on base plates or milled surfaces they must be connected at their ends with bolts or welds. If welds are used the weld lengths must at least equal the maximum width of the member. If bolts or rivets are used they may not be spaced longitudinally more than four diameters on center and the connection must extend for a distance at least equal to $1\frac{1}{2}$ times the maximum width of the member.

The LRFD Specification also requires the use of welded, bolted, or riveted connections between the end ones described in the last paragraph. These must be sufficient to provide for the transfer of calculated stresses. If it is desired to have a close fit over the entire faying surfaces between the components it may be necessary to place the connectors even closer than is required for shear transfer.

When the component of a built-up column consists of an outside plate the LRFD Specification provides specific maximum spacings for fastening. If intermittent welds are used along the edges of the components or if bolts or rivets are provided along all gage lines at each section, their maximum spacing may not be greater than $127/\sqrt{F_y}$ times the thickness of the thinner outside plate or 12 in. Should these fasteners be staggered on each gage line, however, they may not be spaced farther apart on each gage line than $190/\sqrt{F_y}$ times the thickness of the thinner part or 18 in.

Other spacing values are given in LRFD Specification E4 for built-up members made from unpainted weathering steel.

For the discussion to follow the letter a represents the distance between connectors and r_i is the minimum radius of gyration of an individual component of the column. According to test results if the ratio a/r_i is ≤ 50 all the parts of a built-up column will act together as a unit. If the ratio is greater than 50 the LRFD Specification requires that a modified or larger column slenderness ratio $(KL/r)_m$ be used to determine the design stress in place of the value used if the whole cross section is fully effective $(KL/r)_0$. The modified value follows

$$\left(\frac{KL}{r}\right)_m = \sqrt{\left(\frac{KL}{r}\right)_0^2 + \left(\frac{a}{r_i} - 50\right)^2} \qquad \text{(LRFD Formula E4-2)}$$

The preceding equation is only applicable to the buckling axis where connectors are required to resist the shear.

It is to be clearly remembered that the design strength of a built-up column will be reduced if the spacing of the connectors is such that one of the components of the column can buckle before the whole column buckles. Such a situation will be prevented if the ratio a/r_i is kept $\leq$ the governing ratio, that is, the lesser value of $(KL/r)_x$ or $(KL/r)_y$ for the whole member.

For built-up columns whose components are connected at their ends only with snug-tight bolts (described in Chapter 12) the LRFD Specification states that the following slenderness ratio is to be used:

$$\left(\frac{KL}{r}\right)_m = \sqrt{\left(\frac{KL}{r}\right)_0^2 + \left(\frac{a}{r_i}\right)^2} \qquad \text{(LRFD Formula E4-1)}$$

The column tables for double angles given in Part 2 of the LRFD Manual were developed on the basis of the modified slenderness ratio and also upon flexural-torsional buckling as described in Section 6-5.

Example 6-4 illustrates the design of a column consisting of a W section with cover plates welded to its flanges as shown in Fig. 6-5. It is assumed that the plates are connected to the W section at its ends and at intermediate points so the parts act together and $a/r_i \leq 50$. As this type of section is not shown in the column tables of the LRFD Manual it is necessary to use a trial-and-error procedure. An effective slenderness ratio is assumed; $\phi_c F_{cr}$ for that KL/r is determined and divided into the column design load to estimate the total area required. The area of the W section is subtracted from the estimated total area to obtain the estimated cover plate area. A cover plate size is selected to provide the estimated area and then $\phi_c P_n$ is calculated for the whole section, after which it may be necessary to revise the cover plate size and try again.

EXAMPLE 6-4

It is desired to design a column for P_u = 1740 k using A36 steel and KL = 14 ft. A W12×120 (for which $\phi_c P_n$ = 928 k from the Part 2 tables of the Manual) is on hand. Design cover plates to be welded to the W section as shown in Fig. 6-5 to enable the column to support the required load. Assume $a/r_i \leq 50$.

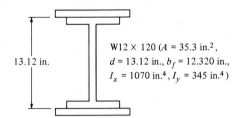

13.12 in.

W12 × 120 (A = 35.3 in.^{2}, d = 13.12 in., b_f = 12.320 in., I_x = 1070 in.^{4}, I_y = 345 in.^{4})

Figure 6-5

Solution

Assume $\dfrac{KL}{r} = 40$

$$\phi_c F_{cr} = 28.13 \text{ ksi}$$

$$A \text{ required} = \frac{1740}{28.13} = 61.86 \text{ in.}^2$$

$$-A \text{ of W12×120} = -35.30$$

$$A \text{ of 2 cover plates} = 26.56 \text{ in.}^2 \text{ or } 13.28 \text{ in.}^2 \text{ each}$$

Try 1 PL $\frac{3}{4}$×18 each flange*

$$A = 35.30 + (2)(\tfrac{3}{4})(18) = 62.3 \text{ in.}^2$$

$$I_x = 1070 + (2)(\tfrac{1}{12})(18)(\tfrac{3}{4})^3 + (2)(\tfrac{3}{4})(18)(6.935)^2$$

$$= 2369 \text{ in.}^4$$

$$I_y = 345 + (2)(\tfrac{1}{12})(\tfrac{3}{4})(18)^3 = 1074 \text{ in.}^4$$

$$r_y = \sqrt{\frac{1074}{62.3}} = 4.15 \text{ in.}$$

$$\frac{KL}{r} = \frac{(12)(14)}{4.15} = 40.48$$

$$\phi_c F_{cr} = 28.07 \text{ ksi}$$

$$\phi_c P_n = (28.07)(62.3) = 1749 \text{ k} > 1740 \text{ k} \qquad \text{OK}$$

Use W12×120 with 1 cover plate $\frac{3}{4}$×18 each flange*

*Many other plate sizes could be selected.

The design of another built-up section is presented in Example 6-5 where a pair of channels are selected for a compression member.

EXAMPLE 6-5

Select a pair of 12-in. channels for the column and load shown in Fig. 6-6 using A36 steel. For connection purposes the back-to-back distance of the channels is to be 12 in.

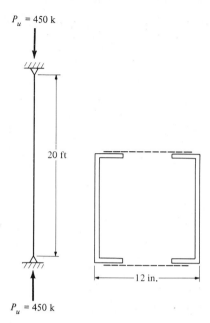

$P_u = 450$ k

20 ft

—12 in.—

$P_u = 450$ k

Figure 6-6

Solution

Assume $\dfrac{KL}{r} = 50$

$$\phi_c F_{cr} = 26.83 \text{ ksi}$$

$$A \text{ required} = \frac{450}{26.83} = 16.77 \text{ in.}^2$$

Try 2C12×30s (For each channel $A = 8.82$ in.^{2}, $I_x = 162$ in.^{4}, $I_y = 5.14$ in.^{4}, $\bar{x} = 0.674$ in.)

$$I_x = (2)(162) = 324 \text{ in.}^4$$

$$I_y = (2)(5.14) + (2)(8.82)(5.326)^2 = 511 \text{ in.}^4$$

$$r_x = \sqrt{\frac{324}{(2)(8.82)}} = 4.29 \text{ in.}$$

$$KL = (1.0)(20) = 20 \text{ ft}$$

$$\frac{KL}{r} = \frac{(12)(20)}{4.29} = 55.94$$

$$\phi_c F_{cr} = 25.95 \text{ ksi}$$

$$\phi_c P_n = (25.95)(17.64) = 457.8 \text{ k} > 450 \text{ k} \qquad \text{OK}$$

Use $2C12 \times 30\text{s}$

6-5 FLEXURAL-TORSIONAL BUCKLING OF COMPRESSION MEMBERS

Axially loaded compression members can theoretically fail in three different fashions: by flexural buckling, by torsional buckling, or by flexural-torsional buckling.

Flexural buckling (also called Euler buckling) is the situation considered up to this point in our column discussions where we have computed slenderness ratios for the principal column axes and determined $\phi_c F_{cr}$ for the highest ratios so obtained. Doubly symmetrical column members (such as W sections) are subject only to flexural buckling and torsional buckling.

As torsional buckling can be very complex it is very desirable to prevent its occurrence. This may be done by careful arrangements of the members and by providing bracing to prevent lateral movement and twisting. If sufficient end supports and intermediate lateral bracing are provided, flexural buckling will always control. The values given in the LRFD column tables for W, M, S, tube and pipe sections are based on flexural buckling.

Open sections such as Ws, Ms, and channels have little torsional strength but box beams have a great deal. Thus if a torsional situation is encountered it may be well to use box sections or to make box sections out of our W sections by adding welded side plates (⊏⊐). Another way in which torsional problems can be reduced is to shorten the lengths of members that are subject to torsion.

For a singly symmetrical section such as a tee or double angle Euler buckling may occur about the x or y axes. For equal-leg single angles Euler buckling may occur about the z axis. For all these sections flexural-torsional buckling is definitely a possibility and may control. (It will always control for unequal-leg single-angle columns.) The values given in the LRFD column load tables for double-angle and structural tee sections were computed for buckling about the weaker of the x or y axes and for flexural-torsional buckling.

Usually symmetrical members such as W sections are used as columns. Torsion will not occur in such sections if the lines of action of the lateral loads pass through their shear centers. The *shear center* is that point in the cross section of a member through which the resultant of the transverse loads must pass so that no torsion will occur. The calculations necessary to locate shear centers are presented in Chapter 10. The shear centers of the commonly used doubly symmetrical sections occur at their centroids. This is not necessarily the case for other sections such as channels and angles. Shear center locations for several types of sections are shown in Fig. 6.7. Also shown in the figure are the coordinates x_0 and y_0 for the shear center of each section with respect to its centroid.

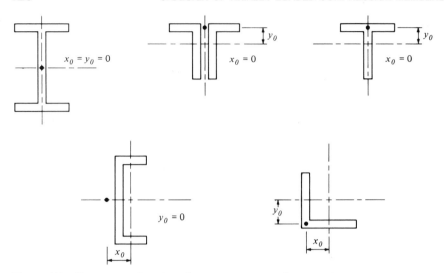

Figure 6-7 Shear center locations for some common column sections.

These values are needed to solve the flexural-torsional formulas, presented later in this section.

Even though loads pass through shear centers torsional buckling may still occur. If you load any section through its shear center no torsion will occur—but one still computes torsional buckling strength for these members, i.e., buckling load does not depend on the nature of the axial or transverse loading, rather it depends on the cross-section properties, column length, and support conditions.

The average designer does not consider the torsional buckling of symmetrical shapes or the flexural-torsional buckling of unsymmetrical shapes. Their usual feeling is that these conditions don't control the critical column loads or at least don't affect them very much. Should we, however, have unsymmetrical columns or even symmetrical columns made up of thin plates we will find that torsional buckling or flexural-torsional buckling may significantly reduce column capacities.

In Appendix E of the LRFD Specification a long list of formulas is presented for computing the flexural-torsional strength of column sections. The values given for column design strengths ($\phi_c P_n$ values) for double angles and tees in Part 2 of the LRFD Manual make use of these formulas. The LRFD does not provide tables for single-angle columns. It is stated in the Manual that this is because of the difficulty of loading such members concentrically. This is a rather puzzling statement because even if double angles or tees are loaded concentrically flexural-torsional buckling may occur. Why then does the Manual not provide similar loading tables for single angles? The LRFD feels that in practice actual eccentricities for single-angle members are rather large and they feel that neglect of these eccentricities may lead to the use of some much underdesigned members. This discussion boils down to the fact that the designer using single-angle struts is going to have to wade through the flexural-torsional formulas and make some allowances for eccentricities of loading. An example problem of this type is presented on pages 2-49 and 2-50 of the Manual.

For flexural-torsion $P_u = \phi_c P_n = \phi_c A g F_{cr}$ with $\phi_c = 0.85$ and F_{cr} to be determined from the formulas to follow from the specification. A list of definitions that are needed for using these formulas is also provided.

If $\lambda_e \sqrt{Q} \le 1.5$

$$F_{cr} = Q(0.658^{Q\lambda_e^2})F_y \qquad \text{(A-E3-2)}$$

If $\lambda_e \sqrt{Q} > 1.5$

$$F_{cr} = \left(\frac{0.877}{\lambda_e^2}\right)F_y \qquad \text{(A-E3-3)}$$

in which $Q = 1.0$ for elements meeting the width-thickness ratios λ_r of LRFD section B5.1 and if not calculated as described in LRFD Appendixes E3 and B5.3.

$$\lambda_e = \sqrt{\frac{F_y}{F_e}} \qquad \text{(A-E3-4)}$$

F_e = critical flexural-torsional elastic buckling stress

For doubly symmetric shapes

$$F_e = \left[\frac{\pi^2 E C_w}{(K_z L)^2} + GJ\right]\frac{1}{I_x + I_y} \qquad \text{(A-E3-5)}$$

For singly symmetric shapes where y is the axis of symmetry

$$F_e = \frac{F_{ey} + F_{ez}}{2H}\left[1 - \sqrt{1 - \frac{4F_{ey}F_{ez}H}{(F_{ey} + F_{ez})^2}}\right] \qquad \text{(A-E3-6)}$$

For unsymmetrical sections F_e is the lowest root of the following cubic equation

$$(F_e - F_{ex})(F_e - F_{ey})(F_e - F_{ez}) - F_e^2(F_e - F_{ey})\left(\frac{x_0}{\bar{r}_0}\right)^2$$

$$- F_e^2(F_e - F_{ex})\left(\frac{y_0}{\bar{r}_0}\right)^2 = 0 \qquad \text{(A-3E-7)}$$

A perhaps more convenient form of LRFD Formula A-E3-7 follows

$$HF_e^3 + \left[\frac{1}{\bar{r}_0^2}(y_0 F_{ex} + x_0^2 F_{ey}) - (F_{ex} + F_{ey} + F_{ez})\right]F_e^2$$

$$+ (F_{ex}F_{ey} + F_{ex}F_{ez} + F_{ey}F_{ez})F_e - F_{ex}F_{ey}F_{ez} = 0$$

K_z = effective length factor for torsional buckling

G = shear modulus (ksi)

C_w = warping constant (in.^{6})

J = torsional constant (in.^{4})

$$\bar{r}_0^2 = x_0^2 + y_0^2 + \frac{I_x + I_y}{A} \qquad \text{(A-E3-8)}$$

$$H = 1 - \left(\frac{x_0^2 + y_0^2}{\bar{r}_0^2}\right) \qquad \text{(A-E3-9)}$$

$$F_{ex} = \frac{\pi^2 E}{(KL/r)_x^2} \qquad \text{(A-E3-10)}$$

$$F_{ey} = \frac{\pi^2 E}{(KL/r)_y^2} \qquad \text{(A-E3-11)}$$

$$F_{ez} = \left[\frac{\pi^2 E C_w}{(K_z L)^2} + GJ\right]\frac{1}{A\bar{r}_0^2} \qquad \text{(A-E3-12)}$$

The values of C_w, J, $\bar{r}_0$, and H are provided for many sections in the "Flexural-Torsional Properties" tables of Part 1 of the Manual.

In Example 6-6 which follows the author has gone through all of these formulas for a double-angle column. The resulting value for $\phi_c P_n$ is shown to coincide with the value given in the Manual.

EXAMPLE 6-6

Determine (a) the flexural buckling strength and (b) the flexural-torsional buckling strength of an 18-ft pinned-end column consisting of two LS $8\times6\times1/2$ long legs back to back with a 3/8-in. gusset plate between them. The angles are made from A36 steel and are connected to each other at 6-ft intervals with two fully tensioned high-strength bolts as shown in Fig. 6-8. The cross section of the member is shown in Fig. 6-9. $G = 11, 200$ ksi, and $K = 1.0$.

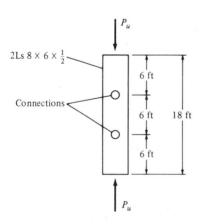

Figure 6-8

Solution

Using two Ls $8\times6\times\frac{1}{2}$ long legs back to back $\frac{3}{8}$-in. gusset plate in between ($A = 13.5$ in.^{2}, $r_x = 2.56$ in., r_z for one $L = 1.30$ in., and $r_y = 2.44$ in.)

(a)   Flexural buckling

$$\left(\frac{KL}{r}\right)_y = \frac{(1.0)(12 \times 18)}{2.44} = 88.52$$

$$\phi_c F_{cr} = 20.26 \text{ ksi}$$

$$\phi_c P_n = (20.26)(13.5) = \underline{\underline{273.5 \text{ k}}}$$

(b)   Flexural-torsional buckling

Cross section of member shown in Fig. 6-9 together with some properties needed in the calculations.

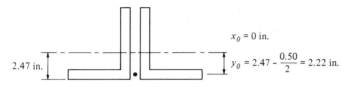

$x_0 = 0$ in.

$y_0 = 2.47 - \dfrac{0.50}{2} = 2.22$ in.

2.47 in.

Figure 6-9

From flexural-torsional properties tables: $J = 0.584$ in.^{4}

$$C_w = 2.280 \text{ in.}^6$$

$$r_0 = 4.18 \text{ in.}$$

$$H = 0.718$$

Checking table values of r_0 and H

$$\bar{r}_0^2 = x_0^2 + y_0^2 + r_x^2 + r_y^2 = 0^2 + 2.22^2 + 2.56^2 + 2.44^2$$

$$\bar{r}_0 = 4.18 \text{ in.}$$

$$H = 1 - \frac{0^2 + 2.22^2}{(4.18)^2} = 0.718$$

Computing values of F_{ey}, F_{ez}, and F_e

$$\frac{a}{r_z} = \frac{72}{1.30} = 55.38 > 50 \therefore \text{ Use } \left(\frac{KL}{r}\right)_m \text{ rather than } \frac{K_y L_y}{r_y}$$

$$\left(\frac{KL}{r}\right)_y = \frac{(1.0)(12 \times 18)}{2.44} = 88.52$$

$$\left(\frac{KL}{r}\right)_m = \sqrt{(88.52)^2 + (55.38 - 50)^2} = 88.68$$

$$F_{ey} = \frac{(\pi^2)(29,000)}{(88.68)^2} = 36.40 \text{ ksi}$$

$$F_{ez} = \left[\frac{(\pi)^2(29,000)(2 \times 2.28)}{(1.0 \times 12 \times 18)^2} + 11,200 \times 2 \times 0.584 \right] \frac{1}{(13.5)(4.18)^2}$$

$$= 55.58 \text{ ksi}$$

$$F_e = \frac{36.40 + 55.58}{(2)(0.718)} \left[1 - \sqrt{1 - \frac{(4)(36.40)(55.58)(0.718)}{(36.40 + 55.58)^2}} \right]$$

$$= 28.20 \text{ ksi}$$

$$\lambda_e = \sqrt{\frac{F_y}{F_e}} = \sqrt{\frac{36}{28.20}} = 1.130 < 1.5$$

$$\phi_c F_{cr} = (0.85)(0.911 \times 36)(0.658)^{1.130^2} = 16.34 \text{ ksi}$$

where F_y is multiplied by $Q_s = 0.911$ from manual page 1–89 as a slender unstiffened element

$$\phi_c P_n = (16.34)(13.5) = \underline{220.6 \text{ k}}$$

which checks LRFD table value page 2-59. This is an 11.5 percent reduction in load from the flexural buckling value.

6-6 LACING AND TIE PLATES

The necessity for built-up compression members in large buildings and bridges has previously been discussed in Chapter 5. When members are built up from more than one section it is necessary for them to be connected or laced together across their open sides. The purpose of lacing or lattice work is to hold the various parts parallel and the correct distance apart and to equalize the stress distribution between the various parts. The student will understand the necessity for lacing if he or she considers a built-up member consisting of several sections (such as the four-angle member of Fig 5-1) which supports a heavy compressive load. Each of the parts will tend to individually buckle laterally unless they are tied together to act as a unit in supporting the load. In addition to lacing it is necessary to have tie plates (also called stay plates or batten plates) as near the ends of the member as possible and at intermediate points if the lacing is interrupted. Parts (a) and (b) of Fig. 6-10 show arrangements of tie plates and lacing. Other possibilities are shown in parts (c) and (d) of the same figure.

The failure of several structures in the past has been attributed to inadequate lacing of built-up compression members. Perhaps the best-known example was the failure of the Quebec Bridge in 1907. Following its collapse the general opinion was that the lattice work of the compression chords was too weak and resulted in failure.

Dimensions of tie plates and lacing are usually controlled by specifications. In their Section E, Part 6 the LRFD Specification states that tie plates shall have

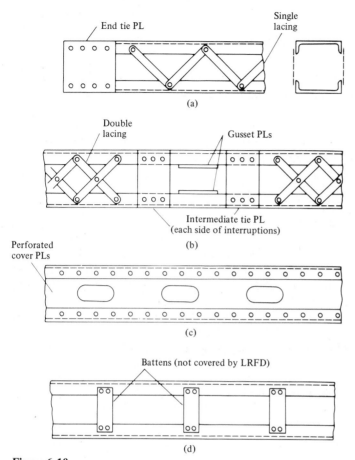

Figure 6-10

a thickness at least equal to one-fiftieth of the distance between the connection lines of welds or other fasteners and shall have a length parallel to the axis of the main member at least equal to the distance between the connection lines.

Lacing usually consists of flat bars but may occasionally consist of angles, perforated cover plates, channels, or other rolled sections. These pieces must be so spaced that the individual parts being connected will not have L/r values between connections which exceed the governing value for the entire built-up member. (The governing value is KL/r for the whole built-up section.) Lacing is assumed to be subjected to a shearing force normal to the member equal to not less than 2 percent of the compression design strength $\phi_c P_n$ of the member. The LRFD column formulas are used to design the lacing in the usual manner. Slenderness ratios are limited to 140 for single lacing and 200 for double lacing. Double lacing or single lacing made with angles should be used if the distance between connection lines is greater than 15 in.

Rather than using tie plates and lacing bars it is permissible to use continuous cover plates over the open sides of built-up sections. Access holes are necessary, and these plates are referred to as perforated cover plates. Stress concentrations and secondary bending stresses are usually neglected but lateral shearing forces must be checked as they are for other types of lattice work. (The unsupported width of such plates at access holes is assumed to contribute to the design strength $\phi_c P_n$ of the member if the conditions as to sizes, width-thickness ratios, etc., described in LRFD specification E4 are met.) Perforated cover plates are attractive to many designers because of several advantages they possess.

1. They are easily fabricated with modern gas cutting methods.
2. Some specifications permit the inclusion of their net areas in the effective section of the main members, provided the holes are made in accordance with their empirical requirements which have been developed on the basis of extensive research.
3. Painting of the members is probably simplified as compared with ordinary lacing bars.

Example 6-7 illustrates the design of lacing and end tie plates for the built-up column of Example 6-5. Bridge specifications are somewhat different in their lacing requirements from the LRFD but the design procedures are much the same.

EXAMPLE 6-7

Using the LRFD Specification design bolted single lacing for the column of Example 6-5. Reference is made to Fig. 6-11. Assume $\frac{3}{4}$-in. bolts.

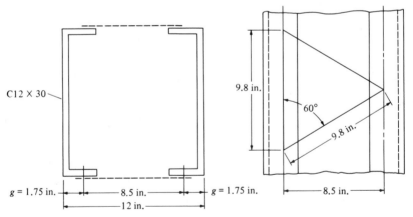

Figure 6-11

Solution

Distance between lines of bolts is 8.5 in. < 15 in.; therefore, single lacing OK.

Assume lacing bars are inclined at 60° with axis of member. Length of channels between lacing connections is 8.5/cos30° = 9.8 in., and L/r of 1 channel between connections is 9.8/0.763 = 12.9 < 56, which is L/r of main member previously determined in Example 6-5.

Force on lacing bar:

$$V_u = \begin{array}{l} 0.02 \text{ times design compressive strength of member} \\ \text{(from Example 6-5)} \end{array}$$

$$V_u = (0.02)(457.8) = 9.16 \text{ k}$$

$$\frac{1}{2} V_u = 4.58 \text{ k} = \text{shearing force on each plane of lacing}$$

Force in bar (with reference to bar dimensions in Fig. 6-7)

$$= \left(\frac{9.8}{8.5}\right)(4.58) = 5.28 \text{ k}$$

Properties of flat bar

$$I = \tfrac{1}{12}bt^3$$

$$A = bt$$

$$r = \sqrt{\frac{\tfrac{1}{12}bt^3}{bt}} = 0.289t$$

Design of bar:

$$\text{Assume } \frac{L}{r} = \text{maximum value of } 140$$

$$\frac{9.8}{0.289t} = 140$$

$$t = 0.242 \text{ in. (try } \tfrac{1}{4}\text{-in. flat bar)}$$

$$\frac{L}{r} = \frac{9.8}{(0.289)(0.250)} = 136$$

$$\phi_c F_{cr} = 11.54 \text{ ksi}$$

$$\text{Area required} = \frac{5.28}{11.54} = 0.458 \text{ in.}^2 \ (1.83 \times \tfrac{1}{4} \text{ needed})$$

Minimum edge distance if $\tfrac{3}{4}$-in. bolt used $= 1\tfrac{1}{4}$ in.

$\therefore$ Minimum length of bar $= 9.8 + (2)(1\tfrac{1}{4}) = 12.3$ in. say 14 in.

$$\underline{\text{Use } \tfrac{1}{4} \times 2\tfrac{1}{2} \times 1 \text{ ft 2 in. bars}}$$

Design of end tie plates:

Minimum length $= 8.5$ in.

Minimum $t = (\tfrac{1}{50})(8.5) = 0.17$ in.

Minimum width $= 8.5 + (2)(1\tfrac{1}{4}) = 11$ in.

$$\underline{\text{Use } \tfrac{3}{16} = 8\tfrac{1}{2} \times 0 \text{ ft 12 in. end tie plates}}$$

PROBLEMS

All of the columns for these problems are assumed to be in frames that are braced against sidesway.

For Probs. 6-1 to 6-3 use a trial-and-error procedure in which a KL/r value is estimated, a value of $\phi_c F_{cr}$ determined from the appropriate LRFD table (3-36 or 3-50 in Part 6 of the Manual), the estimated column area determined, a trial section selected, its P_u computed, and then another size tried if necessary, and so on.

6-1. Select the lightest available W12 section to support the axial compression loads $P_D = 110$ k and $P_L = 125$ k if $KL = 14$ ft and A36 steel is used. (*Ans.* W12×53)

6-2. Select the lightest available W14 section to support the axial loads $P_D = 200$ k and $P_L = 300$ k if $KL = 12$ ft and A36 steel is used.

6-3. Repeat Prob. 6-2 if $F_y = 50$ ksi. (*Ans.* W14×74)

For the solution of Probs. 6-4 to 6-20 take advantage of all available column tables in the Manual, particularly those in Part 2.

6-4. Repeat Prob. 6-1.

6-5. Repeat Prob. 6-2. (*Ans.* W14×90)

6-6. Repeat Prob. 6-3.

6-7. Several building columns are to be designed using A36 steel and the LRFD Specification. Select the lightest available W sections for these columns that are described as follows:
(a) $P_u = 500$ k, $L = 12$ ft, pinned end supports. (*Ans.* W12×65)
(b) $P_u = 450$ k, $L = 14$ ft, fixed end supports. (*Ans.* W12×58 or W8×58)
(c) $P_u = 720$ k, $L = 16$ ft 6 in., fixed at bottom, pinned at top. (*Ans.* W14×90)
(d) $P_u = 1900$ k, $L = 15$ ft, pinned end supports. (*Ans.* W12×252.A W14×233 almost works.)

6-8. Select the lightest W section available in A36 steel to serve as a pinned end column to support the following axial loads: $P_D = 300$ k and P_W due to wind $= 400$ k. Assume KL is 14 ft.

6-9. A W section is to be selected to support an axial compressive load $P_u = 1630$ k. The member, which is to be 24 ft long and is to be pinned top and bottom, has lateral support (pinned) supplied in the weak direction at middepth. Select the lightest W12 or 14 section using A36 steel. (*Ans.* W14×211 or W12×210)

6-10. Repeat Prob. 6-9 if $F_y = 50$ ksi.

6-11. Repeat Prob. 6-9 if lateral support is provided in the weak direction at the one-third points and the total column length is changed to 33 ft. (*Ans.* W14×233 or W12×230)

6-12. A 27-ft column is laterally supported in the weak direction at its middepth. Select the lightest W section that can adequately support the axial gravity loads $P_D = 150$ k and $P_L = 100$ k using A36 steel. Assume all K's are 1.0.

6-13. A 14-ft column is to be built into a wall in such a manner that it will be continuously braced in its weak direction but not in its strong direction. If the member is to consist of A36 steel and is assumed to have pinned ends, select the lightest satisfactory

W12 section available using the LRFD Specification. $P_u = 1000$ k. (*Ans.* W12×120)

6-14. Repeat Prob. 6-13 if $F_y = 50$ ksi.

6-15. A W14 section of A36 steel is to be selected to support the axial compressive loads $P_D = 200$ k and $P_L = 325$ k. The member which is to be 30 ft long is to be fixed top and bottom and is to have lateral support at its one-third points perpendicular to the y axis (pinned). (*Ans.* W14×99)

6-16. Using A36 steel (except $F_y = 46$ ksi for square and rectangular tubing) select the lightest available rolled sections (W, M, S, HP, square, rectangular, or round tubing) that are adequate for the following situations:
(a) $P_u = 300$ k, $L = 14$ ft, pinned ends.
(b) $P_u = 420$ k, $L = 15$ ft, fixed ends.
(c) $P_u = 740$ k, $L = 20$ ft, one end pinned and the other fixed.

6-17. Assuming axial loads only, select W sections for an interior column of the frame shown in the accompanying illustration. Use A36 steel and the LRFD Specification. Each column section can be used for one or two stories before it is spliced, whichever seems advisable. Miscellaneous data: concrete weighs 150 lb/ft³. LL on roof = 30 psf. Roofing = 6 psf. LL on interior floors = 80 psf. Partition load on interior floors = 15 psf. All points pinned. Frames 30 ft on center. (One answer. W14×61 top two stories and W14×132 bottom two stories)

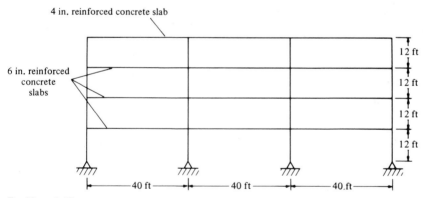

Problem 6-17

6-18. It is desired to design a column for $P_u = 2100$ k using A36 steel with $KL = 12$ ft. A W14×145 is on hand with a good supply of $\frac{3}{4}$-in. thick plates. Design cover plates to be welded to the flanges of the W section to enable the column to support the required load.

6-19. Four 4×4×$\frac{1}{2}$ angles are used to form the member shown in the accompanying illustration. The member is 30 ft long, has pinned ends, and consists of A36 steel. Determine the design compressive strength of the member. Design single lacing and end tie plates assuming connections are made to the angles with $\frac{3}{4}$-in. bolts. (*Ans.* $P_u = 411.6$ k; end tie plates $\frac{9}{32}$×13×1 ft 4 in.; single lacing at 60° $\frac{3}{8}$×2$\frac{1}{2}$×1 ft 5$\frac{1}{2}$ in.)

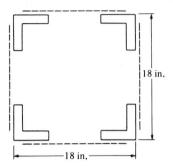

Problem 6-19

6-20. Select a pair of standard channels to support an axial compressive load of $P_u = 700$ k. The member is to be 24 ft long with both ends pinned and is to be arranged as shown in the accompanying illustration. Use A36 steel, and design single lacing and end tie plates assuming $\frac{3}{4}$-in. bolts are to be used for connections. Assume bolts located $2\frac{1}{4}$ in. from back of channels.

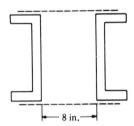

Problem 6-20

Chapter 7

Design of Axially Loaded Compression Members Continued

7-1 FURTHER DISCUSSION OF EFFECTIVE LENGTHS

The subject of effective lengths was introduced in Chapter 5 and some suggested *K* factors were presented in Table 5-1. These factors were developed for columns with certain idealized conditions of end restraint which may be very different from practical design conditions. The table values are usually quite satisfactory for preliminary designs and for situations where sidesway is prevented by bracing.

The effective length of a column is a property of the whole structure of which the column is a part. In many smaller buildings masonry walls provide sufficient lateral support to prevent sidesway and to keep the value of *K* = 1.0 or less. A large percentage of modern buildings, however, particularly the taller ones, have only light curtain walls which provide little or no lateral bracing. The result is sidesway and *K* values larger than 1.0 unless a definite lateral bracing system is used. For buildings without such bracing systems it seems logical to assume that resistance to sidesway is primarily provided by the lateral stiffness of the frame alone.

Perhaps a few explanatory remarks should be made at this time concerning the definition of sidesway as it pertains to effective lengths. In this discussion sidesway refers to a sidesway type of buckling. This not only includes sidesway as used in the analysis of statically indeterminate frames (where the frames deflect laterally due to the presence of lateral loads or unsymmetrical vertical loads or where the frames themselves are unsymmetrical) but also to columns whose ends can move transversely if the columns are loaded until buckling occurs.

Shearson Lehman/American Express Information Services Center, New York City. (Courtesy Owen Steel Company, Inc.)

Theoretical mathematical analyses may be made to determine effective lengths in structures with sidesway. Such a procedure, however, is normally too lengthy and perhaps too difficult for the average designer. The usual procedure is to either use Table 5-1, interpolating between the idealized values contained there as the designer feels appropriate, or to use LRFD Table C-C2.2 (Alignment Chart for Effective Length of Columns in Continuous Frames), which is reproduced here as Fig. 7-1.[1,2] This chart is to be used to estimate the effective lengths of columns in frames where resistance to lateral movement is provided by the stiffness of the members of the frames. To use the chart it is necessary to obtain the sizes of the girders and columns framing into the column in question

---

[1]B.G. Johnson, editor, *Guide to Stability Design for Metal Structures,* 3d ed. (New York: Wiley, 1976), p. 420.

[2]O.G. Julian and L.S. Lawrence, "Notes on J and L Nomograms for Determination of Effective Lengths," unpublished, 1959.

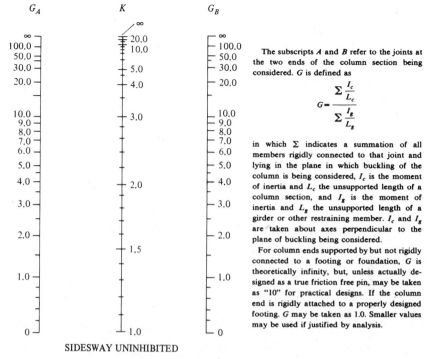

The subscripts A and B refer to the joints at the two ends of the column section being considered. G is defined as

$$G = \frac{\Sigma \dfrac{I_c}{L_c}}{\Sigma \dfrac{I_g}{L_g}}$$

in which Σ indicates a summation of all members rigidly connected to that joint and lying in the plane in which buckling of the column is being considered, I_c is the moment of inertia and L_c the unsupported length of a column section, and I_g is the moment of inertia and L_g the unsupported length of a girder or other restraining member. I_c and I_g are taken about axes perpendicular to the plane of buckling being considered.

For column ends supported by but not rigidly connected to a footing or foundation, G is theoretically infinity, but, unless actually designed as a true friction free pin, may be taken as "10" for practical designs. If the column end is rigidly attached to a properly designed footing. G may be taken as 1.0. Smaller values may be used if justified by analysis.

SIDESWAY UNINHIBITED

Figure 7-1 Alignment chart for effective length of columns in continuous frames.

before its effective length can be determined. In other words, before the chart can be used a trial design has to be made of each of the members.

The Structural Stability Research Council (SSRC) makes several recommendations concerning the use of the alignment charts. Some of these pertain to column supports and are shown in Fig. 7-1. For instance, G (which is the ratio of the sum of the stiffnesses of the columns connected at a joint to the sum of the stiffnesses of the beams and girders connected at the same joint) is theoretically infinite where a column is connected to a footing with a frictionless hinge. Practically it is recommended that G be taken $= 10$ where nonrigid supports are used. For rigid connections of columns to footings G theoretically approaches zero but a value of 1.0 is recommended.

For further improvement of the G values used in the chart the SSRC recommends that girder stiffnesses be multiplied by certain factors when certain conditions are known to exist at the far ends of the girders. If sidesway is permitted and the far end of the girder is hinged the girder stiffness should be multiplied by 0.5.

The effective lengths of each of the columns of a frame are estimated with the alignment chart in Example 7-1. (When sidesway is possible it will be found that the effective lengths are always greater than the actual lengths as is illustrated in this example. When frames are braced in such a manner that sidesway is not possible K will be less than 1.0.) An initial design has provided preliminary sizes for each of the members in the frame of Example 7-1. After the effective lengths are determined each column can be redesigned. Should the sizes change appreciably, new effective lengths can be determined, the column designs repeated, etc. Several tables are used in the solution of this example. These should be self-explanatory after the directions on the alignment chart are examined.

For most buildings the values of K_x and K_y should be examined separately. The reason for such individual study lies in the different possible framing conditions in the two directions. Many multistory frames consist of rigid frames in one direction and conventionally connected frames with sway bracing in the other. In addition the points of lateral support may often be entirely different in the two planes.

EXAMPLE 7-1

Determine the effective lengths of each of the columns of the frame shown in Fig. 7-2 using the alignment chart given in Fig. 7-1. The tentative sizes of each member are given in the figure.

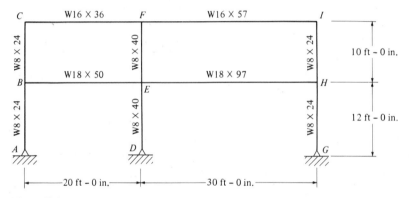

Figure 7-2

Solution

Stiffness factors:

Member	Shape	I	L	I/L
AB	W8×24	82.8	144	0.575
BC	W8×24	82.8	120	0.690
DE	W8×40	146	144	1.014
EF	W8×40	146	120	1.217
GH	W8×24	82.8	144	0.575
HI	W8×24	82.8	120	0.690
BE	W18×50	800	240	3.333
CF	W16×36	448	240	1.867
EH	W18×97	1,750	360	4.861
FI	W16×57	758	360	2.106

G factors for each joint:

Joint	$\Sigma(I_c/L_c)/\Sigma(I_g/L_g)$	G
A	See Fig. 8-5	10.0
B	$\dfrac{0.575 + 0.690}{3.333}$	0.380
C	$\dfrac{0.690}{1.867}$	0.370
D	See Fig. 8-5	10.0
E	$\dfrac{1.014 + 1.217}{3.333 + 4.861}$	0.272
F	$\dfrac{1.217}{1.867 + 2.106}$	0.306
G	See Fig. 8-5	10.0
H	$\dfrac{0.575 + 0.690}{4.861}$	0.260
I	$\dfrac{0.690}{2.106}$	0.328

Column K factors from chart (Fig. 7-1).

Column	G values at column ends		K
AB	10.0	0.380	1.72
BC	0.380	0.370	1.12
DE	10.0	0.272	1.70
EF	0.272	0.306	1.09
GH	10.0	0.260	1.70
HI	0.260	0.328	1.10

There are actually two alignment charts presented in the LRFD Manual (on their page 2-5). The chart shown in Fig. 7-1 of this chapter is for cases where sidesway is permitted while the one of Fig. 7-3 is for frames where sidesway is prevented. For the latter case it will be noted that K must equal 1.0 or less. For better results the girder stiffnesses are again modified for certain end conditions. If the far ends are hinged the stiffnesses are multiplied by 1.5 and by 2.0 if the far ends are fixed.

The alignment chart of Fig. 7-1 for frames with sidesway uninhibited always indicates that $K \geq 1.0$. In fact K factors of 2.0 to 3.0 are common and even larger values are occasionally obtained. To many designers such large factors seem completely unreasonable. If the designer obtains seemingly high K factors he or she should carefully review the numbers used to enter the chart (that is, the G values) as well as the basic assumptions used in preparing the charts. These assumptions are discussed in the next two sections of this chapter.

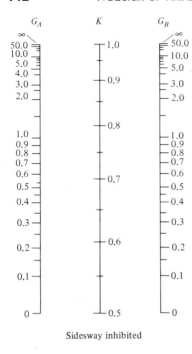

Figure 7-3 Alignment chart for effective length of columns in continuous frames.

7-2 STIFFNESS REDUCTION FACTORS

The alignment charts were developed on the basis of a set of idealized conditions which are seldom if ever completely met in a real structure. A complete list of these assumptions is shown on page 6-152 of the LRFD Manual. A few of them are: column behavior is purely elastic, all columns buckle simultaneously, all members have constant cross sections, all joints are rigid, and so on.

If the actual conditions are different from these assumptions unrealistically high K factors may be obtained from the charts and overconservative designs may result. A large percentage of columns will fail in the inelastic range but the alignment charts were prepared assuming elastic failure. This situation previously discussed in Chapter 5 is illustrated in Fig. 7-4. For such cases the chart K values are too conservative and should be corrected as described in this section.

In the elastic range the stiffness of a column is proportional to EI where $E = 29,000$ ksi, while in the inelastic range its stiffness is more accurately proportional to EI where E is a reduced or tangent modulus.

The buckling strength of columns in framed structures was shown in the alignment charts to be related to

$$G = \frac{\text{column stiffness}}{\text{girder stiffness}} = \frac{\Sigma(EI/L) \text{ columns}}{\Sigma (EI/L) \text{ girders}}$$

If the columns behave elastically the modulus of elasticity will be canceled from the preceding expression for G. If, however, the column behavior is inelastic (that is, if $\lambda < 1.5$) the column stiffness factor will be smaller and will equal

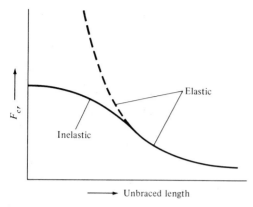

Figure 7-4

$E_T I/L$. As a result the G factor used to enter the alignment chart will be smaller and the K factor selected from the chart will be smaller.

Though the alignment charts were developed for elastic column action they may be used for an inelastic column situation if the G value is multiplied by a correction factor called the *stiffness reduction factor* (SRF). This reduction factor equals the tangent modulus over the elastic modulus (E_T/E) and is approximately equal to $F_{cr\ \text{inelastic}}/F_{cr\ \text{elastic}} \approx P_u/A/F_{cr\ \text{elastic}}$. Values of this correction are shown for various P_u/A values in the LRFD Manual. On their pages 2-6 and 2-7 a direct method for considering inelastic buckling is presented. The steps involved follow:

1. Calculate P_u and select a trial column size.
2. Calculate P_u/A and pick the SRF from Table A in Part 2 of the Manual. (If P_u/A is less than the values given in the table the column is in the elastic range and no reduction needs to be made.)
3. The value of G_{elastic} is computed and multiplied by the SRF and K is picked from the chart.
4. The effective slenderness ratio KL/r is computed, $\phi_c F_{cr}$ obtained from the Manual and multiplied by the column area to obtain P_u. If this value is appreciably different from the value computed in step 1, another trial column size is attempted and the four steps are repeated.

Example 7-2 which follows illustrates these steps for the design of a column in a frame subject to sidesway. *It will be noticed in this example that the author has only considered in-plane behavior and only bending about the x axis.* As a result of inelastic behavior the effective length factor is appreciably reduced.

EXAMPLE 7-2

Select a W12 section for column AB of the frame shown in Fig. 7-5 (a) assuming elastic column behavior and (b) assuming inelastic column behavior. $P_u = 1240k$ and A36 steel is used. The columns above and below

AB are assumed to be approximately the same size as *AB*. Consider only in-plane behavior.

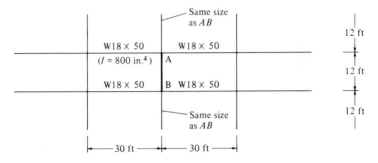

Figure 7-5

Solution

(a) Assuming column in elastic range

Try W12×170 ($A = 50.0$ in.^{2}, $I_x = 1650$ in.^{4}, $r_x = 5.74$ in.)

$$G_A = G_B = \frac{\Sigma(I_c/L_c)}{\Sigma(I_g/L_g)} = \frac{(2)(1650/12)}{(2)(800/30)} = 5.16$$

$K = 2.30$ from Fig. 7-1 alignment chart

$$\frac{KL}{r} = \frac{(2.30)(12 \times 12)}{5.74} = 57.7$$

$\phi_c F_{cr} = 25.68$ ksi

$$P_u = (25.68)(50.0) = 1284 \text{ k} > 1240 \text{ k} \qquad\qquad \text{OK}$$

$$\underline{\text{Use W12×170}}$$

(b) Inelastic solution

Try a lighter section W12×152 ($A = 44.7$ in.^{2},

$$I_x = 1430 \text{ in.}^4, r_x = 5.66 \text{ in.})$$

$$\frac{P_u}{A} = \frac{1240}{44.7} = 27.74 \text{ ksi}$$

SRF = 0.547 from page 2-9 in LRFD Manual

∴ Column is in inelastic range

$$G_A = G_B = \frac{\Sigma(I_c/L_c)}{\Sigma(I_g/L_g)} (\text{SRF})$$

$$= \frac{(2)(1430/12)}{(2)(800/30)} (0.547) = 2.44$$

$K = 1.68$ from Fig. 7.1 alignment chart

$$\frac{KL}{r} = \frac{(1.68)(12 \times 12)}{5.66} = 42.74$$

$\phi_c F_{cr} = 27.79$ ksi

$P_u = (27.79)(44.7) = 1242$ k > 1240 k   OK

Use W12×152

7-3 COLUMNS LEANING ON EACH OTHER FOR IN-PLANE DESIGN

When we have an unbraced frame with beams rigidly attached to columns it is safe to design each individually using the sidesway uninhibited alignment chart to obtain the K factors (which will probably be appreciably larger than 1.0). If this procedure is followed, however, the result may be overly conservative.

A column cannot buckle by sidesway unless all of the columns on that story buckle by sidesway. One of the assumptions on which the alignment chart of Fig. 7-1 was prepared was that all of the columns on the story in question would buckle at the same time. If this assumption is correct the columns cannot support or brace each other because if one gets ready to buckle they all supposedly are ready to buckle.

In some situations, however, certain columns in a frame have some excess buckling strength. If, for instance, the buckling loads of the exterior columns of the unbraced frame of Fig. 7-6 have not been reached when the buckling loads of the interior columns are reached the frame will not buckle. The interior columns in effect will lean against the exterior columns, that is, the exterior columns will brace the interior ones. For this situation shear resistance is provided in the exterior columns which resists the sidesway tendency.[3]

There are many practical situations where some columns have excessive buckling strength. This might happen when the designs of different columns on a particular story are controlled by different loading conditions. For such cases failure of the frame will occur only when the gravity loads are increased sufficiently to offset the extra strength of the lightly loaded columns. As a result the critical loads for the interior columns of Fig. 7-6 are increased and in effect their effective lengths are decreased. In other words if the exterior columns are bracing the interior ones against sidesway the K factors for those interior columns are approaching 1.0. Yura[3] says that the effective length of some of the columns in a frame subject to sidesway can be reduced to 1.0 in this type of situation even though there is no apparent bracing system present.

The net effect of the information presented here is that the total gravity load which an unbraced frame can support equals the sum of the strength of the

[3]J.A. Yura, "The Effective Length of Columns in Unbraced Frames," *Engineering Journal,* AISC, 8, no. 2 (second quarter, 1971), pp. 37–42.

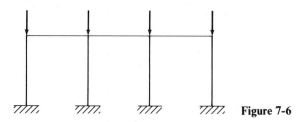

Figure 7-6

individual columns. In other words, the total gravity load that will cause sidesway buckling in a frame can be split up among the columns in any proportion just so long as the maximum load applied to any one column does not exceed the maximum load that column could support if it were braced against sidesway with $K = 1.0$.

For this discussion the unbraced frame of Fig. 7-7(a) is considered. It is assumed that each column has a $K = 2.0$ and will buckle under the loads shown.

When sidesway occurs the frame will lean to one side as shown in part (b) of the figure and $P\Delta$ moments equal to 200Δ and 700Δ will be developed.

Suppose that we load the frame with 200 k on the left-hand column and 500 k on the right-hand column (or 200 k less than we had before). We know that for this situation, which is shown in part (c) of the figure, the frame will not buckle by sidesway until we reach a moment of 700Δ at the right-hand column base. This means that our right-hand column can take an additional moment of 200Δ. Thus as Yura says the right-hand column has a reserve of strength that can be used to brace the left-hand column and prevent its sidesway buckling.

Obviously the left-hand column is now braced against sidesway and sidesway buckling will not occur until the moment at its base reaches 200Δ. Therefore, it can be designed with a K factor less than 2.0 and can support an additional load of 200 k, giving it a total load of 400 k—but this load must not be greater than would be its capacity if it were braced against sidesway with $K = 1.0$. It should be mentioned that the total load the frame can carry is still 900 k as in part (a) of the figure.

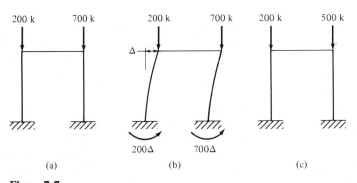

(a)   (b)   (c)

Figure 7-7

The advantage of the frame behavior described here is illustrated in Example 7-3 where it is assumed that the interior columns of a frame are braced against sidesway by the exterior columns. As a result the interior columns are assumed to each have K factors equal to 1.0. They are designed for the factored loads shown (660 k each). Then the K factors are determined for the exterior columns with the sidesway uninhibited chart of Fig. 7-1 and they are each designed for column loads equal to $440 + 660 = 1100$ k.

EXAMPLE 7-3

For the frame of Fig. 7-8, which consists of A36 steel, beams are rigidly connected to the exterior columns while all other connections are simple. The columns are braced top and bottom against sidesway out of the plane of the frame so that $K = 1.0$ in that direction. Sidesway is possible in the plane of the frame. Design the interior columns assuming $K = 1.0$ and the exterior columns with K as determined from the alignment chart and $P_u = 1100$ k. (With this approach to column buckling, the interior columns could carry no load at all since they appear to be unstable under sidesway conditions.)

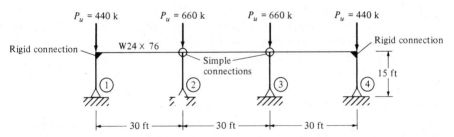

Figure 7-8

Solution

Design of interior columns

Assume $K = 1.0$, $KL = (1.0)(15) = 15$ ft, $P_u = 660$ k

Use W14×90

Design of exterior columns

Out of plane $K_y = 1.0$, $P_u = 440$ k
In plane $P_u = 440 + 660 = 1100$ k, K_x to be determined from alignment chart.

<u>Try W14×159 ($A = 46.7$ in.^{2}, $I_x = 1900$ in.^{4}, $r_x = 6.38$ in.)</u>

$$G_{top} = \frac{1900/15}{2100/30 \times 0.5} = 3.62$$

(noting that girder stiffness is multiplied by 0.5 since sidesway is permitted and far end of girder is hinged.)

$$G_{bottom} = 10$$

$$K_x = 2.40$$

$$\frac{K_x L_x}{r_x} = \frac{(2.40)(12 \times 15)}{6.38} = 67.71$$

$$\phi_c F_{cr} = 24.04 \text{ ksi}$$

$$P_u = (24.04)(46.7) = 1122.7 \text{ k} > 1100 \text{ k} \qquad \text{OK}$$

$$\underline{\underline{\text{Use W14} \times 159}}$$

7-4 BASE PLATES FOR CONCENTRICALLY LOADED COLUMNS

The design compressive stress in a concrete or other type of masonry footing is much smaller than it is in a steel column. When a steel column is supported by a footing it is necessary for the column load to be spread over a sufficient area to keep the footing from being overstressed. Loads from steel columns are transferred through a steel base plate to a fairly large area of the footing below. (It will be noted that a footing performs a related function in that it spreads the load over an even larger area so that the underlying soil will not be overstressed.)

The base plates for steel columns can be welded directly to the columns or they can be fastened by means of some type of bolted or welded lug angles. These connection methods are illustrated in Fig. 7-9. A base plate welded directly to the column is shown in part (a) of the figure. For small columns these plates are probably shop-welded to the columns but for larger columns it may be necessary to ship the plates separately and set them to the correct elevations. The columns

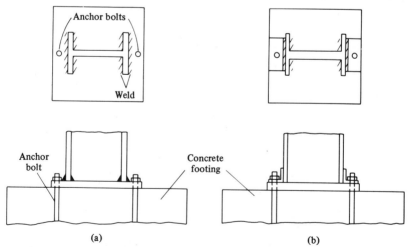

Figure 7-9 Column base plates.

are then set and connected to the footing with anchor bolts which pass through the lug angles which have been shop-welded to the columns. This type of arrangement is shown in part (b) of the figure. Some designers like to use lug angles on both flanges and web.

The lengths and widths of column base plates are usually selected in multiples of even inches and their thicknesses in multiples of $\frac{1}{8}$ in. To make sure that column loads are spread uniformly over their base plates it is essential to have good contact between the two. To accomplish this objective it is necessary to straighten plates thicker than 2 in. up through 4 in. by pressing or milling. Plates thicker than 4 in. need to have all of their bearing surfaces milled except for two situations. If the bottom surfaces of the plates are to be in contact with cement grout to ensure full bearing contact on the foundation they do not have to be milled. Furthermore the top surfaces of plates thicker than 4 in. do not have to be milled if full-penetration welds (to be described in Chapter 14) are used between the columns and base plates.

Initially columns will be considered which support average-size loads. Should the loads be very small so that base plates are very small the design procedure will have to be revised as described later in this section.

The method of design presented is the one recommended by the LRFD Manual. To analyze the base plate shown in Fig. 7-10 the column is assumed to apply a total load equal to P_u to the base plate and this load is assumed to be transmitted uniformly through the base plate to the footing below with a pressure of P_u/A where A is the area of the base plate. The footing will push back with a pressure of P_u/A and will tend to curl up the cantilevered parts of the base plate outside the column as shown in the figure. This pressure also tends to push up the part of the base plate between the flanges of the column.

With reference to Fig. 7-10 the LRFD Manual suggests that maximum moments in a base plate occur at distances approximately $0.80b_f$ and $0.95d$ apart. The bending moment is calculated at each of these sections and the larger value used to determine the plate thickness needed. This method of analysis is only a rough approximation of the true conditions because the actual plate stresses are caused by a combination of bending in two directions.

Plate Area   The design strength of the concrete in bearing beneath the base plate must at least equal the load to be carried. When the base plate covers the entire area of the concrete support (a pedestal) this strength equals ϕ_c (which is 0.60 for bearing on concrete) times the nominal strength of the concrete 0.85 $f'_c A_1$ (where f'_c is the 28-day compression strength of the concrete in ksi and A_1 is the area of the plate).

$$P_u = \phi_c(0.85 f'_c A_1)$$

Should the full area of the concrete support not be covered by the plate the concrete under the bearing plate surrounded by concrete outside the bounds of the plate will be somewhat stronger and its nominal strength is computed by the following expression in which A_2 is the maximum area of the portion of the sup-

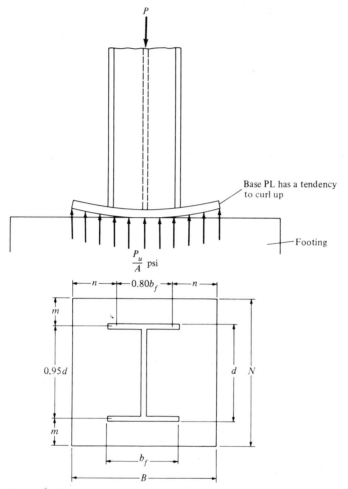

Figure 7-10

porting surface which is geometrically similar to and concentric with the loaded area.

$$P_u = \phi_c(0.85 f'_c A_1) \sqrt{\frac{A_2}{A_1}} \le \phi_c\, 1.7 f'_c A_1$$

Plate Thickness With reference to Fig. 7-10 we take moments in each direction as though the plate is cantilevered out by the dimension m or n. For instance, the moments are $(P_u/A_1)(m)(B)(m/2)$ or $(P_u/A_1)(n)(A)(n/2)$.

It will be noted that if we could make m and n equal we would keep the two moments equal and thus keep the plate thickness to a minimum. This condition can be approached when

$$N \approx \sqrt{A_1} + \Delta$$

where $\Delta = 0.5(0.95\,d - 0.80\,b_f)$ and $B \approx \dfrac{A_1}{N}$

The larger of these two moments is equated to the design bending strength of the plate whose thickness is t_p. The design bending strength $(\phi_b F_y Z)$ is discussed in Chapters 8 and 9. The resulting t_p values are:

$$t_p = m\sqrt{\frac{2\,P_u}{0.9\,F_y BN}}$$

$$t_p = n\sqrt{\frac{2\,P_u}{0.9\,F_y BN}}$$

From these expressions the minimum areas of base plates may be determined with the following results.

Minimum base plate area for the upper limit of concrete bearing strength

$$A_1 = \frac{1}{A_2}\left[\frac{P_u}{0.6(0.85\,f'_c)}\right]^2$$

or

$$A_1 = \frac{P_u}{0.6(1.70\,f'_c)}$$

Furthermore A_1 may not be less than the depth of the column times its flange width.

$$A_1 = b_f d$$

If columns are lightly loaded the load is assumed to be distributed over the cross-hatched area shown in Fig. 7-11. The thickness required for the base plate is determined by taking the bearing on the H-shaped area and calculating the moment for the base plate acting as a cantilever fixed to the web or flange of the column. The length of the cantilever is represented by the letter c in the figure and is measured from the center line of the beam web or flange.

The LRFD Manual then determines for lightly loaded columns the part of P_u applied to the area enclosed by the column $(b_f d)$ and they call this P_0.

$$P_0 = \left(\frac{P_u}{BN}\right)(b_f d)$$

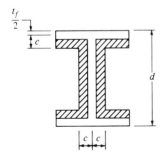

Figure 7-11

They calculate the area of the H-shaped region by dividing P_0 by the permissible pressure

$$A_H = \frac{P_0}{0.6\sqrt{A_2/b_f d}\,f_c'} \le \frac{P_0}{0.6(1.70\,f_c')}$$

Using the complicated dimensions of the H-shaped area, c can be determined from A_H as follows:

$$c = \tfrac{1}{4}[d + b_f - t_f - \sqrt{(d + b_f - t_f)^2 - 4(A_H - t_f b_f)}]$$

Finally the thickness of the base plate for this case is determined by taking moments of the pressure in the cantilever of c length and equating it to the bending strength of the plate of thickness t. The results are

$$t_p = c\sqrt{\frac{2\,P_0}{0.9\,F_y A_H}}$$

The thickness of column base plates is the largest value determined from the following three equations.

$$t_p = m\sqrt{\frac{2\,P_u}{0.9\,F_y BN}} \qquad t_p = n\sqrt{\frac{2\,P_u}{0.9\,F_y BN}} \qquad t_p = c\sqrt{\frac{2\,P_0}{0.9\,F_y A_H}}$$

A base plate is designed in Example 7-4 using the equations described here for A_1 and t_p.

EXAMPLE 7-4

Design a base plate with A36 steel for a W12 × 65 column with a dead working load of 150 k and a working live load of 230 k. The concrete has a 28-day compression strength = 3 ksi. Assume footing size is 9 ft × 9 ft. (Thus A_2 is 108 × 108 = 11,664 in.^{2}.)

Solution

$$P_u = (1.2)(150) + (1.6)(230) = 548 \text{ k}$$

1. Area of base plate

$$A_1 = \frac{1}{A_2}\left(\frac{P_u}{0.6 \times 0.85\,f_c'}\right)^2 = \frac{1}{11,664}\left(\frac{548}{0.6 \times 0.85 \times 3}\right)^2 = 11.00 \text{ in.}^2$$

$$A_1 = \frac{P_u}{0.6 \times 1.7\,f_c'} = \frac{548}{(0.6)(1.7 \times 3)} = 179.1 \text{ in.}^2 \leftarrow$$

$$A_1 = (d)(b_f) = (12.12)(12.00) = 145.44 \text{ in.}^2$$

2. Plate dimensions B and N

$$\Delta = 0.5(0.95\,d - 0.80\,b_f) = (0.5)(0.95 \times 12.12 - 0.80 \times 12.00)$$
$$= 0.957 \text{ in.}$$

$N \approx \sqrt{A_1} + \Delta = \sqrt{179.1} + 0.957 = 14.34$ in.   Use $N = 15$ in.

$B = \dfrac{A_1}{N} = \dfrac{179.1}{15} = 11.94$ in.   Use $B = 14$ in. because of W flange width

3. Drawing sketch of plate (Fig. 7-12)

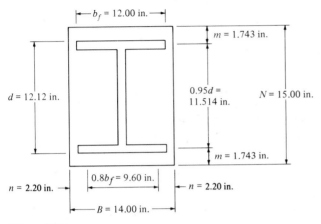

Figure 7-12

4. Computing P_0, the load contributing to the area enclosed by the column

$$P_0 = \dfrac{P_u}{B \times N} b_f d = \dfrac{548}{(14)(15)} (12.00)(12.12) = 379.5 \text{ k}$$

5. Value of H-shaped area

$$A_H = \dfrac{P_0}{0.6(0.85) \sqrt{A_2/b_f d f'_c}} \leq \dfrac{P_0}{(0.6)(1.7 f'_c)}$$

$$A_H = \dfrac{379.5}{(0.6)(0.85) \sqrt{11,664/(12.00 \times 12.12)(3)}} = 27.7 \text{ in.}^2$$

$$A_H = \dfrac{379.5}{(0.6)(1.7 \times 3)} = 124.0 \text{ in.}^2$$

6. The cantilever distance c

$$c = \dfrac{(d + b_f - t_f) - \sqrt{(d + b_f - t_f)^2 - 4(A_H - t_f b_f)}}{4}$$

$$c = \dfrac{(12.12 + 12.00 - 0.605) - \sqrt{(12.12 + 12.00 - 0.605)^2 - 4(30.9 - 0.605 + 12.00)}}{4}$$

$$c = 0.452 \text{ in.}$$

7. Plate thickness

$$t_p = n \sqrt{\frac{2P_u}{0.9F_y BN}} = 2.20 \sqrt{\frac{(2)(548)}{(0.9)(36)(14)(15)}} = 0.883 \text{ in.}$$

$$t_p = c \sqrt{\frac{2P_o}{0.9F_y A_H}} = 0.452 \sqrt{\frac{(2)(379.5)}{(0.9)(36)(27.7)}} = 0.416 \text{ in.}$$

Use PL $\frac{15}{16}$ × 14 × 15

PROBLEMS

7-1. Select a W12 section for column *AB* shown in the accompanying illustration. P_u = 1100 k. Use A36 steel. The columns above and below *AB* are to be approximately the same size as *AB*. Consider only in-plane behavior. (a) Assume elastic column behavior. (b) Assume inelastic column behavior. (*Ans.* W12×152, W12×136)

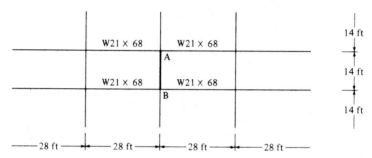

Problem 7-1

7-2. Repeat Prob. 7-1 if F_y is 50 ksi.

7-3. Repeat Prob. 7-1 if a W14 is used. (*Ans.* W14×145, W14×132)

7-4. Select a W14 section for column *CD* shown in the accompanying illustration. Consider only in-plane behavior. (a) Assume elastic column behavior. (b) Assume inelastic column behavior. P_u = 1330 k and A36 steel is to be used. The columns above and below are assumed to be approximately the same size as *CD*.

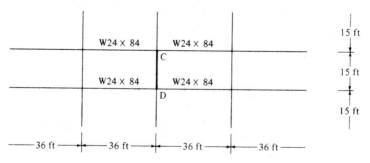

Problem 7-4

7-5. Design W12 columns for the bent shown in the accompanying illustration with A36 steel using the inelastic K-factor procedure. The columns are braced top and bottom against sidesway out of the plane of the frame so that $K = 1.0$ in that direction. Sidesway is possible in the plane of the frame. Design the right-hand column using $K = 1.0$ and the left-hand column with K as determined from the alignment chart and $P_u = 1500$ k. The beam is rigidly connected to the left column but has only a simple connection to the right column. (*Ans.* W12×106, W12×210)

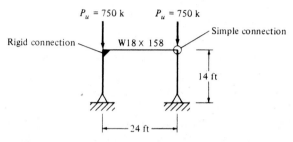

P_u = 750 k   P_u = 750 k   Simple connection   Rigid connection   W18 × 158   14 ft   24 ft

Problem 7-5

7-6. Repeat Prob. 7-5 if the P_u loads are 880 k each.

7-7. The columns for the frames shown in the accompanying illustration are braced top and bottom against sidesway out of the plane of the frame so that $K = 1.0$ in that direction. Sidesway is possible in the plane of the frame. Design the interior column assuming $K = 1.0$ and the exterior columns with K as determined from the alignment chart and $P_u = 1600$ k. Use A36 steel and a W14 section. (*Ans.* W14×193, W14×193)

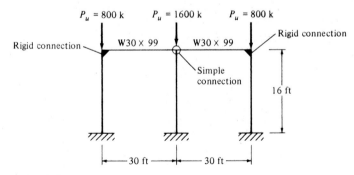

P_u = 800 k   P_u = 1600 k   P_u = 800 k   Rigid connection   Rigid connection   W30 × 99   W30 × 99   Simple connection   16 ft   30 ft   30 ft

Problem 7-7

7-8. Repeat Prob. 7-7 using $F_y = 50$ ksi.

7-9. For the frame shown in the accompanying illustration the beams are rigidly connected to the exterior columns while all other connections are simple. The columns are braced top and bottom against sidesway out of the plane of the frame so that $K = 1.0$ in that direction. Sidesway is possible in the plane of the frame. Design W14 interior columns of A36 steel assuming $K = 1.0$ and W14 exterior columns with K as determined from the alignment chart and $P_u = 1625$ k. (*Ans.* W14×99, W14×233)

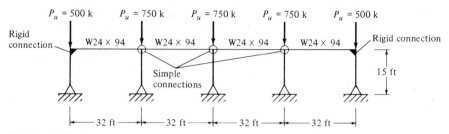

Problem 7-9

7-10. Repeat Prob. 7.9 using $F_y = 50$ ksi and assuming the beam section is a W24×76.

7-11. Design a base plate with A36 steel for a W14×82 column with a dead working load of 120 k and a working live load of 460 k. The concrete 28-day strength is 3 ksi. Footing size is 11 by 11 ft. (*Ans.* PL$1\frac{1}{2}$×15×20)

7-12. Design a base plate with A36 steel for a 12×106 column with a dead working load of 100 k and a live working load of 420 k. The concrete 28-day strength is 4 ksi. Footing size is 12 by 12 ft.

7-13. Repeat Prob. 7.12 if the column is supported by a 28- by 28-in. pedestal. (*Ans.* PL$1\frac{7}{8}$×13×16)

7-14. Design a column base plate of A36 steel for a W14×120 column supporting an axial load $P_u = 960$ k. Footing size is 10 by 10 ft and f'_c is 3 ksi.

Chapter 8

Introduction to Beams

8-1 TYPES OF BEAMS

Beams are usually said to be members that support transverse loads. They are probably thought of as being used in horizontal positions and subjected to gravity or vertical loads; but there are frequent exceptions—rafters, for example.

Among the many types of beams are joists, lintels, spandrels, stringers, and floor beams. *Joists* are the closely spaced beams supporting the floors and roofs of buildings, while *lintels* are the beams over openings in masonry walls such as windows and doors. A *spandrel beam* supports the exterior walls of buildings and perhaps part of the floor and hallway loads. The discovery that steel beams as a part of a structural frame could support masonry walls (together with the development of passenger elevators) is said to have permitted the construction of today's "skyscrapers." *Stringers* are the beams in bridge floors running parallel to the roadway, whereas *floor beams* are the larger beams in many bridge floors which are perpendicular to the roadway of the bridge and are used to transfer the floor loads from the stringers to the supporting girders or trusses. The term *girder* is rather loosely used but usually indicates a large beam and perhaps one into which smaller beams are framed. These and other types of beams are discussed in the sections to follow.

8-2 SECTIONS USED AS BEAMS

The W shapes will normally prove to be the most economical beam sections and they have largely replaced channels and S sections for beam usage. Channels are sometimes used for beams subjected to light loads, such as purlins, and in places

Harrison Avenue Bridge, Beaumont, Tex. (Courtesy of Bethlehem Steel Corporation.)

where clearances available require narrow flanges. They have very little resistance to lateral forces and need to be braced as illustrated by the sag rod problem in Chapter 4. The W shapes have more steel concentrated in their flanges than do S beams and thus have larger moments of inertia and resisting moments for the same weights. They are relatively wide and have appreciable lateral stiffness. (The small amount of space devoted to S beams in the LRFD Manual clearly shows how much their use has decreased from former years. They are today used primarily for special situations as where narrow flange widths are desirable, or where shearing forces are very high, or where the greater flange thickness next to the web may be desirable where lateral bending occurs as perhaps with crane rails.)

Another common type of beam section is the open web joist or bar joist which is discussed at length in Chapter 16. This type of section which is commonly used to support floor and roof slabs is actually a light shop fabricated parallel chord truss. It is particularly economical for long spans and light loads.

8-3 BENDING STRESSES

For an introduction to bending stresses the rectangular beam and stress diagrams of Fig. 8-1 are considered. (For this initial discussion the beam's compression flange is assumed to be fully braced against lateral buckling. Lateral buckling is discussed at length in Chapter 9). If the beam is subjected to some bending moment the stress at any point may be computed with the usual flexure formula $f_b = Mc/I$. *It is to be remembered, however, that this expression is only applicable when the maximum computed stress in the beam is below the elastic limit.* The formula is based on the usual elastic assumptions: stress is proportional to

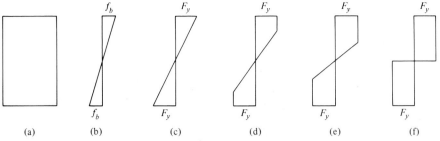

f_b F_y F_y F_y F_y

f_b F_y F_y F_y F_y

(a)   (b)   (c)   (d)   (e)   (f)

Figure 8-1

strain, a plane section before bending remains a plane section after bending, etc. The value of I/c is a constant for a particular section and is known as the *section modulus* (S). The flexure formula may then be written as follows:

$$f_b = \frac{M}{S}$$

where $S = I/c$, the section modulus

Initially when the moment is applied to the beam the stress will vary linearly from the neutral axis to the extreme fibers. This situation is shown in part (b) of Fig. 8-1. If the moment is increased there will continue to be a linear variation of stress until the yield stress is reached in the outermost fibers as shown in part (c) of the figure. The *yield moment* of a cross section is defined as the moment that will just produce the yield stress in the outermost fiber of the section.

If the moment in a ductile steel beam is increased beyond the yield moment the outermost fibers that had previously been stressed to their yield point will continue to have the same stress but will yield, and the duty of providing the necessary additional resisting moment will fall on the fibers nearer to the neutral axis. This process will continue with more and more parts of the beam cross section stressed to the yield point as shown by the stress diagrams of parts (d) and (e) of the figure, until finally a full plastic distribution is approached as shown in part (f). When the stress distribution has reached this stage a *plastic hinge* is said to have formed because no additional moment can be resisted at the section. Any additional moment applied at the section will cause the beam to rotate with little increase in stress.

The *plastic moment* is the moment that will produce full plasticity in a member cross section and create a plastic hinge. The ratio of the plastic moment M_p to the yield moment M_y is called the *shape factor*. The shape factors equals 1.50 for rectangular sections and varies from about 1.10 to 1.20 for standard rolled-beam sections.

8-4 PLASTIC HINGES

This section is devoted to a description of the development of a plastic hinge in the simple beam shown in Fig. 8-2. The load shown is applied to the beam and

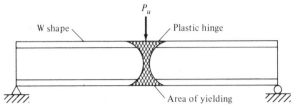

Figure 8-2

increased in magnitude until the yield moment is reached and the outermost fiber is stressed to the yield point. The magnitude of the load is further increased with the result that the outer fibers begin to yield. The yielding spreads out to the other fibers away from the section of maximum moment as indicated in the figure. The length in which this yielding occurs away from the section in question is dependent on the loading conditions and the member cross section. For a concentrated load applied at the center line of a simple beam with a rectangular cross section, yielding in the extreme fibers at the time the plastic hinge is formed will extend for one-third of the span. For a W shape in similar circumstances yielding will extend for approximately one-eighth of the span. During this same period the interior fibers at the section of maximum moment yield gradually until nearly all of them have yielded and a plastic hinge is formed as shown in Fig. 8-2.

Although the effect of a plastic hinge may extend for some distance along the beam it is assumed to be concentrated at one section for analysis purposes. For the calculation of deflections and for the design of bracing, the length over which yielding extends is quite important.

When steel frames are loaded to failure, the points where rotation is concentrated (plastic hinges) become quite visible to the observer before collapse occurs.

8-5 ELASTIC DESIGN

Until recent years almost all steel beams were designed on the basis of the elastic theory. The maximum load that a structure could support was assumed to equal the load that first caused a stress somewhere in the structure to equal the yield stress of the material. The members were designed so that computed bending stresses for service loads did not exceed the yield stress divided by a safety factor (say 1.5 to 2.0). Engineering structures have been designed for many decades by this method with satisfactory results. The design profession, however, has long been aware that ductile members do not fail until a great deal of yielding occurs after the yield stress is first reached. This means that such members have greater margins of safety against collapse than the elastic theory would seem to indicate.

8-6 THE PLASTIC MODULUS

The yield moment M_y equals the yield stress times the elastic modulus. The elastic modulus equals I/c or $bd^2/6$ for a rectangular section and the yield moment

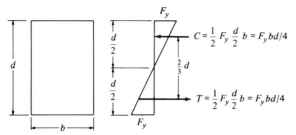

Figure 8-3

equals $F_y \frac{bd^2}{6}$. This same value can be obtained by considering the resisting internal couple shown in Fig. 8-3.

The resisting moment equals T or C times the lever arm between them, as follows:

$$M_y = \left(\frac{F_y bd}{4}\right)\left(\frac{2}{3} d\right) = \frac{F_y bd^2}{6}$$

The elastic section modulus can again be seen to equal $bd^2/6$ for a rectangular beam.

The resisting moment at full plasticity can be determined in a similar manner. The result is the so-called plastic moment, M_p. It is also the nominal moment of the section, M_n. This plastic or nominal moment equals T or C times the lever arm between them. For the rectangular beam of Fig. 8-4 we have

$$M_p = M_n = T\frac{d}{2} = C\frac{d}{2} = \left(F_y \frac{bd}{2}\right)\left(\frac{d}{2}\right) = F_y \frac{bd^2}{4}$$

The plastic moment is said to equal the yield stress times the plastic modulus. From the foregoing expression for a rectangular section, the plastic modulus Z can be seen to equal $bd^2/4$. The shape factor, which equals M_n/M_y, F_yZ/F_yS, or Z/S, is $(bd^2/4)/(bd^2/6) = 1.50$ for a rectangular section. A study of the plastic modulus determined here shows that it equals the statical moment of the tension and compression areas about the neutral axis. Unless the section is symmetrical, the neutral axis for the plastic condition will not be in the same location as for the elastic condition. The total internal compression must equal the total internal tension. As all fibers are considered to have the same stress (F_y) in the plastic condition, the areas above and below the neutral axis must be equal. This

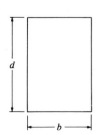
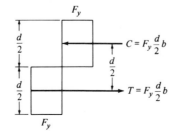

Figure 8-4

situation does not hold for unsymmetrical sections in the elastic condition. Example 8-1 illustrates the calculations necessary to determine the shape factor for a tee beam and the nominal uniform load w_n that the beam can theoretically support.

EXAMPLE 8-1

Determine M_y, M_n, and Z for the steel tee beam shown in Fig. 8-5. Also calculate the shape factor and the nominal uniform load (w_n) which can be placed on the beam for a 12-ft simple span. $F_y = 36$ ksi.

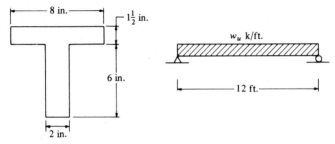

Figure 8-5

Solution

Elastic calculations:

$$A = (8)(1\tfrac{1}{2}) + (6)(2) = 24 \text{ in.}^2$$

$$\bar{y} = \frac{(12)(0.75) + (12)(4.5)}{24} = 2.625 \text{ in. from top flange}$$

$$I = (\tfrac{1}{3})(2)(1.125^3 + 4.875^3) + (\tfrac{1}{12})(8)(1\tfrac{1}{2})^3 + (12)(1.875)^2 = 122.4 \text{ in.}^4$$

$$S = \frac{I}{c} = \frac{122.4}{4.875} = 25.1 \text{ in.}^3$$

$$M_y = F_y S = \frac{(36)(25.1)}{12} = 75.3 \text{ ft-k}$$

Plastic calculations: Neutral axis is at base of flange.

$$Z = (12)(0.75) + (12)(3) = 45 \text{ in.}^3$$

$$M_n = F_y Z = \frac{(36)(45)}{12} = 135 \text{ ft-k}$$

$$\text{Shape factor} = \frac{M_n}{M_y} \quad \text{or} \quad \frac{Z}{S} = \frac{45}{25.1} = 1.79$$

$$M_n = \frac{w_u L^2}{8}$$

$$w_n = \frac{(8)(135)}{(12)^2} = 7.5 \text{ k/ft}$$

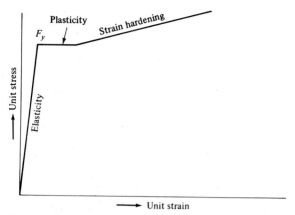

Figure 8-6

The values of the plastic moduli for the standard steel beam sections are tabulated in Part 3 of the LRFD Manual in the "Load Factor Design Selection Table for Shapes Used as Beams" as well as being listed for each shape in the "Dimensions and Properties" section of the Handbook (Part 1). These Z values will frequently be used throughout the text.

8-7 THEORY OF PLASTIC ANALYSIS

The basic plastic theory has been shown to be a major change in the distribution of stresses after the stresses at certain points in a structure reach the yield point. The theory is that those parts of the structure which have been stressed to the yield point cannot resist additional stresses. They instead will yield the amount required to permit the extra load or stresses to be transferred to other parts of the structure where the stresses are below the yield stress and thus in the elastic range and able to resist increased stress. Plasticity can be said to serve the purpose of equalizing stresses in cases of overload.

As early as 1914 the Hungarian, Dr. Gabor Kazinczy, recognized that the ductility of steel permitted a redistribution of stresses in an overloaded statically indeterminate structure.[1] In the United States Prof. J.A. Van den Broek introduced his plastic theory which he called limit design. This theory was published in a paper entitled "Theory of Limit Design" in February 1939 in the *Proceedings of the ASCE*.

For this discussion the stress-strain diagram is assumed to have the idealized shape shown in Fig. 8-6. The yield point and the proportional limit are assumed to occur at the same point for this steel, and the stress-strain diagram is assumed to be a perfectly straight line in the plastic range. Beyond the plastic range there is a range of strain hardening. This latter range could theoretically permit steel members to withstand additional stress, but from a practical stand-

[1]Lynn S. Beedle, *Plastic Design of Steel Frames* (New York: Wiley, 1958), p. 3.

point the strains occurring are so large that they cannot be considered. Furthermore, inelastic buckling will limit the ability of a section to develop a moment greater than M_n even if strain hardening is significant.

8-8 THE COLLAPSE MECHANISM

A statically determinate beam will fail if one plastic hinge develops. To illustrate this fact, the simple beam of constant cross section loaded with a concentrated load at midspan shown in Fig. 8-7 (a) is considered. Should the load be increased until a plastic hinge is developed at the point of maximum moment (underneath the load in this case) an unstable structure will have been created as shown in part (b) of the figure. Any further increase in load will cause collapse. P_n represents the nominal or theoretical maximum load which the beam can support.

For a statically indeterminate structure to fail it is necessary for more than one plastic hinge to form. The number of plastic hinges required for failure of statically indeterminate structures will be shown to vary from structure to structure, but may never be less than two. The fixed-end beam of Fig. 8-8 cannot fail unless the three plastic hinges shown in the figure are developed.

Although a plastic hinge may have formed in a statically indeterminate structure, the load can still be increased without causing failure if the geometry of the structure permits. The plastic hinge will act like a real hinge insofar as increased loading is concerned. As the load is increased there is a redistribution of moment because the plastic hinge can resist no more moment. As more plastic hinges are formed in the structure there will eventually be a sufficient number of them to cause collapse. Actually some additional load can be carried after this time before collapse occurs as the stresses go into the strain hardening range, but the deflections that would occur are too large to be permissible.

The propped beam of Fig. 8-9 is an example of a structure that will fail after two plastic hinges develop. Three hinges are required for collapse but there is a real hinge on the right end. In this beam the largest elastic moment caused by the design concentrated load is at the fixed end. As the magnitude of the load is increased a plastic hinge will form at that point.

The load may be further increased until the moment at some other point (here it will be at the concentrated load) reaches the plastic moment. Additional

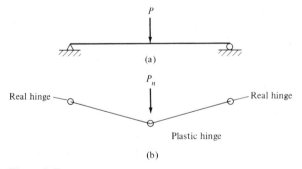

Figure 8-7

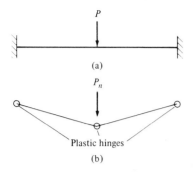

(a)

(b)

Figure 8-8

load will cause the beam to collapse. The arrangement of plastic hinges and perhaps real hinges which permit collapse in a structure is called the *mechanism*. Parts (b) of Figs. 8-7, 8-8, and 8-9 show mechanisms for various beams.

After observing the large number of fixed-end and propped beams used for illustration in this text, the student may form the mistaken idea that he or she will frequently encounter such beams in engineering practice. These types of beams are difficult to find in actual structures but are very convenient to use in illustrative examples. They are particularly convenient for introducing plastic analysis before continuous beams and frames are considered.

8-9 THE VIRTUAL-WORK METHOD

One very satisfactory method used for the plastic analysis of structures is the *virtual-work method*. The structure in question is assumed to be loaded to its nominal capacity, M_n, and is then assumed to deflect through a small additional displacement after the ultimate load is reached. The work performed by the external loads during this displacement is equated to the internal work absorbed by the hinges. For this discussion the *small-angle theory* is used. By this theory the sine of a small angle equals the tangent of that angle and also equals the same angle expressed in radians. In the pages to follow the author uses these values interchangeably because the small displacements considered here produce extremely small rotations or angles.

As a first illustration the uniformly loaded fixed-ended beam of Fig. 8-10 is considered. This beam and its collapse mechanism are shown in the figure.

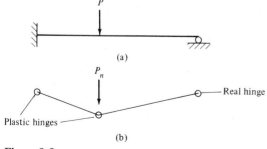

(a)

Real hinge

Plastic hinges

(b)

Figure 8-9

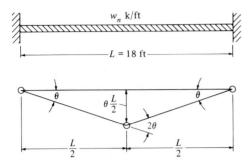

Figure 8-10

Owing to symmetry, the rotations at the end plastic hinges are equal and they are represented by θ in the figure; thus the rotation at the middle plastic hinge will be 2θ.

The work performed by the total external load $(w_n L)$ is equal to $w_n L$ times the average deflection of the mechanism. The average deflection equals one-half the deflection at the center plastic hinge $(\frac{1}{2} \times \theta \times L/2)$. The external work is equated to the internal work absorbed by the hinges or to the sum of M_n at each plastic hinge times the angle through which it works. The resulting expression can be solved for M_n and w_n as follows.

$$M_n(\theta + 2\theta + \theta) = w_n L \left(\frac{1}{2} \times \theta \times \frac{L}{2} \right)$$

$$M_n = \frac{w_n L^2}{16}$$

$$w_n = \frac{16 M_n}{L^2}$$

For the 18-ft span used in Fig. 8-10 these values become

$$M_n = \frac{(w_n)(18)^2}{16} = 20.25 w_n$$

$$w_n = \frac{M_n}{20.25}$$

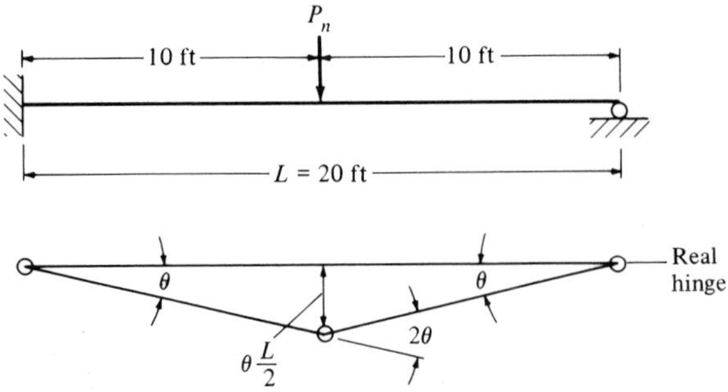

Figure 8-11

Plastic analysis can be handled in a similar manner for the propped beam of Fig. 8-11. There the collapse mechanism is shown and the end rotations (which are equal to each other) are assumed to equal θ.

The work performed by the external load P_n as it moves through the distance $\theta L/2$ is equated to the internal work performed by the plastic moments at the hinges, noting that there is no moment at the real hinge on the right end of the beam.

$$M_n(\theta + 2\theta) = P_n\left(\theta\frac{L}{2}\right)$$

$$M_n = \frac{P_nL}{6} \quad \text{(or 3.33 } P_n \text{ for the 20-ft beam shown)}$$

$$P_n = \frac{6M_n}{L} \quad \text{(or 0.3 } M_n \text{ for the 20-ft beam shown)}$$

The fixed-end beam of Fig. 8-12 together with its collapse mechanism and assumed angle rotations is next considered. From this figure the values of M_n and P_n can be determined by virtual work as follows.

$$M_n(2\theta + 3\theta + \theta) = P_n\left(2\theta \times \frac{L}{3}\right)$$

$$M_n = \frac{P_nL}{9} \quad \text{(or 3.33 } P_n \text{ for this beam)}$$

$$P_n = \frac{9M_n}{L} \quad \text{(or 0.3 } M_n \text{ for this beam)}$$

A person beginning the study of plastic analysis needs to learn to think of all the possible ways in which a particular structure might collapse. Such a habit is of the greatest importance when one begins to analyze more complex structures. In this light the plastic analysis of the propped beam of Fig. 8-13 is considered by the virtual-work method. The beam with its two concentrated loads is shown together with four possible collapse mechanisms and the necessary calculations. It is true that the mechanisms of parts (b), (d), and (e) of the figure do

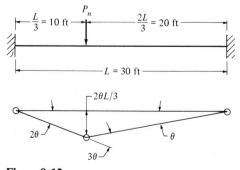

Figure 8-12

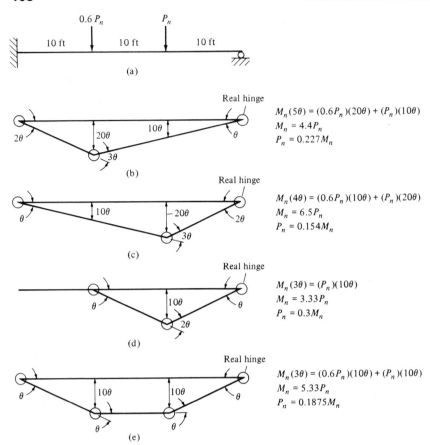

Figure 8-13

not control, but such a fact is not obvious to the average student until he or she makes the virtual-work calculations for each case. Actually the mechanism of part (e) is based on the assumption that the plastic moment is reached at both the concentrated loads simultaneously (a situation that might very well occur).

The value for which the collapse load is the smallest in terms of M_n is the correct value (or the value where M_n is the greatest in terms of P_n). For this beam the second plastic hinge forms at the P_n concentrated load and P_n equals 0.154 M_n.

8-10 LOCATION OF PLASTIC HINGE FOR UNIFORM LOADINGS

There was no difficulty in locating the plastic hinge for the uniformly loaded fixed-end beam, but for other beams with uniform loads, such as propped or continuous beams, the problem may be rather difficult. For this discussion the uniformly loaded propped beam of Fig. 8-14 (a) is considered.

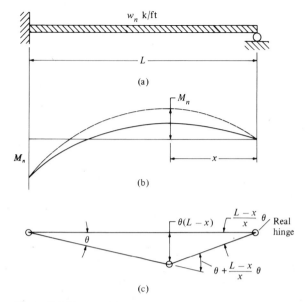

Figure 8-14

The elastic moment diagram for this beam is shown in part (b) of the figure. As the uniform load is increased in magnitude, a plastic hinge will first form at the fixed end. At this time the beam will, in effect, be a "simple" beam with a plastic hinge on one end and a real hinge on the other. Subsequent increases in the load will cause the moment to change as represented by the dotted line in part (b) of the figure. This process will continue until the moment at some other point (a distance x from the right support in the figure) reaches M_n and creates another plastic hinge.

The virtual-work expression for the collapse mechanism of this beam shown in part (c) of Fig. 8-14 is written as follows.

$$M_n\left(\theta + \theta + \frac{L-x}{x}\theta\right) = (w_n L)(\theta)(L-x)\left(\frac{1}{2}\right)$$

Solving this equation for M_n, taking $dM_n/dx = 0$, the value of x can be calculated to equal $0.414L$. This value is also applicable to uniformly loaded end spans of continuous beams with simple end supports as will be illustrated in the next section.

The beam and its collapse mechanism are redrawn in Fig. 8-15 and the following expression for the plastic moment is written using the virtual-work procedure.

$$M_n(\theta + 2.414\theta) = (w_n L)(0.586\theta L)(\tfrac{1}{2})$$

$$M_n = 0.0858 w_n L^2$$

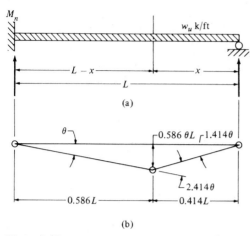

(a)

(b)

Figure 8-15

8-11 CONTINUOUS BEAMS

Continuous beams are very common in engineering structures. Their continuity causes analysis to be rather complicated in the elastic theory, and even though one of the complex "exact" methods is used for analysis, the resulting stress distribution is not nearly so accurate as is usually assumed.

Plastic analysis is applicable to continuous structures as it is to one-span structures. The resulting values definitely give a more realistic picture of the limiting strength of a structure than can be obtained by elastic analysis. Continuous statically indeterminate beams can be handled by the virtual-work procedure as they were for the single-span statically indeterminate beams. As an introduction to continuous beams Examples 8-2 and 8-3 are presented to illustrate two of the more elementary cases.

Here it is assumed that if any or all of a structure collapses failure has occurred. Thus in the continuous beams to follow virtual-work expressions are written separately for each span. From the resulting expressions it is possible to determine the limiting or maximum loads which the beams can support.

EXAMPLE 8-2
A W18×55 (Z_x = 112 in.³) has been selected for the beam shown in Fig. 8-16. Using A36 steel and assuming full lateral support determine the value of w_n.

Solution

$$M_n = F_y Z = \frac{(36)(112)}{12} = 336 \text{ ft-k}$$

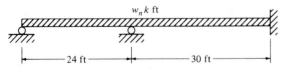

Figure 8-16

Drawing the (collapse) mechanisms for the two spans

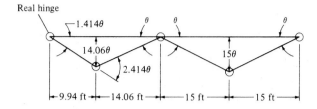

Left-hand span

$$(M_n)(3.414\theta) = (24w_n)(\tfrac{1}{2})(14.06\theta)$$

$$w_n = 0.0202 \, M_n = (0.0202)(336) = 6.8 \text{ klf}$$

Right-hand span

$$(M_n)(4\theta) = (30w_n)(\tfrac{1}{2})(15\theta)$$

$$w_n = 0.0178 \, M_n = (0.0178)(336) = \underline{\underline{5.97 \text{ klf}}} \leftarrow$$

Additional spans have little effect on the amount of work involved in the plastic analysis procedure. The same cannot be said for elastic analysis. Example 8-3 illustrates the analysis of a three-span beam which is loaded with a concentrated load on each span. The student from his knowledge of elastic analysis can see that plastic hinges will initially form at the first interior supports and then at the center lines of the end spans, at which time each end span will have a collapse mechanism.

EXAMPLE 8-3

Using a W21×44 ($Z_x = 95.4$ in.^{3}) consisting of A36 steel determine the value of P_n for the beam of Fig. 8-17.

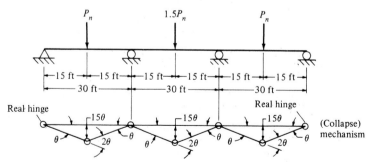

Figure 8-17

Solution

$$M_n = F_y Z = \frac{(36)(95.4)}{12} = 286.2 \text{ ft-k}$$

For first and third spans

$$(P_n)(15\theta) = M_n(3\theta)$$
$$P_n = 0.2M_n = (0.2)(286.2) = 57.2 \text{ k}$$

For center span

$$(1.5P_n)(15\theta) = (M_n)(4\theta)$$
$$P_n = 0.178\,M_n = (0.178)(286.2) = 50.9 \text{ k}$$

Problems

8-1 to 8-10. Find the values of S, Z, and the shape factor about the x axes for the sec-
tions shown in the accompanying illustrations.

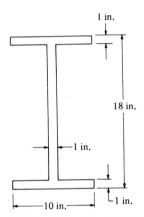

Problem 8-1. (*Ans.* 198.7, 234, 1.18)

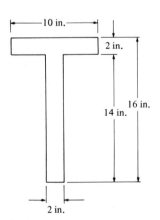

Problem 8-2.

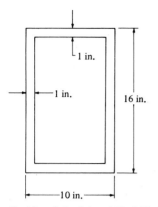

Problem 8-3. (*Ans.* 198, 248, 1.25)

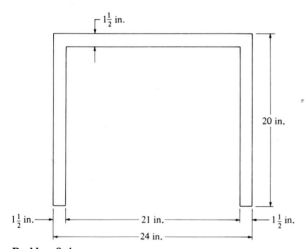

Problem 8-4.

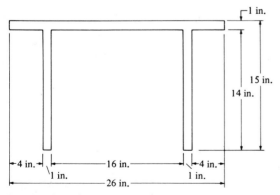

Problem 8-5. (*Ans.* 114.8, 208.5, 1.82)

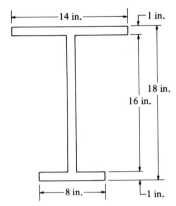

Problem 8-6.

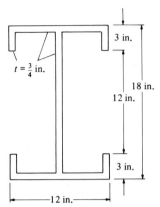

Problem 8-7. (*Ans.* 218.4, 254.4, 1.16)

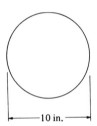

Problem 8-8.

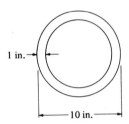

Problem 8-9. (*Ans.* 58.0, 81.3, 1.40)

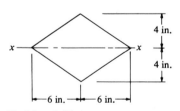

Problem 8-10.

8-11 to 8-17. Determine the values of S, Z, and the shape factor about the x axes for the situations described. Use the web and flange dimensions given in the LRFD Manual for making these calculations.

8-11. A W30×173. (*Ans.* 533.9, 600.1, 1.124)

8-12. A W24×104 with one $\frac{3}{4}$×16 in. PL on each flange.

8-13. Two 8×6×$\frac{5}{8}$-in. Ls long legs vertical and back to back. (*Ans.* 19.7, 35.8, 1.82)

8-14. Two C10×30s back to back.

8-15. Four 8×8×$\frac{3}{4}$ in. Ls arranged as shown in the accompanying illustration. (*Ans.* 64.5, 104.2, 1.62)

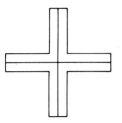

Problem 8-15

8-16. The section of Prob. 5-4(b).

8-17. The section of Prob. 5-6(b). (*Ans.* 205.5, 244.8, 1.19)

8-18. Rework Prob. 8-2 considering the *y* axis.

8-19. Rework Prob. 8-12 considering the *y* axis. (*Ans.* 96.4, 158.4, 1.64)

8-20. Rework Prob. 8-14 considering the *y* axis.

8-21 to 8-34. *Using the given sections all of A36 steel and the plastic theory determine the values of P_n and w_n as indicated.*

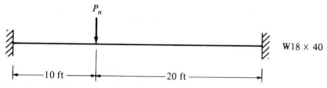

Problem 8-21   (*Ans.* $P_n = 70.6$ k)

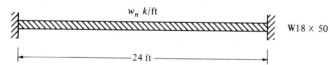

Problem 8-22

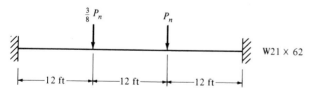

Problem 8-23   (*Ans.* $P_n = 90.9$ k)

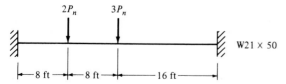

Problem 8-24

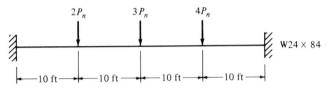

Problem 8-25   (*Ans. P_n = 22.4 k*)

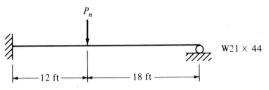

Problem 8-26

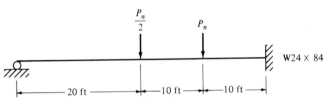

Problem 8-27   (*Ans. P_n = 100.8 k*)

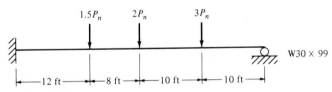

Problem 8-28

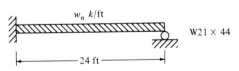

Problem 8-29   (*Ans. w_n = 5.79 k/ft*)

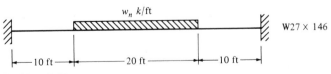

Problem 8-30

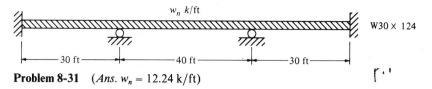

Problem 8-31   (*Ans.* $w_n = 12.24$ k/ft)

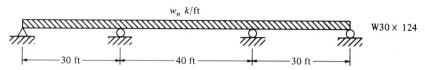

Problem 8-32

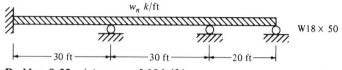

Problem 8-33   (*Ans.* $w_n = 5.39$ k/ft)

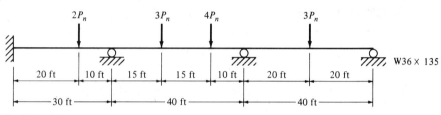

Problem 8-34

Chapter 9

Design of Beams for Moments

9-1 INTRODUCTION

In this chapter the buckling moments of a series of compact ductile steel beams with different lateral bracing situations are considered.

1. First the beams will be assumed to have continuous lateral bracing for their compression flanges.
2. Next the beams will be assumed to be braced laterally at short intervals.
3. Then the beams will be assumed to be braced at larger and larger intervals.

In Fig. 9-1 a typical curve showing the nominal resisting or buckling moments of one of these beams with varying unbraced lengths is shown.

An examination of Fig. 9-1 will show that the beams have three distinct ranges or zones of buckling depending on their lateral bracing situation. If we have continuous or closely spaced lateral bracing, the beams will buckle plastically and fall into what is classified as Zone 1 buckling. As the distance between lateral bracing is increased further, the beams will begin to fail inelastically at smaller moments and fall into Zone 2. Finally, with even larger unbraced lengths, the beams will fail elastically and fall into Zone 3. A brief discussion of these three types of buckling is presented in this section while the remainder of the chapter is devoted to a detailed discussion of each type, together with a series of numerical examples.

Plastic buckling (Zone 1) If we were to take a compact beam whose compression flange is continuously braced laterally, we would find that we could load

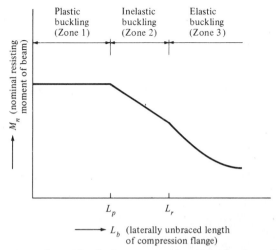

Figure 9-1 Nominal moment as function of unbraced length of compression flange.

it until its full plastic moment M_p is reached; further loading then produces a redistribution of moments as was described in Chapter 8. In other words, the moments in these beams can reach M_p and then develop a rotation capacity sufficient for moment redistribution.

If we now take one of these compact beams and provide closely spaced intermittent lateral bracing for its compression flanges, we will find that we can

Bridge over Allegheny River at Kittanning, PA. (Courtesy American Bridge Company.)

still load it until the plastic moment plus moment redistribution is achieved if the spacing between the bracing does not exceed a certain value called L_p herein. (The value of L_p is dependent on the dimensions of the beam cross section and on its yield stress).

Inelastic buckling (Zone 2)   If we now further increase the spacing between points of lateral bracing, the section may be loaded until some but not all of the compression fibers are stressed to F_y. The section will have insufficient rotation

150 Federal Street, Boston, Mass. (Courtesy Owen Steel Company, Inc.)

capacity to permit full moment redistribution and thus will not permit plastic analysis. In other words, in this zone, we can bend the member until the yield strain is reached in some but not all of its compression elements before buckling occurs. This is referred to as inelastic buckling.

As we increase the unbraced length we will find that the moment the section resists, will decrease until finally it will buckle before the yield stress is anywhere reached. The maximum unbraced length at which we can still reach F_y at one point is the end of the inelastic range. It's shown as L_r in Fig. 9-1; its value is dependent upon the properties of the beam cross section as well as on the yield and residual stresses of the beam. At this point, as soon as we have a moment which theoretically causes the yield stress to be reached anywhere (actually it's less than F_y because of residual stresses), the section will buckle.

Elastic buckling (Zone 3)   If the unbraced length is greater than L_r the section will buckle elastically before the yield stress is reached anywhere. As the unbraced length is further increased, the buckling moment becomes smaller and smaller. As the moment is increased in such a beam, the beam will deflect more and more transversely until a critical moment value M_{cr} is reached. At this time the beam cross section will twist and the compression flange will move laterally. The moment M_{cr} is provided by the torsional resistance and the warping resistance of the beam as will be discussed in Section 9-7.

9-2 PLASTIC BUCKLING, ZONE 1

In this section and the next two beam formulas for plastic buckling (Zone 1) are presented while in Sections 9-5 through 9-7 formulas are presented for inelastic buckling (Zone 2) and elastic buckling (Zone 3). After seeing some of these expressions the reader may become quite concerned that he or she is going to spend an enormous amount of time in formula substitution. This is not generally true, however, as the values needed are tabulated and graphed in simple form in the LRFD Manual.

If the unbraced length L_b of the compression flange of a compact I- or C-shaped section including hybrid members does not exceed L_p (if elastic analysis is being used) or L_{pd} (if plastic analysis is being used) then the member's bending strength about its major axis may be determined as follows:

$$M_n = M_p = F_y Z$$
$$M_u = \phi_b M_n = \phi_b F_y Z \qquad \text{with } \phi_b = 0.9$$

For elastic analysis L_b may not exceed the value L_p to follow if M_n is to equal $F_y Z$.

$$L_p = \frac{300 r_y}{\sqrt{F_{yf}}}$$

For plastic analysis L_b (which is defined as the laterally unbraced length of the compression flange at plastic hinge locations associated with failure mechanisms) may not exceed the value L_{pd} to follow if M_n is to equal $F_y Z$.

$$L_{pd} = \frac{3600 + 2200\,(M_1/M_p)}{F_y} r_y$$

In this expression M_1 is the smaller moment at the end of the unbraced length of the beam and the ratio M_1/M_p is positive when the moments cause the member to be bent in double curvature ($\searrow\!\curvearrowright$) and negative if they bend it in single curvatures ($\searrow\,\diagup$). Only steels with F_y values (F_y is the specified minimum yield stress of the compression flange) of 65 ksi or less may be considered. Higher-strength steels may not be ductile.

There is no limit of the unbraced length for circular or square cross sections or for I-shaped beams bent about their minor axes. (If I-shaped sections are bent about their minor or y axes they will not buckle before the full plastic moment M_p about the y axis is developed.) Section F of the LRFD Specification provides other values for L_p and L_{pd} for solid rectangular bars and symmetric box beams.

For these sections to be compact the width-thickness ratios of the flanges and webs of I- and C-shaped sections are limited to the maximum values to follow which are taken from Table B5.1 of the LRFD Specification.
For flanges

$$\lambda_p = \frac{b_f}{2t_f} \le \frac{65}{\sqrt{F_y}}$$

For webs

$$\lambda_p = \frac{h_c}{t_w} \le \frac{640}{\sqrt{F_y}}$$

In this last expression h_c is the distance from the web toe of the fillet in the top of the web to the web toe of the fillet in the bottom of the web (that is, twice the distance from the neutral axis to the inside face of the compression flange less the fillet or corner radius).

9-3 DESIGN OF BEAMS, ZONE 1

Included in the items that need to be considered in beam design are moments, shears, deflections, crippling, lateral bracing for the compression flanges, fatigue, and others. Beams will probably be selected that provide sufficient design moment capacities ($\phi_b M_n$) and then checked to see if any of the other items are critical. The factored moment will be computed and a section having that much design moment capacity will be initially selected from the LRFD Manual.

A table is given in Section 3 of the LRFD Manual entitled "Load Factor Design Selection Table for Shapes Used as Beams." From this table steel shapes having sufficient plastic moduli to resist certain moments can quickly be selected. Two important items should be remembered in selecting shapes. These are:

1. These steel sections cost so many cents per pound and it is therefore desirable to select the lightest possible shape having the required plastic modulus (assuming that the resulting section is one which will reasonably fit into the structure). The table has the sections arranged in various groups having certain ranges of plastic moduli. The heavily typed section at the top of each group is the lightest section in that group and the others are arranged successively in the order of their plastic moduli. Normally the deeper sections will have the lightest weights giving the required plastic moduli, and they will be generally selected unless their depth causes a problem in obtaining the desired headroom, in which case a shallower but heavier section will be selected.

2. The plastic moduli values in the table are given about the horizontal axes for beams in their upright positions. If a beam is to be turned on its side the proper plastic modulus can be found in the tables giving dimensions and properties of shapes in Part 1 of the LRFD Manual. A W shape turned on its side may only be from 10 to 30 percent as strong as one in the upright position when subjected to gravity loads. In the same manner, the strength of a wood joist with the actual dimensions 2×10 in. turned on its side would only be 20 percent as strong as in the upright position.

The examples to follow illustrate the design of compact steel beams whose compression flanges have full lateral support or bracing thus permitting plastic analysis. For the selection of such sections the designer may enter the tables either with the required plastic modulus or with the factored design moment (if $F_y = 36$ or 50 ksi).

Beam Weight Estimates   In each of the examples to follow the weight of the beam is included in the calculation of the bending moment to be resisted, as the beam must support itself as well as the external loads. The estimates of beam weight are very close here because the author was able to perform a little preliminary paperwork before making his estimates. The beginning student is not expected to be able to glance at a problem and estimate exactly the weight of the beam required. A very simple method is available, however, with which the student can quickly and accurately estimate beam weights. He or she can calculate the maximum bending moment not counting the effect of the beam weight and then pick a section from the LRFD table. Then the weight of this section or a little bit more (since the beam's weight will increase the moment somewhat) can be used as the estimated beam weight. The results will almost always be very close to the weight of the member selected in the final design.

EXAMPLE 9-1

Select a beam section for the span and loading shown in Fig. 9-2, assuming full lateral support is provided for the compression flange by the floor slab above (that is, $L_b = 0$). Use A36 steel with $F_y = 36$ ksi.

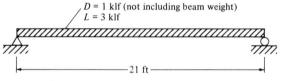

D = 1 klf (not including beam weight)
L = 3 klf

21 ft

Figure 9-2

Solution

Assume beam weight = 55 lb/ft

$$w_u = (1.2)(1.055) + (1.6)(3) = 6.07 \text{ klf}$$

$$M_u = \frac{(6.07)(21)^2}{8} = 334.6 \text{ ft-k}$$

$$Z_x \text{ required} = \frac{M_u}{\phi_b F_y} = \frac{(12)(334.6)}{(0.9)(36)} = 123.9 \text{ in.}^3$$

Use W24×55

EXAMPLE 9-2

The 5-in. reinforced-concrete slab shown in Fig. 9-3 is to be supported with steel W sections 8 ft 0 in. on centers. The beams, which will span 20 ft, are assumed to be simply supported. If the concrete slab is designed to support a live load of 100 psf, determine the lightest steel section required to support the slab. It is assumed that the compression flange of the beam will be fully supported laterally by the concrete slab. The concrete weighs 150 lb/ft³. Use A36 steel.

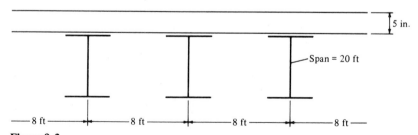

5 in.

Span = 20 ft

8 ft   8 ft   8 ft   8 ft

Figure 9-3

Solution

$$\text{Dead loads:} \quad \text{Slab} = (\tfrac{5}{12})(150)(8) = 500 \text{ lb/ft}$$

$$\text{Estimated beam wt} = \underline{\quad 26 \quad}$$

$$\text{Total} = 526 \text{ lb/ft}$$

$$w_u = (1.2)(526) + (1.6)(8 \times 100) = 1911 \text{ lb/ft} = 1.911 \text{ klf}$$

$$M_u = \frac{(1.911)(20)^2}{8} = 95.55 \text{ ft-k}$$

$$Z_x \text{ required} = \frac{(12)(95.55)}{(0.9)(36)} = 35.4 \text{ in.}^3$$

$$\text{Use W12} \times 26$$

Holes in Beams It is often necessary to have holes in steel beams. They are obviously required for the installation of bolts and sometimes for pipes, conduits, ducts, etc. If at all possible these latter types of holes should be completely avoided. When absolutely necessary they should be placed through the web if the shear is small and through the flange if the moment is small. Cutting a hole through the web of a beam does not reduce its section modulus greatly or its resisting moment; but, as will be described in Section 10-2, a large hole in the web tremendously reduces the shearing strength of a steel section. When large holes are put in beam webs, extra plates are sometimes connected to the webs around the holes to serve as reinforcing against possible web buckling. An example design for such reinforcing is presented by Kussman and Copper.[1]

The presence of holes of any type in a beam certainly does not make it stronger and in all probability weakens it somewhat. The effect of holes has been a subject which has been argued back and forth for many years. The questions, "Is the neutral axis affected by the presence of holes?" and "Is it necessary to subtract holes from the compression flange which are going to be plugged with bolts?" are frequently asked.

The theory that the neutral axis might move from its normal position to the theoretical position of its net section when holes are present is rather questionable. Tests seem to show that flange holes for bolts do not appreciably change the location of the neutral axis. It is logical to assume that the location of the neutral axis will not follow the exact theoretical variation with its abrupt changes in position at rivet or bolt holes as shown in part (b) of Fig. 9-4. A more reasonable change in neutral axis location is shown in part (c) of this figure where it is assumed to have a more gradual variation in position.

It is interesting to note that flexure tests of steel beams seem to show that their failure is based on the strength of the compression flange even though there may be bolt holes in the tension flange. The presence of these holes does not seem

[1]R. L. Kussman and P. B. Cooper, "Design Example for Beams with Web Openings," *Engineering Journal*, second quarter 1976, 13, no. 2 (New York: AISC), pp. 48–56.

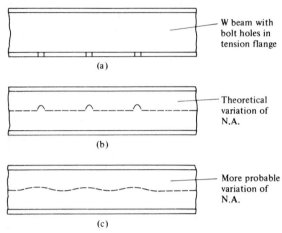

Figure 9-4

to be as serious as might be thought particularly as compared with holes in a pure tension member. These tests show little difference in the strengths of beams with no holes and in beams with holes up to 15 percent of the gross area of either flange.

The LRFD Specification (B1) does not require the subtraction of holes in either flange provided the hole area in any one flange does not exceed 15 percent of the gross area of that flange, and then the deduction is only for the area in excess of 15 percent. Furthermore the LRFD does not make a distinction between holes in the compression and tension flanges. Although the 15 percent value is permitted by the LRFD, some specifications (notably the bridge ones) and a good many designers have not adopted the idea and follow the more conservative practice of deducting all holes.

The usual practice is to subtract the same area of holes from both flanges whether they are present or not. For a section with two holes in the tension flange only, the properties of the section would be computed based on the subtraction of two holes from the tension flange and two holes from the compression flange. Example 9-3 illustrates this method. Again the author has made a few preliminary calculations in estimating the member size.

EXAMPLE 9-3

Select the lightest section available for the beam of Fig. 9-5 assuming that it will be necessary to punch holes for two 1-in. bolts in the tension flange at the point of maximum moment. Use A36 steel and assume full lateral bracing for the compression flange.

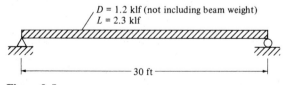

Figure 9-5

Solution

Assume beam weight = 84 lb/ft

$$w_u = (1.2)(1.284) + (1.6)(2.3) = 5.22 \text{ klf}$$

$$M_u = \frac{(5.22)(30)^2}{8} = 587.3 \text{ ft-k}$$

$$\text{Net } Z_x \text{ required} = \frac{(12)(587.3)}{(0.9)(36)} = 217.5 \text{ in.}^3$$

It appears from the table that we should use a W24×84, but to account for the holes the author estimated that a little larger section would be required and he thus assumed a W27×84 for the following calculations.

Try a W27×84 (see Fig. 9-6 for sketch with dimensions).

Assuming two holes in each flange, Z of holes about N.A. is calculated.

Area of 2 holes in each flange = $(2)(1\frac{1}{8})(0.640) = 1.44 \text{ in.}^2$

−15 percent of flange area as per Specification B1 of LRFD

$$= -(0.15)(9.96)(0.640) = \underline{-0.956}$$

Area to be subtracted in Z_{net} calculation = 0.484 in.²

Calculation of Z_{net} by taking hole area from both flanges

$$Z_{\text{net}} = 244 - (0.484)(13.035)(2) = 231.4 \text{ in.}^3 > 217.5 \text{ in.}^3 \qquad \text{OK}$$

$$\underline{\underline{\text{Use W27×84}}}$$

Should a hole be present in only one side of a flange of a W section there will be no axis of symmetry for the net section of the shape. A correct theoretical solution of the problem would be very complex. Rather than going through such a lengthy process over a fairly minor point it seems logical to assume holes in both sides of the flange. The results obtained will probably be just as satisfactory as those obtained by a laborious theoretical method.

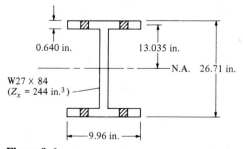

0.640 in.

13.035 in.

N.A. 26.71 in.

W27 × 84
(Z_x = 244 in.³)

9.96 in.

Figure 9-6

9-4 LATERAL SUPPORT OF BEAMS

Probably the large majority of steel beams are used in such a manner that their compression flanges are restrained against lateral buckling. (Unfortunately, however, the percentage has not been quite as high as the design profession has assumed.) The upper flanges of beams used to support concrete building and bridge floors are often incorporated in these concrete floors. For situations of this type where the compression flanges are restrained against lateral buckling the beams will fall into Zone 1.

Should the compression flange of a beam be without lateral support for some distance it will have a stress situation similar to that existing in columns. As is well known the longer and slenderer a column becomes the greater becomes the danger of its buckling for the same loading condition. When the compression flange of a beam is long enough and slender enough it may quite possibly buckle unless lateral support is provided.

There are many factors affecting the amount of stress which will cause buckling in the compression flange of a beam. Some of these factors are the properties of the material, the spacing and types of lateral support provided, residual stresses in the sections, the types of end support or restraints, the loading conditions, etc.

The tension in the other flange of a beam tends to keep that flange straight and restrain the compression flange from buckling; but as the bending moment is increased the tendency of the compression flange to buckle may become large enough to overcome the tensile restraint. When the compression flange does begin to buckle, twisting or torsion will occur, and the smaller the torsional strength of the beam the more rapid will be the failure. The W, S, and channel shapes so frequently used for beam sections do not have a great deal of resistance to lateral buckling and the resulting torsion. Some other shapes, notably the built-up box shapes, are tremendously stronger. These types of members have a great deal more torsional resistance than the W, S, and plate girder sections. Tests have shown that they will not buckle laterally until the strains developed are well in the plastic range.

Some judgment needs to be used in deciding what does and what does not constitute satisfactory lateral support for a steel beam. Perhaps the most common question asked by practicing steel designers is "What is lateral support?" A beam that is wholly encased in concrete or that has its compression flange incorporated in a concrete slab is certainly well supported laterally. When a concrete slab rests on the top flange of a beam, the engineer must study the situation carefully before he counts on friction to provide full lateral support. Perhaps if the loads on the slab are fairly well fixed in position, they will contribute to the friction and it may be reasonable to assume full lateral support. If on the other hand there is much movement of the loads and appreciable vibration, the friction may well be reduced and full lateral support not be assumed. Such situations occur in bridges due to traffic, and in buildings with vibrating machinery such as printing presses.

Should lateral support of the compression flange not be provided by a floor slab, it is possible that such support may be provided with connecting beams or

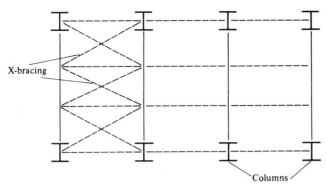

X-bracing

Columns

Figure 9-7

with special members inserted for that purpose. Beams that frame into the sides of the beam or girder in question and are connected to the compression flange can usually be counted on to provide full lateral support at the connection. If the connection is made primarily to the tensile flange, little lateral support is provided to the compression flange. Before support is assumed from these beams the designer should note if they themselves are prevented from moving. The series of beams represented with horizontal dotted lines in Fig. 9-7 provide questionable lateral support for the main beams between columns. For a situation of this type some system of x-bracing may be desirable in one of the bays. Such a system is shown in Fig. 9-7. This one system will provide sufficient lateral support for the beams for several bays.

The corrugated sheet-metal roofs which are usually connected to the purlins with metal straps probably furnish only partial lateral support. A similar situation exists when wood flooring is bolted to supporting steel beams. At this time the student quite naturally asks, "If only partial support is available, what am I to consider to be the distance between points of lateral support?" The answer to this question is for him or her to use his or her judgment. As an illustration, a wood floor is assumed to be bolted every 4 ft to the supporting steel beams in such a manner that it is thought only partial lateral support is provided at those points. After studying the situation the engineer might well decide that the equivalent of full lateral support at 8-ft intervals is provided. Such a decision seems to be within the spirit of the specifications.

9-5 INTRODUCTION TO INELASTIC BUCKLING, ZONE 2

If intermittent lateral bracing is supplied for the compression flange of a beam section such that the member can be bent until the yield strain is reached in some but not all of its compression elements before lateral buckling occurs we have inelastic buckling. In other words the bracing is insufficient to permit the member to reach a full plastic strain distribution before buckling occurs.

Because of the presence of residual stresses (discussed in Section 5-2) yielding will begin in a section at applied stresses equal to $F_{yw} - F_r$, where F_{yw} is the yield stress of the web and F_r equals the compressive residual stress assumed equal to 10 ksi for rolled shapes and 16.5 ksi for welded shapes. It should be noted that the definition of plastic moment $F_y Z$ in Zone 1 is not affected by residual stresses because the sum of the compressive residual stresses equals the sum of the tensile residual stresses in the section and the net effect is theoretically zero.

If the unbraced length, L_b, of a compact I- or C-shaped section is larger than L_p the beam will fail inelastically unless L_b is greater than a distance L_r (to be discussed) beyond which the beam will fail elastically before F_y is reached (thus falling into Zone 3).

Bending Coefficients   In the formulas presented in these next few sections for inelastic and elastic buckling we will use a term C_b. This is a moment coefficient that is included in the formulas to account for the effect of different moment gradients on lateral-torsional buckling. In other words lateral buckling may be appreciably affected by the end restraint and loading conditions of the member.

As an illustration the reader can see that the moment in the unbraced beam of part (a) of Fig. 9-8 causes a worse compression flange situation than does the moment in the unbraced beam of part (b). For one reason the upper flange of the beam in part (a) is in compression for its entire length while in (b) the length of the "column," that is, the length of the upper flange which is in compression, is much less (thus a much shorter "column").

For the simply supported beam of part (a) of the figure C_b is taken as 1.0 while for the beam of part (b) it is taken as larger than 1.0. The basic moment capacity equations for Zones 2 and 3 were developed for laterally unbraced beams subject to single curvature with $C_b = 1.0$. Frequently beams are not bent in single curvature with the result that they can resist more moment. We have seen this in Fig. 9-8. To handle this situation the LRFD Specification provides moment or C_b coefficients larger than 1.0 which are to be multiplied by the computed M_n values. The results are higher moment capacities. The designer who conservatively says "I'll always use $C_b = 1.0$" is missing out on the possibility of significant savings in steel weight for some situations. *When using C_b values the*

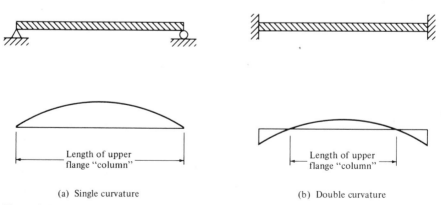

(a) Single curvature   (b) Double curvature

Figure 9-8

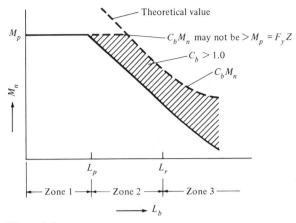

Figure 9-9

designer should clearly understand that the moment capacity obtained by multiplying M_n by C_b may not be larger than the plastic M_n of Zone 1 which is equal to $F_y Z$. This situation is illustrated in Fig. 9-9.

The value of C_b is determined from the expression to follow in which M_1 is the smaller and M_2 the larger of the bending moments at the ends of the unbraced length taken about the strong axis of the member. Should the moment at any point within the unbraced length be larger than the end moments, C_b shall be taken as 1.0. The ratio M_1/M_2 is considered positive if M_1 and M_2 have the same sign (reverse curvature bending) and negative if they have opposite signs (single curvature bending).

$$C_b = 1.75 + 1.05 \left(\frac{M_1}{M_2}\right) + 0.3 \left(\frac{M_1}{M_2}\right)^2 < 2.3$$

C_b is equal to 1.0 for unbraced cantilever beams and for beams that have a moment over an appreciable part of their unbraced span equal to or larger than the larger of the segment's end moments. Some typical values of C_b are shown in Fig. 9-10 for different beam and loading conditions. Table 7 in Part 6 of the LRFD Manual presents calculated values of C_b for different M_1/M_2 ratios.

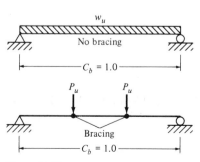
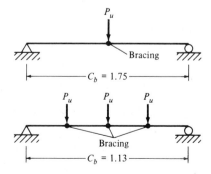

Figure 9-10

9-6 MOMENT CAPACITIES, ZONE 2

As the unbraced length of the compression flange of a beam is increased beyond L_p the moment capacity of the section will become smaller and smaller. Finally at an unbraced length L_r the section will buckle elastically as soon as the yield stress is reached. Owing to the rolling operation, however, there is a residual stress in the section equal to F_r. Thus the elastically computed stress caused by bending can only reach $F_{yw} - F_r$. Assuming $C_b = 1.0$ the permissible moment capacity for a compact I- or C-shaped section bent about its x axis may be determined as follows if $L_b = L_r$.

$$M_u = \phi_b M_r = \phi_b S_x (F_{yw} - F_r)$$

L_r is a function of several of the section's properties such as its cross-sectional area, modulus of elasticity, yield stress, and warping and torsional properties. The very complex formulas needed for its computation are given in the LRFD Specification (F2) and space is not taken to show them here. Fortunately numerical values have been determined for sections normally used as beams and are given in the "Load Factor Design Selection Table."

Going backward from an unbraced length of L_r toward an unbraced length L_p we can see that buckling does not occur when the yield stress is first reached. We are in the inelastic range (Zone 2) where there is some penetration of the yield stress into the section from the extreme fibers. For these cases when the unbraced length falls between L_p and L_r the moment capacity will fall approximately on a straight line between $M_u = \phi_b F_y Z$ at L_p and $\phi_b S_x (F_{yw} - F_r)$ at L_r. For intermediate values of the unbraced length the moment capacity may be determined by proportions or by substituting into the expression at the end of this paragraph. If C_b is larger than 1.0 the section will resist additional moment but not more than $\phi_b F_y Z = \phi_b M_p$

$$\phi_b M_n = C_b [\phi_b M_p - BF(L_b - L_p)] \le \phi_b M_p$$

In which BF is a factor given in the "Load Factor Design Selection Table" for each section which enables us to do the proportioning with a simple formula.

Example 9-4 illustrates the determination of the moment capacities of a section with L_b between L_p and L_r while Example 9-5 demonstrates the design of a beam in the same range.

EXAMPLE 9-4

Determine the moment capacity of a W24×62 with $F_y = 36$ ksi and again with $F_y = 50$ ksi if $L_b = 8.0$ ft and $C_b = 1.0$.

Solution if $F_y = 36$ ksi

From the Load Factor Design Selection Table for a W24×62

$$L_p = 5.8 \text{ ft}$$

$$L_r = 17.2 \text{ ft}$$

$$\phi_b M_r = 255 \text{ ft-k}$$

$$\phi_b M_p = 413 \text{ ft-k}$$

$$BF = 13.8 \text{ k}$$

Since $L_b > L_p < L_r$ the section is in Zone 2 for inelastic buckling and $\phi_b M_n$ can be determined as follows:

$$\phi_b M_n = C_b[\phi_b M_p - BF(L_b - L_p)]$$
$$= 1.0[413 - (13.8)(8.0 - 5.8)] = 382.6 \text{ ft-k}$$

Or directly by proportions

$$\phi_b M_n = 255 + \left(\frac{17.2 - 8.0}{17.2 - 5.8}\right)(413 - 255) = 382.5 \text{ ft-k}$$

Solution if $F_y = 50$ ksi

From the Load Factor Design Selection Table

$$L_p = 4.9 \text{ ft}$$

$$L_r = 13.3 \text{ ft}$$

$$\phi_b M_r = 393 \text{ ft-k}$$

$$\phi_b M_p = 574 \text{ ft-k}$$

$$BF = 21.4 \text{ k}$$

Since $L_b > L_p < L_r$

$$\phi_b M_n = 1.0[574 - (21.4)(8.0 - 4.9)] = 507.7 \text{ ft-k}$$

EXAMPLE 9-5

Select the lightest available section for a factored moment of 290 ft-k if $L_b = 10.0$ ft. Use A36 steel and assume $C_b = 1.0$.

Solution

Enter the Load Factor Design Selection Table and notice that $\phi_b M_p$ for a W21×50 is 297 ft-k but L_p is 5.4 ft $< L_b$ of 10.0 ft. Also $\phi_b M_r = 184$ ft = k and $BF = 10.5$ k.

$$\therefore \phi_b M_n = 1.0[297 - (10.5)(10.0 - 5.4)]$$
$$= 248.7 \text{ ft-k} < 290 \text{ ft-k} \qquad \text{NG}$$

Moving up in the Load Factor Design Selection Table try a W24×55 (with $L_p = 5.6$ ft, $L_r = 16.6$ ft, $\phi_b M_r = 222$ ft-k, $\phi_b M_p = 362$ ft-k, and $BF = 12.7$ k).

$$\phi_b M_n = 1.0[362 - (12.7)(10.0 - 5.6)]$$
$$= 306.1 \text{ ft-k} > 290 \text{ ft-k} \qquad \text{OK}$$

$$\text{Use W24×55}$$

Note: A much easier solution is presented in the next section.

9-7 ELASTIC BUCKLING, ZONE 3

If the unbraced length of the compression flange of a beam section is greater than L_r the section will buckle elastically before the yield stress is reached anywhere in the section. In Section F1.4 of the LRFD Specification the classic equation for determining this flexural-torsional buckling moment called M_{cr} is presented. This expression follows

$$M_{cr} = C_b \frac{\pi}{L_s} \sqrt{EI_y GJ + \left(\frac{\pi E}{L_b}\right)^2 I_y C_w}$$

In this equation G is the shear modulus of elasticity of the steel = 11,200 ksi, J is a torsional constant (in.^{4}), while C_w is the warping constant (in.^{6}). The values of J and C_w are provided for rolled sections in the tables entitled "Torsion Properties" in Part 1 of the LRFD Manual.

This expression is applicable to compact doubly symmetric I-shaped members, Channel sections loaded in the plane of their webs and for I-shaped singly symmetric sections with their compression flanges larger than their tension ones (⊥). Expressions are also given in Sections F1.4 and F1.5 of the LRFD Specification for M_{cr} in the elastic range for other sections such as solid rectangular bars, symmetric box sections, tees, and double angles.

Example 9-6 illustrates the computation of $\phi_b M_{cr}$ for an elastic buckling situation.

EXAMPLE 9-6

Compute $M_u = \phi_b M_{cr}$ for a W18×97 consisting of A36 steel if the unbraced length L_b is 44 ft. Assume $C_b = 1.0$.

Solution

Noting L_b = 44 ft > L_r = 38.1 ft from Load Factor Design Selection Table (Part 3 LRFD Manual).

The following values for the W18×97 are also obtained from the Manual: I_y = 201 in.^{4}, J = 5.86 in.^{4}, and C_{w-} = 15,800 in.^{6}

$$\phi_b M_n = \phi M_{cr} = M_u =$$

$$(0.9)(1.0)\left(\frac{\pi}{12 \times 44}\right)\sqrt{(29 \times 10^3)(201)(11,200)(5.86) + \left(\frac{\pi 29 \times 10^3}{44 \times 12}\right)^2 (201)(15,800)}$$

$$= 3698.9 \text{ in.k} = 308.2 \text{ ft-k}$$

(This value may be checked with the LRFD Manual charts described in the next few paragraphs. There we will be able to read $\phi M_n = M_u = 308$ ft-k.)

The LRFD Manual (page 3–6) also presents the elastic buckling equation in an alternate form as follows:

$$M_{cr} = \frac{C_b S_x X_1 \sqrt{2}}{L_b/r_y}\sqrt{1 + \frac{X_1^2 X_2}{2(L_b/r_y)^2}}$$

in which

$$X_1 = \frac{\pi}{S_x}\sqrt{\frac{EGJA}{2}}$$

$$X_2 = 4\frac{C_w}{I_y}\left(\frac{S_x}{GJ}\right)^2$$

Values of X_1 and X_2 are shown for W shapes in the "Properties for W Shapes" section of Part 1 of the Manual.

Fortunately the values of $\phi_b M_{cr} = \phi_b M_n$ for the sections normally used as beams are computed for different unbraced lengths and the results plotted as curves in Part 3 of the LRFD Manual. The values not only cover unbraced lengths in the elastic range but also the inelastic range enabling us to very easily handle the problems presented in the last section as well as the ones in this section which fall in Zone 3. The moments are plotted for F_y values of 36 ksi and 50 ksi and for $C_b = 1.0$.

The curve for a typical W section is shown in Fig. 9-11. For each of the shapes L_p is indicated with a solid circle (●) while L_r is shown with a hollow circle (○).

The charts were developed without regard to such things as shear, deflection, etc., items that may occasionally control the design as described in Chapter 10. The curves extend to unbraced lengths equal to 30 times the section depths. As such they cover almost all of the unbraced lengths encountered in practice. If C_b is greater than 1.0 the values given will be magnified somewhat as illustrated in Fig. 9-9.

To select a member it is only necessary to enter the chart with the unbraced length L_b and the factored design moment M_u. For an illustration let's assume $F_y = 36$ ksi and assume that we wish to select a beam with $L_b = 20.0$ ft for a moment $M_u = 590$ ft-k. We enter the charts in Part 3 entitled "Beam Design Moments" and go through the pages where $F_y = 36$ ksi until we find in the left-hand column $\phi_b M_n$ equal to 590. We proceed up from the bottom of the chart for an unbraced length $= 20.0$ ft until we intersect a horizontal line from the 590.

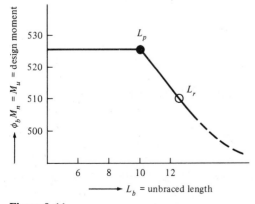

Figure 9-11

Any section to the right and above this intersection point (↗) will have a greater unbraced length and a greater moment capacity.

Moving up and to the right we first encounter a W21×101. In this area of the charts this section is shown with a dashed line. This section will provide the necessary moment capacity but the dashed line indicates that it is in an uneconomical range. If we proceed farther upward and to the right the first solid line that we encounter will represent the lightest available section. In this case it's a W30×99. Another illustration of the use of these charts is presented in Example 9-7.

EXAMPLE 9-7

Using A36 steel select the lightest available section for the beam of Fig. 9-12 which has lateral bracing provided for its compression flange only at its ends.

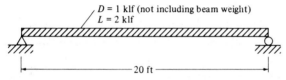

$D = 1$ klf (not including beam weight)
$L = 2$ klf

20 ft

Figure 9-12

Solution

$$\text{Assume beam weight} = 60 \text{ lb/ft}$$

$$w_u = (1.2)(1.060) + (1.6)(2) = 4.472 \text{ klf}$$

$$M_u = \frac{(4.472)(20)^2}{8} = 223.6 \text{ ft-k}$$

$$\text{Noting } C_b = 1.0$$

Entering the Beam Design Moments charts with $L_b = 20$ ft and $M_u = 223.6$ ft-k.

$$\underline{\text{Use W18×60}}$$

For the example problem that follows C_b is greater than 1.0. For such a situation the reader should look back to Fig. 9-9. There he or she will see that the design moment strength of a section can go to $\phi_b C_b M_n$ when $C_b > 1.0$ but may under no circumstances exceed $\phi_b M_p = \phi F_y Z$.

To handle such a problem we calculate an effective moment as follows (the numbers being taken from Example 9-8):

$$M_{\text{effective}} = \frac{M_u}{C_b} = \frac{868.7}{1.75} = 496.4 \text{ ft-k}$$

Then we enter the charts with our unbraced length of 17 ft and with $M_{\text{effective}} = 496.4$ ft-k and there select a W27×84. We must, however, check to see that our M_u does not exceed $\phi_b F_y Z$ for the section. In this case it does and we must keep going until we find the lightest section which has a $\phi_b M_n \geq 496.4$ ft-k at $L_b = 17$ ft and yet which has a $\phi_b F_y Z \geq 868.7$ ft-k.

EXAMPLE 9-8

Using A36 steel select the lightest available section for the situation shown in Fig. 9-13. Bracing is provided only at the ends and center line of the member and thus $C_b = 1.75$.

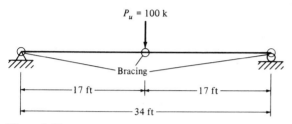

Figure 9-13

Solution

$$\text{Assume beam weight} = 108 \text{ lb/ft}$$

$$w_u = (1.2)(0.108) = 0.1296 \text{ klf}$$

$$M_u = \frac{(100)(34)}{4} + \frac{(0.1296)(34)^2}{8} = 868.7 \text{ ft-k}$$

Entering the Beam Design Moments charts with $M_{u\,\text{effective}} = 868.7/1.75 = 496.4$ ft-k and $L_b = 17$ ft we find a W24×84 will do.

But $\phi_b M_n = 868.7$ ft-k may not exceed $\phi_b F_y Z$ of the section but it does for a W24×84, as can be seen in the Load Factor Design Selection Table, and so we select a larger section from that table.

Use W30×108

9-8 NONCOMPACT SECTIONS

There is one section listed in the Load Factor Design Selection Table that is noncompact when F_y is 36 ksi. It is the W6×15 and the LRFD Manual indicates that it is noncompact with an asterisk*. There are six noncompact sections in the same table when $F_y = 50$ ksi. These sections which are listed with a dagger[†] are the W14×99, W14×90, W12×65, W10×12, W6×15, and the W8×10.

The LRFD Specification Appendix F1.7 provides formulas for the reduced values L'_p and $\phi_b M'_n$ for these sections. These values are tabulated in the Load Factor Design Selection Table and are shown as L_p and $\phi_b M_n$.

The designer will have little trouble with noncompact sections when F_y is 36 or 50 ksi. He or she, however, should be very careful for cases where F_y values are greater than 50 ksi and will have to use the complicated formulas of the Appendix.

PROBLEMS

9-1 to 9-14. *Select the most economical sections using A36 steel unless otherwise specified and assuming full lateral bracing for the compression flanges. Working or service loads are given for each case and beam weights are not included.*

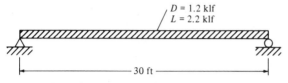

$D = 1.2$ klf
$L = 2.2$ klf

30 ft

Problem 9-1   (*Ans.* W24×84)

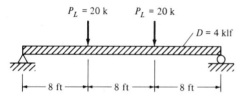

$P_L = 20$ k $P_L = 20$ k

$D = 4$ klf

8 ft   8 ft   8 ft

Problem 9-2

9-3. Repeat Prob. 9-2 using $F_y = 50$ ksi. (*Ans.* W24×68)

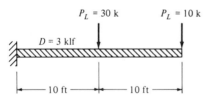

$P_L = 30$ k $P_L = 10$ k

$D = 3$ klf

10 ft   10 ft

Problem 9-4

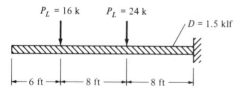

$P_L = 16$ k $P_L = 24$ k

$D = 1.5$ klf

6 ft   8 ft   8 ft

Problem 9-5   (*Ans.* W33×130)

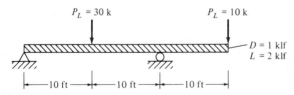

$P_L = 30$ k $P_L = 10$ k

$D = 1$ klf
$L = 2$ klf

10 ft   10 ft   10 ft

Problem 9-6

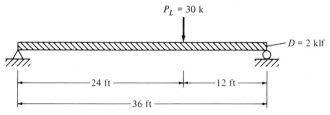

$P_L = 30$ k

$D = 2$ klf

24 ft

12 ft

36 ft

Problem 9-7 (*Ans.* W27×94)

9-8. The accompanying illustration shows the arrangement of beams and girders which are used to support a 6-in. reinforced-concrete floor for a small industrial building. Design the beams and girders assuming they are simply supported. Assume full lateral support and a live load of 120 psf. Concrete weight is 150 lb/ft³.

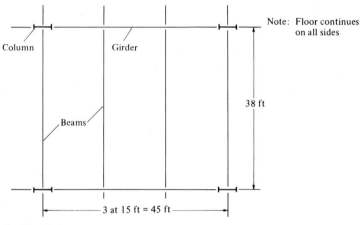

Column

Girder

Beams

Note: Floor continues on all sides

38 ft

3 at 15 ft = 45 ft

Problem 9-8

9-9. A beam consists of a W16×40 with a 1/2×12-in. cover plate welded to each flange. Determine the design uniform load w_u the member can support in addition to its own weight for a 30-ft simple span. (*Ans.* $w_u = 4.03$ k/ft)

9-10. The member shown consists of A36 steel. Determine the maximum service live load that can be placed on the beam if in addition to its own weight it is supporting a service dead load of 1 klf. The member is used for a 24-ft simple span.

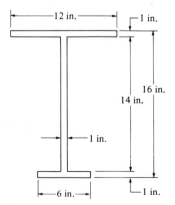

12 in.

1 in.

16 in.

14 in.

1 in.

6 in.

1 in.

Problem 9-10

9-11. Select a section for a 30-ft simple span to support a service dead uniform load of 2 klf and a live serice uniform load of 3 klf if two holes for 1-in. bolts are assumed present in each flange at the section of maximum moment. Follow LRFD Specification B1 as to the 15 percent rule. (*Ans.* W30×108)

9-12. Rework Prob. 9-1 assuming two holes for $1\frac{1}{4}$-in. bolts pass through each flange at the point of maximum moment. Use the LRFD 15 percent rule.

9-13. The section shown in the accompanying illustration has two 1-in. bolts passing through each flange. Find the design or factored load w_u which the section can support in addition to its own weight for a 30-ft simple span if it consists of a steel with $F_y = 50$ ksi. Ignore the 15 percent rule of LRFD specification B1. (*Ans.* 12.17 k/ft)

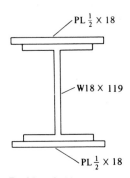

Problem 9-13

9-14. A 40-ft simple span beam is to support two movable 20-k loads a distance of 12 ft apart. Assuming a dead load of 1.2 klf in addition to the beam weight select an A36 steel section to resist the largest possible moment.

9-15 to 9-28. *For these problems different values of L_b are given. Dead loads do not include beam weights.*

9-15. Determine the design moment strength $\phi_b M_n$ of a W21×57 for simple spans of 5, 12, and 25 ft if lateral bracing for the compression flanges is provided at the ends only. A36 steel. (*Ans.* 348, 275.7, 128.1 ft-k)

9-16. Using A36 steel select the lightest satisfactory section if bracing is provided at the ends only.

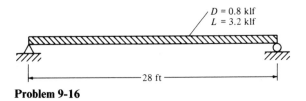

Problem 9-16

9-17. Select the lightest available section if $F_y = 50$ ksi. Lateral bracing is provided at the ends only. (*Ans.* W18×86)

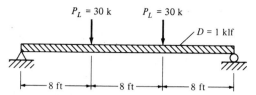

Problem 9-17

9-18. Repeat Prob. 9.17 if lateral bracing is supplied at each of the concentrated loads as well as at the ends of the span.

9-19. A W36×150 consisting of A36 steel is used for a simple span of 30 ft and has lateral bracing provided at its ends only. If the only dead load present is the beam weight what is the largest service concentrated live load which can be placed at the beam center line? (*Ans.* 80.9 k)

9-20. Repeat Prob. 9.19 if lateral bracing is supplied at beam ends and at the concentrated load.

9-21. If F_y is 36 ksi select the lightest section for the situation shown in the accompanying illustration if lateral bracing is supplied at the fixed end only. (*Ans.* W21×62)

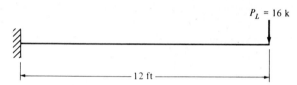

Problem 9-21

9-22. Using a steel with $F_y = 50$ ksi what uniformly distributed service live load can a simply supported W33×241 carry for a span of 40 ft when (a) the compression flange is braced laterally for its full length and when (b) lateral bracing is supplied at the ends only.

9-23. What service live uniform load can be placed on a W14×109 ($F_y = 50$ ksi) which has lateral bracing provided at its ends only? The beam is simply supported and has a span of 30 ft. The dead uniform service load is 1 klf plus the beam weight. (*Ans.* 2.54 klf)

9-24. A W33×130 is used to support the loads shown in the accompanying illustration. Using A36 steel and neglecting the dead weight of the beam, find if the beam is overloaded.
 (**a**) If lateral support is provided for the entire span.
 (**b**) If lateral support is provided at the beam ends only.

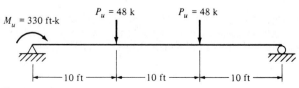

Problem 9-24

9-25. Redesign the beam of Prob. 9-24 if lateral support is provided at the ends and at the concentrated loads. (*Ans.* W27×94)

9-26. The W30×99 shown has lateral support supplied at its center line as well as at its ends. If F_y is 50 ksi determine the maximum permissible service load P. Neglect beam weight.

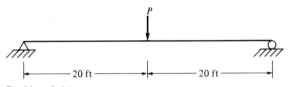

Problem 9-26

9-27. Repeat Prob. 9-26 using A36 steel if a dead uniform service load of 0.8 klf not including the beam weight is placed on the beam for its full length and if lateral bracing is provided at the ends and center line of the beam. (*Ans.* 39.1 k)

9-28. Using A36 steel select the lightest available section for the situation shown in the accompanying illustration if bracing is provided only at the ends and center line.

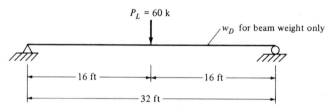

Problem 9-28

9-29. A W33×221 is required for a certain span but, because of a strike in the steel mills, cannot be obtained on time; however, an extra W33×118 of sufficient length is available along with a good supply of plates. Select $\frac{7}{8}$-in.-thick cover plates to be welded to the flanges of the W33×118 to provide a satisfactory substitute for the original beam. Use A36 steel and assume full lateral bracing is to be provided for the compression flange. (*Ans.* $\frac{7}{8}$×16 PL)

9-30. A W30×173 has been specified for use on a certain job. By mistake a W30×124 was shipped to the field. This beam must be erected today. Assuming that 1-in.-thick plates are obtainable immediately, select cover plates to be welded to the flanges to obtain the necessary section. Use A36 steel and assume full lateral bracing is supplied for the compression flange.

9-31. Repeat Prob. 9-29 if the original beam was to have been of a steel with $F_y = 50$ ksi and the substitute material is available only in A36 steel. Assume also that only $1\frac{1}{2}$-in. plates are available. (*Ans.* $1\frac{1}{2}$×16 PL)

9-32. Design a beam of A36 steel with a depth no greater than 12.00 in. to support a uniform load of $w_u = 12.5$ klf (includes beam wt) for a 20-ft simple span. The member is to have full lateral bracing for its compression flange.

Chapter 10

Design of Beams Continued

10-1 DESIGN OF CONTINUOUS BEAMS

Section A5 of the LRFD Specification permits the design of beams analyzed either on the basis of elastic analysis with factored loads or by plastic analysis with the same ultimate loads. Plastic analysis is permitted only for sections with yield stresses no greater than 65 ksi.

Both theory and tests show clearly that continuous ductile steel members meeting the requirements for compact sections with sufficient lateral bracing supplied for their compression flanges have the desirable ability of being able to redistribute moments caused by overloads. If plastic analysis is used this advantage is automatically included in the analysis.

If elastic analysis is used the LRFD handles the redistribution by a rule of thumb that approximates the real plastic behavior. The LRFD Specification (A5) states that for continuous compact sections the design *may* be made on the basis of nine-tenths of the maximum negative moments caused by gravity loads which are maximum at points of support if the positive moments are increased by one-tenth of the average negative moments at the adjacent supports. (*The 0.9 factor is applicable only to gravity loads and not to lateral loads such as those caused by wind and earthquake.* The factor can also be applied to columns that have axial stresses less than $0.15F_y$.) This moment reduction does not apply to members consisting of A514 steel, hybrid girders or to moments produced by loading on cantilevers.

Example 10-1 illustrates the design of a three-span continuous beam analyzed by (a) plastic analysis and (b) elastic analysis.

EXAMPLE 10-1

The beam shown in Fig. 10-1 is assumed to consist of A36 steel. (a) Select the lightest W section available using plastic analysis and assuming full lateral support is provided for its compression flanges. (b) Design the beam using elastic analysis with the factored loads and the 0.9 rule and assuming full lateral support is provided for its compression flanges. (c) Design the beam using elastic analysis with the factored loads and the 0.9 rule and assuming full lateral support is provided for the upper flange *but* only for the lower flange at support points.

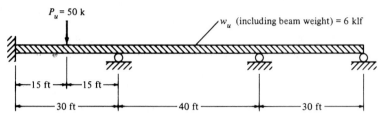

Figure 10-1

Solution

(a)   Plastic analysis and design

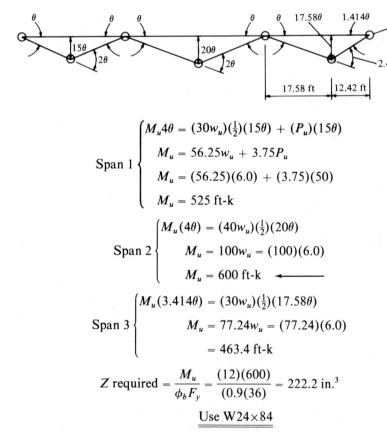

$$\text{Span 1} \begin{cases} M_u 4\theta = (30w_u)(\tfrac{1}{2})(15\theta) + (P_u)(15\theta) \\ M_u = 56.25w_u + 3.75P_u \\ M_u = (56.25)(6.0) + (3.75)(50) \\ M_u = 525 \text{ ft-k} \end{cases}$$

$$\text{Span 2} \begin{cases} M_u(4\theta) = (40w_u)(\tfrac{1}{2})(20\theta) \\ M_u = 100w_u = (100)(6.0) \\ M_u = 600 \text{ ft-k} \quad \longleftarrow \end{cases}$$

$$\text{Span 3} \begin{cases} M_u(3.414\theta) = (30w_u)(\tfrac{1}{2})(17.58\theta) \\ M_u = 77.24w_u = (77.24)(6.0) \\ = 463.4 \text{ ft-k} \end{cases}$$

$$Z \text{ required} = \frac{M_u}{\phi_b F_y} = \frac{(12)(600)}{(0.9)(36)} = 222.2 \text{ in.}^3$$

Use W24×84

(b) Elastic analysis and design—full lateral bracing both flanges

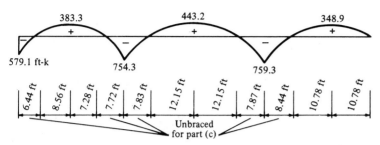

Figure 10-2

Maximum negative moment for design

$$= (0.9)(759.3) = -683.4 \text{ ft-k} \leftarrow$$

Maximum positive moment for design

$$= 443.2 + \frac{1}{10}\left(\frac{754.3 + 759.3}{10}\right) = +518.9 \text{ ft-k}$$

$$Z \text{ required} = \frac{(12)(683.4)}{(0.9)(36)} = 253.1 \text{ in.}^3$$

Use W24×94

(c) Elastic design—full lateral support supplied for top flange but only at support points for bottom flange

Referring to the moment diagram for the part (b) solution (Fig. 10-2) we see that the greatest unbraced length in the negative moment region is 8.44 ft and the reduced M_u is 683.4 ft-k.

From the Load Factor Design Selection Table with $M_u = 683.4$ ft-k we pick a W24×94. Its $L_p = 8.6$ ft > 8.44 ft.

Use W24×94

10-2 SHEAR

For this discussion the beam of Fig. 10-3(a) is considered. As the member bends shear stresses occur because of the changes in length of its longitudinal fibers. For positive bending the lower fibers are stretched and the upper fibers are shortened while somewhere in between there is a neutral axis where the fibers do not change in length. Owing to these varying deformations a particular fiber has a tendency to slip on the fiber above or below.

If a wooden beam was made by stacking boards on top of each other and not connecting them they would obviously tend to take the shape shown in part (b) of

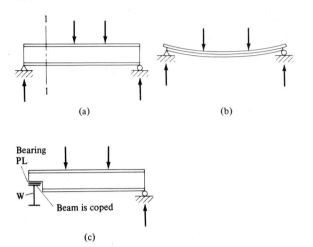

Figure 10-3

the figure. The student may have observed short heavily loaded timber beams with large transverse shears which split along horizontal planes.

This presentation may be entirely misleading in seeming to completely separate horizontal and vertical shears. In reality horizontal and vertical shear at any point are the same and may not be separated. Furthermore, one cannot occur without the other.

Generally shear is not a problem in steel beams because the webs of rolled shapes are capable of resisting rather large shearing forces. Perhaps it is well, however, to list here the most common situations where shear might be excessive. These are as follows.

1. Should large concentrated loads be placed near beam supports they will cause large external shears without corresponding increases in bending moments. A fairly common example of this type occurs in tall buildings where on a particular floor the upper columns are offset with respect to the columns below. The loads from the upper columns applied to the beams on the floor level in question will be quite large if there are many stories above.

2. Probably the most common shear problem occurs where two members (as a beam and a column) are rigidly connected together so their webs lie in a common plane. This situation frequently occurs at the junction of columns and beams (or rafters) in rigid frame structures.

3. Where beams are notched or coped as shown in Fig. 10-3(c) shear can be a problem. For this case shear forces must be calculated for the remaining beam depth. A similar discussion can be made where holes are cut in beam webs for duct work or other items.

4. Theoretically, very heavily loaded short beams can have excessive shears but practically this does not occur too often unless it is like case 1.

Combined welded and bolted joint, Transamerica Pyramid, San Francisco, Calif. (Courtesy of Kaiser Steel Corporation.)

5. Shear may very well be a problem even for ordinary loadings when very thin webs are used as in plate girders or in light gage cold formed steel members.

From his or her study of mechanics of materials the student is familiar with the shear stress formula $f_v = VQ/Ib$ where V is the external shear, Q is the statical moment of that portion of the section lying outside (either above or below) the line on which f_v is desired taken about the neutral axis, and b is the width of the section where the unit shearing stress is desired.

Figure 10.4(a) shows the variation in shear stresses across the cross section of an I-shaped member while part (b) of the same figure shows the shear stress variation in a member with a rectangular cross section. It can be seen in part (a) of the figure that the shear in I-shaped sections is primarily resisted by the web.

If the load is increased on an I-shaped section until the bending yield stress is reached in the flange, the flange will be unable to resist shear stress and it will be carried in the web. If the moment is further increased the bending yield stress will penetrate farther down into the web and the area of web that can resist shear will be further reduced. Rather than assuming the nominal shear stress is resisted by part of the web the LRFD Specification assumes that a reduced shear stress is resisted by the entire web area. This web area, A_w, is equal to the overall depth of the member, h, times the web thickness, t_w.

In the shear strength expressions to follow F_{yw} is the specified minimum yield stress of the web while k is a web plate buckling coefficient defined in

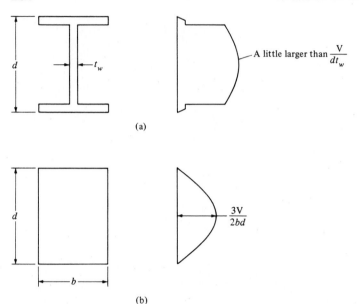

(a)

(b)

Figure 10-4

LRFD Specification F2 and ϕ_v is equal to 0.9. Different expressions are given for different h/t_w ratios where shear failures would be plastic, inelastic, or elastic.

1. <u>Web yielding.</u> All W and channel shapes in the manual fall into this classification.

   If $\dfrac{h}{t_w} \leq 187 \sqrt{\dfrac{k}{F_{yw}}} = 70$ for A36 steel

   $\phi_v V_n = \phi_v 0.6 \, F_{yw} A_w = 19.4 \, dt_w$ in kips for A36 steel

2. <u>Inelastic buckling of web</u>

   If $187 \sqrt{\dfrac{k}{F_{yw}}} < \dfrac{h}{t_w} \leq 234 \sqrt{\dfrac{k}{F_{yw}}} = 87$ for A36 steel

   $\phi_v V_n = \phi_v 0.6 F_{yw} A_w \dfrac{187 \sqrt{k/F_{yw}}}{h/t_w} = 19.4 \, dt_w$ in kips for A36 steel

3. <u>Elastic buckling of web</u>

   If $\dfrac{h}{t_w} > 234 \sqrt{\dfrac{k}{F_{yw}}} = 87$ for A36 steel

   $\phi_v V_n = \phi_v A_w \dfrac{26,400 \, k}{(h/t_w)^2} = \dfrac{118,800}{(h/t_w)^2} \, dt_w$ in kips for A36 steel

In Example 10-2 to follow a beam is checked as to its shear strength.

EXAMPLE 10-2

A W24×55 ($d = 23.57$ in., $t_w = 0.395$ in.) consisting of A36 steel is used for the beam and loading of Fig. 10-5. Check its adequacy in shear.

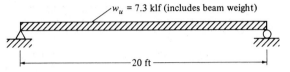

$w_u = 7.3$ klf (includes beam weight)

|← 20 ft →|

Figure 10-5

Solution

$$V_u = (10)(7.3) = 73 \text{ k}$$

$$\phi_v V_n = 19.4 d t_w = (19.4)(23.57)(0.395)$$

$$= 180.6 \text{ k} > 73 \text{ k} \qquad \text{OK}$$

Note. This same value can be taken from the tables entitled "Beams W Shapes" in Part 3 of the LRFD Manual.

10-3 DEFLECTIONS

The deflections of steel beams are usually limited to certain maximum values. Among the several excellent reasons for deflection limitations are the following.

1. Excessive deflections may damage other materials attached to or supported by the beam in question. Plaster cracks caused by large ceiling joist deflections are one example.
2. The appearance of structures is often damaged by excessive deflections.
3. Extreme deflections do not inspire confidence in the persons using a structure although it may be completely safe from a strength standpoint.
4. It may be necessary for several different beams supporting the same loads to deflect equal amounts.

Standard American practice for buildings has been to limit service live-load deflections to approximately $\frac{1}{360}$ of the span length. This deflection is supposedly the largest value that ceiling joists can deflect without causing cracks in underlying plaster. The $\frac{1}{360}$ deflection is only one of many maximum deflection values in use because of different loading situations, different engineers, and different specifications. For situations where precise and delicate machinery is supported maximum deflections may be limited to 1/1500 or 1/2000 of the span lengths. The 1983 AASHTO Specifications limit deflections in steel beams and girders due to live load and impact to $\frac{1}{800}$ of the span. (For bridges in urban areas which are partly used by pedestrians the AASHTO *recommends* a maximum value equal to 1/1000 of the span lengths.)

The LRFD Specification does not specify exact maximum permissible deflections. There are so many different materials, types of structures, and loadings that no one single set of deflection limitations is acceptable for all cases. Thus limitations must be set by the individual designer on the basis of his or her experience and judgment.

Before substituting blindly into a formula that will give the deflection of a beam for a certain loading condition the student should thoroughly understand the theoretical methods of calculating deflections. These methods include the moment area, conjugate beam, and virtual work procedures. From these methods various expressions can be determined such as the common one given at the end of this paragraph for the center line deflection of a uniformly loaded simple beam.

$$\Delta = \frac{5wL^4}{384EI}$$

To use deflection expressions such as this one the reader must be very careful to use consistent units. Example 10-3 illustrates the application of the preceding expression. The author has changed all units to pounds and inches. Thus the uniform load given in the problem as so many kips per foot is changed to so many lb/in.

EXAMPLE 10-3

In Example 9-1 a W24×55 ($I_x = 1350$ in.) was selected for a 21-ft simple span to support a service dead load of 1 klf and a service live load of 3 klf. Is the center line deflection of this section satisfactory for the service live load if the maximum permissible value is 1/360 of the span?

Solution

$$\Delta_c = \frac{5wL^4}{384EI} = \frac{(5)(3000/12)(12 \times 21)^4}{(384)(29 \times 10^6)(1350)} = 0.335 \text{ in.}$$

$$< \left(\frac{1}{360}\right)(12 \times 21) = 0.70 \text{ in.} \qquad\qquad \text{OK}$$

In Part 3 (page 3-24) of the LRFD Manual the following simple formula for determining maximum beam deflections for *W, M, HP, S, C*, and *MC* sections for several different loading conditions is presented.

$$\Delta = \frac{ML^2}{C_1 I_x}$$

In this expression *M* is the service load moment in ft-k based on a uniformly distributed load, C_1 is a constant whose value can be determined from Fig. 10-6, *L* is the span length (ft) and I_x is the moment of inertia (in.⁴).

If we want to use this LRFD expression for the beam of Example 10-3, C_1 from part (a) of Fig. 10.6 would be 161, the center line bending moment is

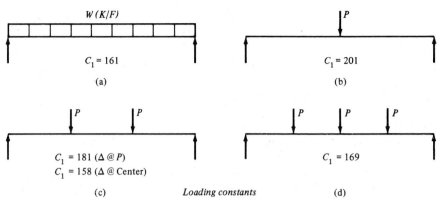

Figure 10-6

$wL^2/8 = (3)(21)^2/8 = 165.375$ ft-k and the center line deflection for the live service load would be

$$\Delta_{\mathbb{C}} = \frac{(165.375)(21)^2}{(161)(1350)} = 0.336 \text{ in.}$$

Some specifications handle the deflection problem by requiring certain minimum depth-span ratios. For example, the AASHTO suggests the depth-span ratio be limited to a minimum value of $1/25$. A shallower section is permitted but it should have sufficient stiffness to prevent a deflection greater than would have occurred if the $1/25$ ratio had been used. The LRFD Manual in their Table 3-1 presents recommended span/depth ratios for simply supported uniformly loaded beams with different service dead to live load (or to total load) ratios which if followed will effectively limit deflections.

A steel beam can be cold-bent or cambered an amount equal to the deflection caused by dead load or the deflection caused by dead load plus some percentage of the live load. Approximately 25 percent of the camber so produced is elastic and will disappear when the cambering operation is completed. Detailed information for particular shapes is given in Part 1 of the LRFD Manual in the section entitled "Standard Mill Practice." It should be remembered that a beam which is bent upward looks much stronger and safer than one which sags downward (even a very small distance).

It is usually more economical to select heavier beams with larger moments of inertia to limit deflections than it is to camber sections. Cambering is something of a nuisance to steel fabricators and it may very well introduce additional problems to them. Frequently when we camber beams we are going to have to adjust our detailed dimensions for proper fitting of members. As a result of these problems and the extra costs to the fabricator (and thus to owners) it is usually more economical to use larger beam sections than it is to use cambering. Furthermore if a higher-strength steel is being used in the structure it is probably also economical to switch to A36 steel for the members that deflect excessively.

Deflections may very well control the sizes of beams for longer spans or for short ones where deflection limitations are severe. To assist the designer in selecting sections where deflections may control, the LRFD Manual includes in its Part 3 a table entitled "Moment of Inertia Selection Table" in which the I_x values are given in numerically descending order for all of the sections normally used as beams. In this table the sections are arranged in groups with the lightest section in each group printed in black type. Example 10-4 presents the design of a beam where deflection controls the design.

EXAMPLE 10-4

Select the lightest available section with $F_y = 36$ ksi to support a service dead load of 1.2 klf and a service live load of 3 klf for a 30-ft simple span. The section is to have full lateral bracing for its compression flange and the maximum total service load deflection is not to exceed 1/1500 the span length.

Solution

After some scratch work assume beam weight = 194 lb/ft

$$w_u = (1.2)(1.394) + (1.6)(3) = 6.473 \text{ klf}$$

$$M_u = \frac{(6.473)(30)^2}{8} = 728.2 \text{ ft-k}$$

$$Z \text{ required} = \frac{(12)(728.2)}{(0.9)(36)} = 269.7 \text{ in.}^3$$

<u>Try W27×94 ($I_x = 3270$ in.^{4})</u>

$$\text{Maximum permissible } \Delta = \left(\frac{1}{1500}\right)(12 \times 30) = 0.24 \text{ in.}$$

$$\text{Actual } \Delta_{\mathbb{C}} = \frac{ML^2}{C_1 I_x} = \frac{\dfrac{(4.394)(30)^2}{8}(30)^2}{(161)(3270)}$$

$$= 0.845 \text{ in.} > 0.24 \text{ in.} \qquad\qquad \text{NG}$$

Maximum I_x required to limit deflection to 0.24 in.

$$= \left(\frac{0.845}{0.24}\right)(3270) = 11,513 \text{ in.}^4$$

From Moment of Inertia Selection Table

<u>Use W36×194</u>

Ponding

If water on a flat roof accumulates faster than it runs off, the result is called *ponding* because the increased load causes the roof to deflect into a dish shape that can hold more water, which causes greater deflections, etc. This process continues until equilibrium is reached or until collapse occurs. Ponding is a serious matter as illustrated by the large number of flat roof failures which occur every year in the United States.

Ponding will occur on almost any flat roof to a certain degree even though roof drains are present. Drains may very well be used but they may be inadequate during severe storms or they may become stopped up and, furthermore, they are often placed along the beam lines which are actually the high points of the roof. The best method of preventing ponding is to have an appreciable slope on the roof ($\frac{1}{4}$ in./ft or more) together with good drainage facilities. It has been estimated that probably two-thirds of the flat roofs in the United States have slopes less than this value which is the minimum recommended by the National Roofing Contractors Association (NRCA). It costs approximately 3 to 6 percent more to construct a roof with this desired slope than building with no slope.[1]

When a very large flat roof (perhaps an acre or more) is being considered the effect of wind on water depth may be quite important. The problem of ponding will logically occur during a heavy rainstorm. Such a storm will frequently be accompanied by heavy winds. When a large quantity of water is present on the roof a strong wind may very well push a great deal of water to one end, creating a dangerous depth of water as regards the load in pounds per square foot applied to the roof. For such situations *scuppers* are sometimes used. These are large holes or tubes in the walls or parapets which enable water above a certain depth to quickly drain off the roof.

Ponding failures will be prevented if the roof system (consisting of the roof deck and supporting beams and girders) has sufficient stiffness. The LRFD Specification (K2) describes a minimum stiffness to be achieved if ponding failures are to be prevented. If this minimum stiffness is not provided it is necessary to make other investigations to be sure that a ponding failure is not possible.

Theoretical calculations for ponding are very complicated. The LRFD requirements are based on work by F. J. Marino[2] in which he considered the interaction of a two-way system of secondary or submembers supported by a primary system of main members or girders.

10-4 WEBS AND FLANGES WITH CONCENTRATED LOADS

When steel members have concentrated loads applied which are perpendicular to one flange and symmetric to the web their flanges and webs must have sufficient flange and web design strength in the areas of flange bending, web yielding, web crippling, and sidesway web buckling. Should a member have concentrated loads applied to both flanges it must have a sufficient web design strength in the areas of web yielding, web crippling, and column web buckling. In this section formulas for determining strengths in these areas are presented.

[1]Gary Van Ryzin, "Roof Design: Avoid Ponding by Sloping to Drain," *Civil Engineering* (New York: American Society of Civil Engineers, January 1980), pp. 77–81.
[2]F. J. Marino, "Ponding of Two-Way Roof System," *Engineering Journal* (New York: AISC, July 1966), pp. 93–100.

Local Flange Bending The nominal tensile load that may be applied through a plate welded to the flange of a W section is to be determined by the expression to follow in which F_{yf} is the specified minimum yield stress of the flange (ksi) and t_f is the flange thickness (in.).

$$R_n = 6.25t_f^2 F_{yf}$$

$$\phi = 0.90 \qquad \qquad \text{(LRFD Formula K1-1)}$$

It is not necessary to check this formula if the length of loading across the beam flange is less than 0.15 times the flange width b_f.

Local Web Yielding The nominal strength of the web of a beam at the web toe of the fillet when a concentrated load or reaction is applied is to be determined by one of the following two expressions in which k is the distance from the outer edge of the flange to the web toe of the fillet, N is the bearing length (in.) of the force, F_{yw} is the specified minimum yield stress (ksi) of the web, and t_w is the thickness of the web.

If the force is a concentrated load or reaction that causes tension or compression and is applied at a distance greater than the member depth from the end of the member

$$R_n = (5k + N)F_{yw}t_w$$

$$\phi = 1.0 \qquad \qquad \text{(LRFD Formula K1-2)}$$

If the force is a concentrated load or reaction applied at or near the end of the member

$$R_n = (2.5k + N)F_{yw}t_w$$

$$\phi = 1.0 \qquad \qquad \text{(LRFD Formula K1-3)}$$

Reference to Fig. 10-7 clearly shows where these expressions were obtained. The nominal strength R_n equals the length over which the force is

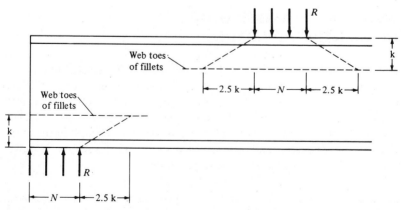

Figure 10-7 Local web yielding.

assumed to be spread when it reaches the web toe of the fillet times the web thickness times the yield stress of the web.

Web Crippling   Should concentrated loads be applied to a member with an unstiffened web the nominal web crippling strength of the web is to be determined by the appropriate equation of the two which follow (in which d is the overall depth of the member). If web stiffeners are provided and extend for at least half of the web depth web crippling does not need to be checked.

If the concentrated load is applied at a distance not less than $d/2$ from the end of the member

$$R_n = 135t_w^2 \left[1 + 3\left(\frac{N}{d}\right)\left(\frac{t_w}{t_f}\right)^{1.5}\right]\sqrt{\frac{F_{yw}t_f}{t_w}}$$

$\phi = 0.75$ (LRFD Formula K1-4)

If the concentrated load is applied at a distance less than $d/2$ from the end of the member

$$R_n = 68t_w^2 \left[1 + 3\left(\frac{N}{d}\right)\left(\frac{t_w}{t_f}\right)^{1.5}\right]\sqrt{\frac{F_{yw}t_f}{t_w}}$$

$\phi = 0.75$ (LRFD Formula K1-5)

Sidesway Web Buckling   Should compressive loads be applied to laterally braced compression flanges the web will be put in compression and the tension flange may buckle as shown in Fig. 10-8.

It has been found that sidesway web buckling will not occur if the flanges are restrained against rotation with $(d_c/t_w)/(\ell/b_f) > 2.3$ or if $(d_c/t_w)/(\ell/b_f) > 1.7$ when flange rotation is not restrained. In these expressions d_c is the web depth between the web toes of the fillets, that is, $d - 2k$, and ℓ is the largest laterally unbraced length along either flange at the point of the load.

It is also possible to prevent sidesway web buckling with properly designed lateral bracing or stiffeners at the load point. The LRFD Commentary suggests that local bracing for *both* flanges be designed for 1 percent of the magnitude of the concentrated load applied at the point. If stiffeners are used they must extend from the point of load for at least one-half of the member depth and should be

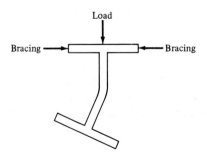

Load

Bracing ———→   ←——— Bracing

Figure 10-8   Sidesway web buckling.

Girders for all-welded Connecticut expressway bridge. (Courtesy of the Lincoln Electric Company.)

designed to carry the full load. Flange rotation must be prevented if the stiffeners are to be effective.

Should members not be restrained against relative movement by stiffeners or lateral bracing and be subject to concentrated compressive loads their strength may be determined as follows:

When the loaded flange is braced against rotation and $(d_c/t_w)/(\ell/b_f)$ is less than 2.3

$$R_n = \frac{12,000\, t_w^3}{h}\left[1 + 0.4\left(\frac{d_c/t_w}{\ell/b_f}\right)^3\right]$$

$$\phi = 0.85$$

(LRFD Formula K1-6)

When the loaded flange is not restrained against rotation and $(d_c/t_w)/(\ell/b_f)$ is less than 1.7

$$R_n = \frac{12,000\, t_w^3}{h}\left[0.4\left(\frac{d_c/t_w}{\ell/b_f}\right)^3\right]$$

$$\phi = 0.85$$

(LRFD Formula K1-7)

EXAMPLE 10-5

Select a W section for moment with A36 steel for the beam shown in Fig. 10-9. Lateral bracing will be provided at the ends and concentrated loads. Determine the minimum bearing length required for web yielding and web crippling at the reactions. Service loads are shown in the figure.

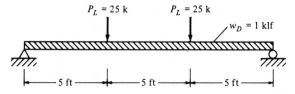

Figure 10-9

Solution

Assume beam weight = 44 lb/ft

$$M_u = \frac{(1.2)(1.044)(15)^2}{8} + (1.6)(25)(5) = 235.2 \text{ ft-k}$$

$$Z \text{ required} = \frac{(12)(235.2)}{(0.9)(36)} = 87.1 \text{ in.}^3$$

Use W21×44 (L_p = 5.3 ft > L_b of 5.0 ft, d = 20.66 in., t_w = 0.350 in., t_f = 0.450 in., k = $1\frac{3}{16}$ in.)

Required bearing length to prevent web yielding at support

$$R_u = (1.2)(1.044)\left(\frac{15}{2}\right) + (1.6)(25) = 49.4 \text{ k}$$

$$R_n = \frac{R_u}{\phi} = \frac{49.4}{1.0} = 49.4 \text{ k}$$

$$R_n = (2.5\,k + N)F_{yw}t_w$$

$$49.4 = (2.5 \times 1\tfrac{3}{16} + N)(36)(0.350)$$

$$N = 0.95 \text{ in.}$$

Required bearing length to prevent web crippling at support

$$R_n = \frac{R_u}{\phi} = \frac{49.4}{0.75} = 65.9 \text{ k}$$

$$65.9 = (68)(0.350)^2\left[1 + 3\left(\frac{N}{20.66}\right)\left(\frac{0.350}{0.450}\right)^{1.5}\right]\sqrt{\frac{(36)(0.450)}{0.350}}$$

$$N = 1.63 \text{ in.} \qquad \underline{\text{Use 4 in. minimum bearing length at support}}$$

Note In Part 3 of the LRFD Manual sets of tables entitled "Beams W Shapes," "Beams S Shapes," and so on are presented with which the calculations for this beam and the other beams in this chapter can be greatly simplified. The definition of the terms in these tables is given on pages 3–25 through 3–29 in the Manual together with several useful example problems.

For instance, for the sections normally used as beams and with yield stresses of 36 and 50 ksi the values $\phi_b M_p$, $\phi_v V_n$, L_p, and L_u are provided along with the quantities ϕR_1, ϕR_2, $\phi_r R_3$, and $\phi_r R_4$. The latter values are useful in making the calculations necessary to check web yielding and web crippling. If we have a W21×44 ($t_w = 0.350$ in., $k = 1\frac{3}{16}$ in; $f_{yw} = 36$ ksi) bearing on a length N of 3.5 in. we can compute ϕR_n for web yielding as follows:

$$\phi R_n = \phi(2.5k + N)F_{yw}t_w$$

$$= (1.0)(2.5 \times 1\tfrac{3}{16} + 3.5)(36)(0.350) = 81.5 \text{ k}$$

Or we can use the table values for ϕR_1 and ϕR_2

$$\phi R_n = \phi R_1 + \phi R_2 N = 37.4 + (12.6)(3.5) = 81.5 \text{ k}$$

10-5 UNSYMMETRICAL BENDING

From mechanics of materials it is remembered that each beam cross section has a pair of mutually perpendicular axes known as the principal axes for which the product of inertia is zero. Bending that occurs about any axis other than one of the principal axes is said to be unsymmetrical bending. When the external loads are not in a plane with either of the principal axes or when loads are simultaneously applied to the beam from two or more directions unsymmetrical bending is the result.

If a load is not perpendicular to one of the principal axes it may be broken into components which are perpendicular to those axes and the moments about each axis, M_{ux} and M_{uy}, determined as shown in Fig. 10-10.

When a section has one axis of symmetry that axis is one of the principal axes and the calculations necessary for determining the moments are quite simple. For this reason unsymmetrical bending is not difficult to handle in the usual beam section, which is probably a W, S, M, or C. Each of these sections has at least one axis of symmetry and the calculations are appreciably reduced. A fur-

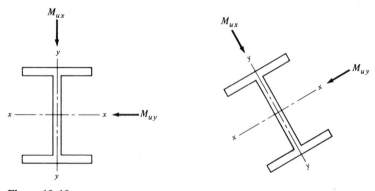

Figure 10-10

ther simplifying factor is that the loads are usually gravity loads and probably perpendicular to the x axis.

Among the beams that must resist unsymmetrical bending are crane girders in industrial buildings and purlins for ordinary roof trusses. The x axes of purlins are parallel to the sloping roof surface while the large percentage of their loads (roofing, snow, etc.) are gravity loads. These loads do not lie in a plane with either of the principal axes of the inclined purlins and the result is unsymmetrical bending. Wind loads are generally considered to act perpendicular to the roof surface and thus perpendicular to the x axes of the purlins with the result that they are not considered to cause unsymmetrical bending. The x axes of crane girders are usually horizontal but the girders are subjected to lateral thrust loads from the moving cranes as well as to gravity loads.

To check the adquacy of members bent about both axes simultaneously the LRFD provides an equation in Section H1 of their Specification. The equation to follow is for combined bending and axial tension or compression if $P_u/\phi P_n <$ 0.2.

$$\frac{P_u}{2\phi P_n} + \left(\frac{M_{ux}}{\phi_b M_{nx}} + \frac{M_{uy}}{\phi_b M_{ny}}\right) \le 1.0 \qquad \text{(LRFD Formula H1-1b)}$$

Since for the problem discussed here P_u is equal to zero the formula reduces to

$$\frac{M_{ux}}{\phi_b M_{nx}} + \frac{M_{uy}}{\phi_b M_{ny}} \le 1.0$$

This is an interaction or percentage equation. If M_{ux} is say 75 percent of the moment it could resist if bent about the x axis only $(\phi_b M_{nx})$ then M_{uy} can be no greater than 25 percent of what it could resist if bent about the y axis only $(\phi_b M_{ny})$.

Examples 10-6 and 10-7 illustrate the design of beams subjected to unsymmetrical bending. To illustrate the trial-and-error nature of the problem the author did not do quite as much scratchwork as in some of his earlier examples. The first design problems of this type which the student attempts may quite well take several trials. Consideration needs to be given to the question of lateral support for the compression flange. Should the lateral support be of questionable nature the engineer should reduce the design moment resistance by means of one of the expressions previously given for that purpose.

EXAMPLE 10-6

A certain beam in its upright position is estimated to have a vertical bending moment M_{ux} = 160 ft-k and a lateral bending moment M_{uy} = 35 ft-k. These moments include the effect of the estimated beam weight. The loads are assumed to pass through the centroid of the section. Select a W24 shape of A36 steel that can resist these moments assuming full lateral support for the compression flange.

Solution

Try W24×62 ($\phi_b M_p = \phi_b M_{nx} = 413$ ft-k, $Z_y = 15.7$ in.³)

$$\phi_b M_{ny} = \frac{(0.9)(36)(15.7)}{12} = 42.39 \text{ ft-k}$$

$$\frac{M_{ux}}{\phi_b M_{nx}} + \frac{M_{uy}}{\phi_b M_{ny}} = \frac{160}{413} + \frac{35}{42.39} = 1.21 > 1.00 \qquad\qquad \text{NG}$$

Try W24×68 ($\phi_b M_p = \phi_b M_{nx} = 478$ ft-k, $Z_y = 24.5$ in.³)

$$\phi_b M_{ny} = \frac{(0.9)(36)(24.5)}{12} = 66.15 \text{ ft-k}$$

$$\frac{160}{478} + \frac{35}{66.15} = 0.864 < 1.00 \qquad\qquad \text{OK}$$

Use W24×68

It will be noted in the solution for Example 10-6 that though the procedure used will yield a section that will adequately support the moments given the selection of the absolutely lightest section listed in the LRFD Manual could be quite lengthy because of the two variables Z_x and Z_y which affect the size. If we have a large M_{ux} and a small M_{uy} the most economical section will probably be quite deep and rather narrow whereas if we have a large M_{uy} in proportion to M_{ux} the most economical section may be rather wide and shallow.

10-6 DESIGN OF PURLINS

To avoid bending in the top chords of roof trusses, it is theoretically desirable to place purlins only at panel points. For large trusses, however, it is more economical to space them at closer intervals. If this practice is not followed for large trusses the purlin sizes may become so large as to be impractical. When intermediate purlins are used the top chords of the truss should be designed for bending as well as axial stress as described in Chapter 11. Purlins are usually spaced from 2 to 6 ft apart depending on loading conditions, while their most desirable depth-to-span ratios are probably in the neighborhood of 1/24. Channels or S sections are the most frequently used sections but on some occasions other shapes may be convenient.

As previously described in Chapter 4, the channel and S sections are very weak about their web axes and sag rods may be necessary to reduce the span lengths for bending about those axes. Sag rods, in effect, make the purlins continuous sections for their y axes and the moments about these axes are greatly reduced, as shown in Fig. 10-11. These moment diagrams are developed on the assumption that the changes in length of the sag rods are negligible. It is further assumed that the purlins are simply supported at the trusses. This assumption is on the conservative side since they are often continuous over two or more trusses

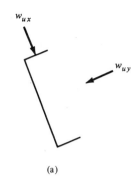

(a)

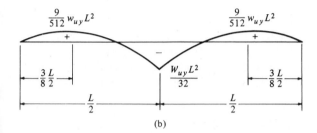

(b)

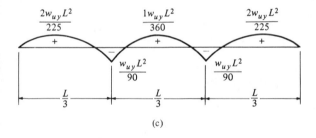

(c)

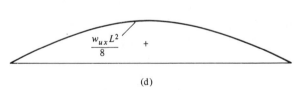

(d)

Figure 10-11 (a) Channel purlin, (b) Moment about web axis of purlins—sag rods at midspan. (c) Moment about web axis of purlins—sag rods at one-third points. (d) Moment about X axis of purlin.

and appreciable continuity may be achieved at their splices. The student can easily reproduce these diagrams from his or her knowledge of the moment distribution or other analysis methods. In the diagrams L is the distance between trusses, w_{uy} is the load component *perpendicular* to the web axis of the purlin, and w_{ux} is the load component *parallel* to the web axis.

If sag rods were not used the maximum moment about the web axis of a purlin would be $w_{uy}L^2/8$. When sag rods are used at midspan this moment is reduced to a maximum of $w_{uy}L^2/32$ (a 75 percent reduction) and when used at one-third points is reduced to a maximum of $w_{uy}L^2/90$ (a 91 percent reduction). In the example problem to follow (10-7), sag rods are used at the midpoints and the purlins are designed for a moment of $w_{uy}L^2/8$ parallel to the web axis and $w_{uy}L^2/32$ perpendicular to the web axis.

In addition to being of advantage in reducing moments about the web axes of purlins, sag rods can serve other useful purposes. First, they can provide lateral support for the purlins; second, they are useful in keeping the purlins in proper alignment during erection until the roof deck is installed and connected to the purlins.

EXAMPLE 10-7

Select a W6 purlin for the roof shown in Fig. 10-12. The trusses are 15 ft 0 in. on center and sag rods are used at the midpoints between trusses. Full lateral support is assumed to be supplied from the roof above. Use A36 steel and the LRFD Specification and assume there is no torsion. Thus the full value of Z_y is to be used. Loads are as follows in terms of pounds per square foot of roof surface.

$$\text{Snow} = 30 \text{ psf}$$

$$\text{Roofing} = 6 \text{ psf}$$

$$\text{Estimated purlin weight} = 3 \text{ psf}$$

$$\text{Wind pressure} = 15 \text{ psf} \perp \text{to roof surface}$$

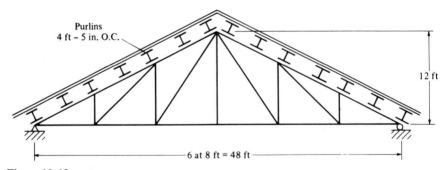

Figure 10-12

Solution

Computing the load w_{ux} to be supported per foot by the purlins (which are 4.42 ft on center).

$$w_{ux} = 1.2D + 1.6S + 0.8W$$

$$= (1.2)(6 + 3)(4.42)\left(\frac{2}{\sqrt{5}}\right) + (1.6)(30)(4.42)\left(\frac{2}{\sqrt{5}}\right)$$

$$+ (0.8)(15)(4.42) = 285.6 \text{ lb/ft} \leftarrow$$

$$w_{ux} = 1.2D + 1.3W + 0.5S$$

$$= (1.2)(6 + 3)(4.42)\left(\frac{2}{\sqrt{5}}\right) + (1.3)(15)(4.42)$$

$$+ (0.5)(30)(4.42)\left(\frac{2}{\sqrt{5}}\right) = 188.2 \text{ lb/ft} \leftarrow$$

$$w_{uy} = (1.2D + 1.6S)\frac{1}{\sqrt{5}}$$

$$= [(1.2)(6 + 3)(4.42) + (1.6)(30)(4.42)]\frac{1}{\sqrt{5}}$$

$$= 116.5 \text{ lb/ft}$$

$$M_{ux} = \frac{(0.2856)(15)^2}{8} = 8.03 \text{ ft-k}$$

$$M_{uy} = \frac{(0.1165)(15)^2}{32} = 0.819 \text{ ft-k}$$

Try W6×9 ($\phi_b M_p = 16.8$ ft-k, $Z_y = 1.73$ in.^{3})

$$\frac{M_{ux}}{\phi_b M_{nx}} + \frac{M_{uy}}{\phi_b M_{ny}} \leq 1.0 \quad \text{since} \quad \frac{P_u}{\phi P_n} < 0.2$$

$$\frac{8.03}{16.8} + \frac{0.819}{(0.9)(36)(1.73)/12} = 0.653 < 1.0 \qquad \text{OK}$$

$$\underline{\underline{\text{Use W6×9}}}$$

10-7 THE SHEAR CENTER

The shear center is defined as the point on the cross section of a beam through which the resultant of the transverse loads must pass so that the stresses in the beam may be calculated only from the theories of pure bending and transverse shear. Should the resultant pass through this point, it is unnecessary to analyze the beam for torsional moments. For a beam with two axes of symmetry the shear center will fall at the intersection of the two axes, thus coinciding with the centroid of the section. For a beam with one axis of symmetry the shear center will fall somewhere on that axis but not necessarily at the centroid of the section. This surprising statement means that to avoid torsion in some beams the lines of action of the loads should not pass through the centroids of the sections.

The shear center is of particular importance for beams whose cross sections are composed of thin parts which provide considerable bending resistance but little resistance to torsion. Many common structural members such as the W, S, and C sections, angles, and various beams made up of thin plates (as in aircraft construction) fall into this class and the problem has wide application.

The location of shear centers for several open sections are shown in Fig. 10-13. Sections such as these are relatively weak in torsion and for them and similar shapes the location of the resultant of the external loads can be a very serious matter. A previous discussion has indicated that the addition of one or more webs to these sections so they are changed into box shapes greatly increases their torsional resistance.

The average designer probably does not take the time to go through the sometimes tedious computations involved in locating the shear center and calcu- lating the effect of twisting. He or she may instead just ignore the situation, or may make a rough estimate of the effect of torsion on the bending stresses. One very poor estimte used on many occasions is that of reducing $\phi_b M_{ny}$ by 50 percent for use in the interaction equation.

Shear centers can be located quickly for beams with open cross sections and relatively thin webs. For other beams the shear centers can probably be found but only with considerable difficulty. The term *shear flow* is often used when speak- ing of thin-wall members although there is really no flowing involved. It refers to the shear per inch of the cross section and equals the unit shearing stress times the thickness of the member. (The unit shearing stress has been determined by the expression VQ/bI and the shear flow can be determined by VQ/I if the shearing stress is assumed to be constant across the thickness of the section.) The shear flow acts parallel to the sides of each element of a member.

The channel section of Fig. 10-14(a) will be considered for this discussion. In this figure the shear flow is shown with the small arrows and in part (b) the values are totaled for each component of the shape and labeled H and V. The two H values are in equilibrium horizontally and the internal V value balances the external shear at the section. Although the horizontal and vertical forces are in equilibrium, the same cannot be said for the moment forces unless the lines of action of the resultant of the external forces passes through a certain point called the shear center. The horizontal H forces in part (b) of the figure can be seen to form a couple. The moment produced by this couple must be opposed by an equal

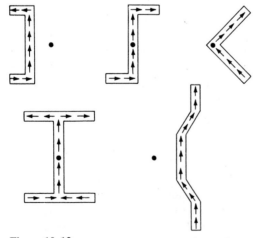

Figure 10-13

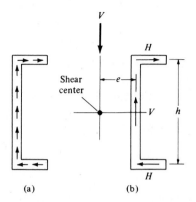

(a)   (b)   **Figure 10-14**

and opposite moment which can only be produced by the two V values. The location of the shear center is a problem in equilibirum; therefore moments should be taken about a point that eliminates the largest number of forces possible.

From this information the following equation can be written from which the shear center can be located.

$$Ve = Hh \quad \text{(moments taken about c.g. of web)}$$

Examples 10-8 and 10-9 illustrate the calculations involved in locating the shear center for two shapes. It will be noted that the location of the shear center is independent of the value of the external shear.

EXAMPLE 10-8

The channel section shown in Fig. 10-15(a) is subjected to an external shear of V. Locate the shear center.

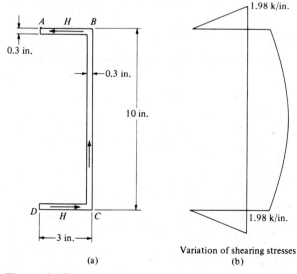

Variation of shearing stresses
(b)

(a)

Figure 10-15

Solution

Properties of section:

$$I_x = (\tfrac{1}{12})(3)(10)^3 - (\tfrac{1}{12})(2.7)(9.4)^3 = 63 \text{ in.}^4$$

$$f_v \text{ at } B = \frac{(V)(2.85 \times 0.3 \times 4.85)}{63} = 0.06582 \ V/\text{in.}$$

$$\text{Total } H = (\tfrac{1}{2})(2.85)(0.06582V) = 0.0938V$$

Location of shear center:

$$Ve = Hh$$

$$Ve = (0.0938V)(9.7)$$

$$e = 0.91 \text{ in. from } \mathcal{E} \text{ of web}$$

The student should clearly understand that shear stress variation across the corners, where the webs and flanges join, cannot be determined correctly with the mechanics of materials expression (VQ/bI or VQ/I for shear flow) and cannot be determined too well even after a complicated study with the theory of elasticity. The author as an approximation in the two examples presented here has assumed that shear flow continues up to the middle of the corners on the same pattern (straight-line variation for horizontal members and parabolic for others). The values of Q are computed for the corresponding dimensions. Other assumptions could have been made such as assuming a shear flow variation continuing vertically for the full depth of webs and only for the protruding parts of flanges horizontally, or vice versa. It is rather disturbing to find that whichever assumption is used the values do not check out perfectly. For instance, in Example 10-9 which follows the sum of the vertical shear flow values does not check out very well with the external shear.

EXAMPLE 10-9

The open section of Fig. 10-16 is subjected to an external shear of V. Locate the shear center.

Solution

Properties of section:

$$I_x = (\tfrac{1}{12})(0.25)(16)^3 + (2)(3.75 \times 0.25)(4.87)^2 = 129.8 \text{ in.}^4$$

Values of shear flow (labeled q):

$$q_A = 0$$

$$q_B = \frac{(V)(3.12 \times 0.25 \times 6.44)}{129.8} = 0.0387V/\text{in.}$$

$$q_C = q_B + \frac{(V)(3.75 \times 0.25 \times 4.87)}{129.8} = 0.0739V/\text{in.}$$

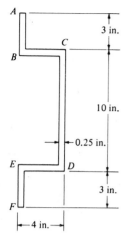

Figure 10-16

$$q_{\mathfrak{E}} = q_C + \frac{(V)(4.87 \times 0.25 \times 2.44)}{129.8} = 0.0968 V/\text{in.}$$

These shear flow values are shown in Fig. 10-17(a) and the summation for each part of the member is given in part (b) of the figure.

Taking moments about the center line of CD

$$-(0.211V)(9.75) + (2)(0.0807V)(3.75) + Ve = 0$$

$$e = 1.45 \text{ in.}$$

The theory of the shear center is a very useful one in design but it has certain limitations which should be clearly understood. For instance, the approximate analysis given in this section is only valid for thin sections. In addition steel beams often have variable cross sections along their spans

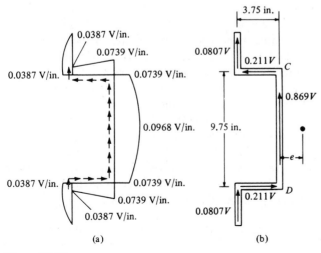

(a)   (b)

Figure 10-17

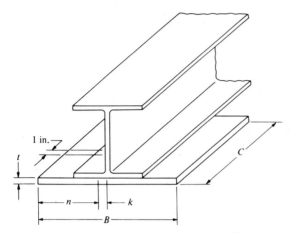

Figure 10-18

with the result that the loci of the shear centers are not straight lines along those spans. The result is that if the resultant of the loads passes through the shear center at one cross section it might very well not do so at other cross sections.

When designers are faced with the application of loads to thin-walled sections such that twisting of those sections will be a problem they will usually provide some means by which the twisting can be constrained. They may specify special bracing at close intervals, or attachments to flooring or roofing, or other similar devices. Should such solutions not be feasible they will probably consider selecting sections with greater torsional stiffnesses. Two references on this topic are given here.[3,4]

10-8 BEAM-BEARING PLATES

When the ends of beams are supported by direct bearing on concrete or other masonry construction it is frequently necessary to distribute the beam reactions over the masonry by means of beam-bearing plates. The reaction is assumed to be spread uniformly through the bearing plate to the masonry and the masonry is assumed to push up against the plate with a uniform pressure equal to the factored reaction R_u over the area of the plate A_1. This pressure tends to curl up the plate and the bottom flange of the beam. The LRFD Manual recommends that the bearing plate be considered to take the entire bending moment produced and that the critical section for moment be assumed to be a distance k from the center line of the beam (see Fig. 10-18). The distance k is the same as the distance from the outer face of the flange to the web toe of the fillet given in the tables for each section (or it equals the flange thickness plus the fillet radius).

[3]C.G. Salmon and J.E. Johnson, *Steel Structures, Design and Behavior,* 2d ed. (New York: Harper & Row, 1980), pp. 672–675.
[4]AISC Engineering Staff, Torsional Analysis of Steel Members (Chicago: AISC, 1983).

The determination of the true pressure distribution in a beam-bearing plate is a very formidable task, and the uniform pressure distribution assumption is usually made. This assumption is probably on the conservative side as the pressure is usually larger at the center of the beam than at the edges. The outer edges of the plate and flange tend to bend upward and the center of the beam tends to go down, concentrating the pressure there.

The required thickness of a 1-in.-wide strip of plate can be determined as follows:

$$M_u = \frac{R_u}{A_1} n \frac{n}{2} = \frac{R_u n^2}{2A_1}$$

Z of 1 in. wide piece of plate of t thickness $= (1)\left(\frac{t}{2}\right)\left(\frac{t}{4}\right)(2) = \frac{t^2}{4}$

$$M_u = \phi_b F_y Z \qquad \text{with} \qquad \phi_b = 0.9$$

$$\frac{R_u n^2}{2A_1} = (0.9)(F_y)\left(\frac{t^2}{4}\right)$$

$$t = \sqrt{\frac{2.22 R_u n^2}{A_1 F_y}}$$

If the plate extends for the full width of the wall or other support parallel to the beam the area of the plate A_1 is determined by dividing the factored reaction R_u by the permissible ultimate pressure on the concrete or other masonry beneath the plate. For concrete this would normally be $\phi_c 0.85f'_c$ where ϕ_c for bearing on concrete is 0.60 and f'_c is the 28-day compressive strength of the concrete in ksi.

$$A_1 = \frac{R_u}{\phi_c 0.85f'_c}$$

If the plate does not extend for the full width of the support A_1 is to be determined by the following expression according to Section J9 of the LRFD Specification where A_2 is the maximum area of the portion of the supporting surface that is geometrically similar to and concentric with the loaded area.

$$A_1 = \frac{1}{A_2}\left(\frac{R_u}{\phi_c 0.85f'_c}\right)^2$$

After A_1 is determined its length (parallel to the beam) and its width are selected. The length may not be less than the N required to prevent web yielding or web crippling of the beam nor may it be less than about $3\frac{1}{2}$ or 4 in. for practical construction reasons. It may not be greater than the thickness of the wall or other support and it may actually have to be less than that thickness particularly at exterior walls to prevent the steel from being exposed.

Example 10-10 illustrates the calculations involved in designing a beam bearing plate. Notice that the width and the length of the plate are desirably taken to the nearest full inch.

EXAMPLE 10-10

A W18×71 beam ($d = 18.47$ in., $t_w = 0.495$ in., $b_f = 7.635$ in., $t_f = 0.810$ in., $k = 1\frac{1}{2}$ in.) has one of its ends supported by a reinforced-concrete wall with $f'_c = 3$ ksi. Design a bearing plate for the beam with A36 steel. The factored end reaction is 120 k and the maximum length of end bearing $\perp$ to the wall is the full wall thickness 8.0 in.

Solution

$$\text{Area required} = A_1 = \frac{R_u}{\phi_c 0.85 f'_c} = \frac{120}{(0.60)(0.85)(3)} = 78.43 \text{ in.}^2$$

Minimum length of bearing $\perp$ to wall required for web yielding

$$R_n = \frac{R_u}{\phi} = \frac{120}{1.0} = 120 \text{ k}$$

$$120 = (2.5 \times 1.50 + N)(36)(0.495)$$

$$N = 2.98 \text{ in.}$$

Minimum length of bearing $\perp$ to wall required for web crippling

$$R_n = \frac{R_u}{\phi} = \frac{120}{0.75} = 160 \text{ k}$$

$$160 = (68)(0.495)^2 \left[1 + 3 \left(\frac{N}{18.47} \right) \left(\frac{0.495}{0.810} \right)^{1.5} \right] \sqrt{\frac{(36)(0.810)}{0.495}}$$

$$N = 3.24 \text{ in.} \leftarrow$$

Try 8×10 in. PL

$$n = 5 - 1\frac{1}{2} = 3.5 \text{ in.}$$

$$t = \sqrt{\frac{2.22 \, R_u n^2}{A_1 F_y}} = \sqrt{\frac{(2.22)(120)(3.5)^2}{(80.0)(36)}} = 1.06 \text{ in.}$$

Use $1\frac{1}{8} \times 8 \times 10$ in. PL

On some occasions the beam flanges alone probably provide sufficient bearing area, but bearing plates are nevertheless recommended as they are useful in erection and ensure an even bearing surface for the beam. They can be placed separately from the beams and carefully leveled to the proper elevations. When the ends of steel beams are enclosed by the concrete or masonry walls it is considered desirable to use some type of wall anchor to prevent the beam from moving longitudinally with respect to the wall. The usual anchor consists of a bent steel bar passing through the web of the beam and running parallel to the wall. These are called government anchors and details of their sizes are given on page 5-163 of the LRFD Manual. Occasionally clip angles attached to the web are used instead of government anchors. Should longitudinal loads of considerable size be anticipated, regular vertical anchor bolts may be used at the beam ends.

If we were to check to see if the flange thickness alone is sufficient, we would have

$$t = \sqrt{\frac{2.22 R_u n^2}{A_1 F_y}} = \sqrt{\frac{(2.22)(120)(3.5)^2}{(8 \times 7.635)(36)}} = 1.221 \text{ in.} > t_f \text{ of } W 18 \times 71$$

$$\therefore \text{ Flanges not sufficient alone}$$

PROBLEMS

10-1 to 10-5. *Considering moment only and assuming full lateral support for the compression flanges select the lightest A36 sections available. The loads shown include the effect of the beam weights. Use elastic analysis, factored loads, and the 0.9 rule.*

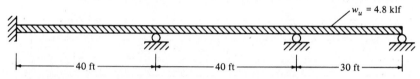

Problem 10-1   (*Ans.* W27 × 94)

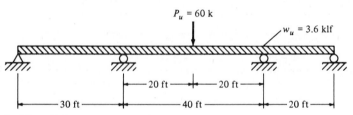

Problem 10-2

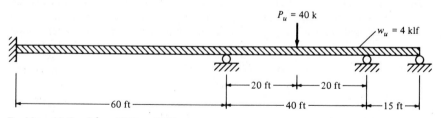

Problem 10-3   (*Ans.* W33 × 130)

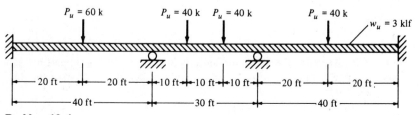

Problem 10-4

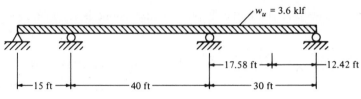

Problem 10-5   (*Ans.* W21 × 68)

10-6. Repeat Prob. 10-1 using plastic analysis.

10-7. Repeat Prob. 10-3 using plastic analysis. (*Ans.* W30×108)

10-8. Repeat Prob. 10-5 using plastic analysis.

10-9. Three methods of supporting a roof are shown in the accompanying illustration. Using an elastic analysis with factored loads, A36 steel, and assuming full lateral support in each case select the lightest section if a dead uniform service load (not including beam weight) of 1 klf and a live uniform service load of 1.2 klf is to be supported. Consider moment only. (*Ans.* W24×76, W21×62, W21×62)

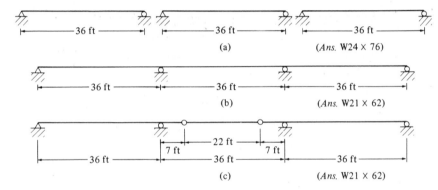

Problem 10-9

10-10. Using A36 steel select the lightest available section for the span and loading shown in the accompanying illustration. Consider moment and shear only and neglect beam weight in all calculations. The member is assumed to have full lateral bracing.

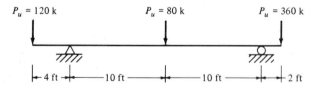

Problem 10-10

10-11. If a fully braced W14×34 section consisting of A36 steel is used for a simple span of 6 ft 6 in. determine the maximum uniform load w_u that it can support in addition to its own weight. Use elastic analysis and consider shear and moment only. (*Ans.* 23.74 klf)

10-12. A W16×40 consisting of A36 steel is used as a simple beam for a span of 6 ft. If it has full lateral support determine the maximum uniform load w_u which it can support in addition to its own weight. Use an elastic analysis and consider shear and moment only.

10-13. A fully braced W36×245 consisting of A36 steel is used as a simple beam for a span of 16 ft. Considering moment and shear only determine the maximum uniform load w_u it can support in addition to its own weight using an elastic analysis. (*Ans.* 69.70 klf)

10-14. A 30-ft simply supported beam is to support a moving concentrated load P_u = 60 k. Using A36 steel select the most economical section considering moment and shear only. Use an elastic analysis and neglect beam weight.

10-15. A 40-ft simple beam which supports a service concentrated load P_L = 30 k at midspan is laterally unbraced except at its ends and center line. If the maximum permissible center line deflection under service loads equals 1/1000 of the span select the most economical W section of A36 steel considering moment, shear, and deflection. Neglect beam weight. (*Ans.* W33×118)

10-16. Design a beam for a 24-ft simple span to support the working uniform loads w_D = 1.2 klf (includes beam weight) and w_L = 2.1 klf. The maximum permissible deflection under working loads is 1/1200 of the span. Use A36 steel and consider moment, shear, and deflection. The beam is to be braced laterally for its full length.

10-17. Select the lightest available W section of A36 steel for the span and service loads shown. The beam will have full lateral support for its compression flange. Its maximum service load center-line deflection may not exceed 1/1500 of the span under working loads. Consider moment, shear, and deflection only. (*Ans.* W36×135)

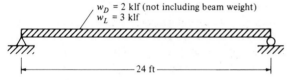

w_D = 2 klf (not including beam weight)
w_L = 3 klf

24 ft

Problem 10-17

10-18. Select the lightest section available if F_y = 50 ksi for the span and working loads shown if the section is to be fully braced laterally and have a maximum service load deflection of 1/800 of its span length. Neglect beam weight in all calculations.

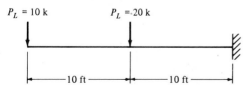

P_L = 10 k P_L = 20 k

10 ft   10 ft

Problem 10-18

10-19. If the maximum permissible service load deflection for the fully braced beam shown is 1/1200 of the span select the lightest available section using A36 steel. Consider moment, shear, and deflections. Working loads are shown. (*Ans.* W36×150)

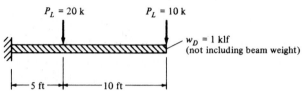

$P_L = 20$ k $P_L = 10$ k

$w_D = 1$ klf
(not including beam weight)

5 ft   10 ft

Problem 10-19

10-20. Select the lightest available W section (A36 steel) for the working live loads and span shown in the accompanying illustration. Lateral support is provided only at the 12-ft points and a maximum deflection (under working loads) equal to 1/1500 of the 24-ft span is permitted. Neglect beam weight. Consider moment, shear, and deflections.

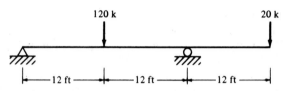

120 k   20 k

12 ft   12 ft   12 ft

Problem 10-20

10-21. Select the lightest available W30 section consisting of A36 steel to resist a gravity moment $M_{ux} = 400$ ft-k and a lateral bending moment of $M_{uy} = 100$ ft-k. The section is assumed to have full lateral support. (*Ans.* W30×132)

10-22. The 20-ft simple beam shown has full lateral support for its compression flange and consists of A36 steel. The beam supports a gravity service dead load of 1.2 klf (includes beam weight) and a gravity live load of 3.4 klf. The loads are assumed to act through the c.g. of the section. Select the lightest available W36 section.

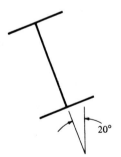

20°

Problem 10-22

10-23. A simply supported W24×146 consisting of A36 steel is supporting a 300-k service live load as shown in the figure. If the length of bearing at the left support is 8 in. and at the concentrated load is 12 in. check the beam for shear, web yielding, and web crippling. (*Ans.* All items except web crippling at concentrated load NG)

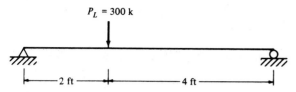

Problem 10-23

10-24. A 32-ft beam with full lateral support for its compression flange is supporting a moving service concentrated load of 32 k. Using A36 steel select the lightest section available for moment. Then check to see if the section is satisfactory in shear and compute the minimum length of bearing required at the supports from the standpoint of web yielding and web crippling.

10-25. Design a steel bearing plate from A36 steel for a W21×68 beam with an end reaction $R_u = 100$ k. The beam will bear on a reinforced-concrete wall with $f'_c = 4$ ksi. In a direction perpendicular to the wall the bearing plate may not be longer than 8 in. (*Ans.* PL $1\frac{1}{8}$×6×0 ft 9 in.)

10-26. Design a steel bearing plate of A36 steel for a W30×116 beam supported by a reinforced-concrete wall with $f'_c = 3$ ksi. The maximum beam reaction R_u is 170 k. Assume the width of the plate perpendicular to the wall is 8 in.

10-27. Repeat Prob. 10-26 using a steel with $F_y = 50$ ksi. (*Ans.* $1\frac{1}{2}$×8×1 ft 2 in.)

Chapter 11

Bending and Axial Force

11-1 OCCURRENCE

Structural members that are subjected to a combination of bending and axial force are far more common than the student may realize. This section is devoted to listing a few of the more obvious cases. Columns that are part of a steel building frame must nearly always resist sizable bending moments in addition to the usual compressive loads. It is almost impossible to erect and center loads exactly on columns even in a testing lab and in an actual building the student can see that it is even more impossible. Even if building loads could be perfectly centered at one time they would not stay in one place. Furthermore, columns may be initially crooked or have other flaws with the result that lateral bending is produced. The beams framing into columns are quite commonly supported with framing angles or brackets on the sides of the columns. These eccentrically applied loads produce moments. Wind and other lateral loads cause columns to bend laterally, and the columns in rigid frame buildings are subjected to moments even when the frame is supporting gravity loads alone. The members of bridge portals must resist combined forces as do building columns. Among the causes of the combined forces are heavy lateral wind loads, vertical traffic loads whether symmetrical or not, and the centrifugal effect of traffic on curved bridges.

The previous experience of the student has probably been to assume truss members axially loaded only. Purlins for roof trusses, however, are frequently placed in between truss joints, causing the top chords to bend. Similarly the bottom chords may be bent by the hanging of light fixtures, ductwork, and other items between the truss joints. All horizontal and inclined truss members have moments caused by their own weights, while all truss members whether vertical or not are subjected to secondary bending forces. Secondary forces are developed

Inryco Building, Chicago, Ill. (Courtesy of I. R. Construction Products Co.)

because the members are not connected with frictionless pins as assumed in the usual analysis, the member centers of gravities or those of their connectors do not exactly coincide at the joints, etc.

Moments in tension members are not as serious as those in compression members because tension tends to reduce lateral deflections while compression increases them. Increased lateral deflection in turn results in larger moments, which result in larger lateral deflections, etc. It is hoped that members in such situations are quite stiff so as to keep the additional lateral deflections from becoming excessive.

11-2 MEMBERS SUBJECT TO BENDING AND AXIAL TENSION

In Section H1 of the LRFD Specification the following interaction equations are given for symmetric shapes subjected simultaneously to bending and axial tensile forces. These equations are also applicable to members subjected to bending and axial compression forces as will be described in Sections 11-6 and 11-7.

$$\text{If } \frac{P_u}{\phi_t P_n} \geq 0.2$$

$$\frac{P_u}{\phi_t P_n} + \frac{8}{9}\left(\frac{M_{ux}}{\phi_b M_{nx}} + \frac{M_{uy}}{\phi_b M_{ny}}\right) \leq 1.0 \qquad \text{(LRFD Formula H1-1a)}$$

If $\dfrac{P_u}{\phi_t P_n} < 0.2$

$$\frac{P_u}{2\phi_t P_n} + \left(\frac{M_{ux}}{\phi_b M_{nx}} + \frac{M_{uy}}{\phi_b M_{ny}}\right) \le 1.0 \qquad \text{(LRFD Formula H1-1b)}$$

The terms in these equations have previously been defined where P_u and M_u are the required tensile and flexural strengths, P_n and M_n are the nominal tensile and flexural strengths, and the resistance factors ϕ_t and ϕ_b are determined as in previous chapters. Usually only a first-order analysis is made for members subject to bending and axial tension. The analyst, however, may and is encouraged to make a second-order analysis for these members and use the results in his or her designs. Examples 11-1 and 11-2 illustrate the application of these interaction expressions to members subjected simultaneously to bending and axial tension.

EXAMPLE 11-1

A W12×35 tension member with no holes consisting of A36 steel is subjected to a factored tensile force P_u of 60 k and a factored bending moment M_{uy} of 25 ft-k. Is the member satisfactory if $L_b < L_p$?

Solution. Using a W12×35 ($A = 10.3$ in.^{3}, $Z_y = 11.5$ in.^{3})

$$\phi_t P_n = \phi_t F_y A_g = (0.9)(36)(10.3) = 333.7 \text{ k}$$

$$\frac{P_u}{\phi_t P_n} = \frac{60}{333.7} = 0.18 < 0.2$$

$$\therefore \text{ Use LRFD Formula H1-1b}$$

$$\phi_b M_{ny} = \phi_b F_y Z_y = \frac{(0.9)(36)(11.5)}{12} = 31.05 \text{ ft-k}$$

$$\frac{P_u}{2\phi_t P_n} + \left(\frac{M_{ux}}{\phi_b M_{nx}} + \frac{M_{uy}}{\phi_b M_{ny}}\right)$$

$$= \frac{60}{(2)(333.7)} + \left(0 + \frac{25}{31.05}\right) = 0.895 < 1.0 \qquad \text{OK}$$

EXAMPLE 11-2

A W10×30 tensile member with no holes consisting of A36 steel and with $L_b = 12.0$ ft is subjected to a factored tensile force $P_u = 100$ k and to the factored moments $M_{ux} = 70$ ft-k and $M_{uy} = 0$. If $C_b = 1.0$, is the member satisfactory?

Solution. Using a W10×30 ($A = 8.84$ in.^{2}, $L_p = 5.7$ ft, $L_r = 20.3$ ft)

$$\phi_t P_n = (0.9)(36)(8.84) = 286.4 \text{ k}$$

$$\frac{P_u}{\phi_t P_n} = \frac{100}{286.4} = 0.349 > 0.2$$

$$\therefore \text{ Use LRFD Formula H1-1a)}$$

Noting $L_b > L_p$, from Load Factor Design Selection Table

$$\phi_b M_p = 98.8 \text{ ft-k}$$

$$\phi_b M_r = 63.2 \text{ ft-k}$$

$$\text{BF} = 2.44$$

$$\phi_b M_n = C_b[\phi_b M_p - \text{BF}(L_b - L_p)]$$

$$= 1.0[98.8 - 2.44(12.0 - 5.7)] = 83.43 \text{ ft-k}$$

$$\frac{P_u}{\phi_t P_n} + \frac{8}{9}\left(\frac{M_{ux}}{\phi_b M_{nx}} + \frac{M_{uy}}{\phi_b M_{ny}}\right) = \frac{100}{286.4} + \frac{8}{9}\left(\frac{70}{83.43} + 0\right)$$

$$= 1.095 > 1.0 \qquad \text{NG}$$

11-3 FIRST-ORDER AND SECOND-ORDER MOMENTS

If a particular member is acted upon by both moments and axial compression loads the results will be increased bending moments and increased lateral deflections. When we analyze a frame by one of the common elastic methods the results are called primary or first-order forces and moments. Even if the frame is braced laterally there will be some secondary moments due to lateral bending of the columns. The column of Fig. 11-1 is assumed to be braced against sidesway but it will bend laterally by some amount δ as shown and a secondary moment $P_u\delta$ will be produced. This moment will cause some additional lateral deflection which will cause some additional moment and so on.

In the LRFD Specification the moment M_1 is assumed to equal M_{nt} (the moment resulting from gravity loads) plus the moment due to the lateral deflection $P_u\delta$. To estimate the sum of these two values the LRFD provides a magnification factor B_1 which is ≥ 1.0 and is to be multiplied by M_{nt} (the formula for B_1 is presented in Section 11-4).

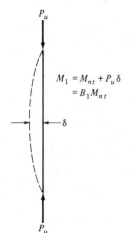

$$M_1 = M_{nt} + P_u\delta$$
$$= B_1 M_{nt}$$

Figure 11-1   Column in a frame braced against sidesway.

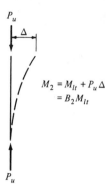

P_u

Δ

$M_2 = M_{lt} + P_u\Delta$
$\quad = B_2 M_{lt}$

P_u

Figure 11-2   Column in an unbraced frame.

If a frame is subject to sidesway where the ends of the columns can move laterally with respect to each other additional secondary moments will result. In Fig. 11-2 the secondary moment produced due to sidesway is equal to $P_u\Delta$.

The moment M_2 is assumed by the LRFD Specification to equal M_{lt} (which is the moment due to the lateral loads) plus the moment due to $P_u\Delta$. The LRFD Specification gives a value for $B_2 \geq 1.0$ (shown in Section 11-4) which is to be multiplied by M_{lt} to estimate the total of these two moments.

The total estimated first-order and second-order moment in a member subjected to axial compression plus bending can then be determined from the expression to follow

$$M_u = B_1 M_{nt} + B_2 M_{lt}$$

Instead of using the LRFD empirical procedure described here the designer may and is encouraged to use a theoretical second-order analysis provided he or she meets certain requirements of Sections C1 and C2 of the Specification. These requirements pertain to axial deformations, maximum axial forces permitted in members, bracing, K factors, and so on.

11-4 MAGNIFICATION FACTORS

The magnification factors are B_1 and B_2. With B_1 the analyst attempts to estimate the $P_u\delta$ effect for a column whether the frame is braced or unbraced against sidesway. With B_2 he or she attempts to estimate the $P_u\Delta$ effect in unbraced frames.

This moment estimation is applicable only when the connections are fully restrained or unrestrained. The LRFD Manual indicates that the determination of secondary moments in between these two classifications for partially restrained moment connections is beyond the scope of their specification. The terms fully restrained and partially restrained are discussed at length in Chapter 15.

In the expression for B_1 that follows C_m is a term that will be defined in Section 11-5, P_u is the required axial strength of the column, while P_e is the Euler buckling strength $A_g F_y/\lambda_c^2$ and is a function of KL in the plane of bending with $KL = 1.0$, and $\lambda_c = (KL/r\pi)\sqrt{F_y/E}$. In Table 9 of the LRFD Specification

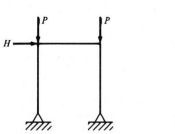
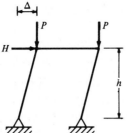

Figure 11-3

Appendix the value of P_e/A_g can be obtained directly. Multiplying it by A_g provides the value of P_e. Notice that B_1 cannot be less than 1.0.

$$B_1 = \frac{C_m}{1 - P_u/P_e} \geq 1.0$$

The designer may use either of two expressions given by the LRFD for B_2. In the first of these expressions ΣP_u represents the required axial strength of all the columns on the floor in question, Δ_{oh}/L represents the story drift index, and ΣH is the sum of all the story horizontal forces producing Δ_{oh}.

$$B_2 = \frac{1}{1 - \Sigma P_u(\Delta_{oh}/\Sigma HL)} \qquad \text{(LRFD Formula H1-5)}$$

$$B_2 = \frac{1}{1 - \Sigma P_u/\Sigma P_e} \qquad \text{(LRFD Formula H1-6)}$$

From a practical standpoint in calculating ΣP_u and ΣP_e we will probably just calculate the values for the columns in that one frame or that single line of columns perpendicular to the wind.

The horizontal deflection of a multistory building due to wind or seismic load is called drift. It is represented by Δ in Fig. 11-3.

Drift is measured by a *drift index* Δ/h where Δ is the deflection and h is the height or distance to the lower level. For the comfort of the occupants of the building the index is usually limited at working loads to a value between 0.0015 and 0.003 and at ultimate load to about 0.004.

Of the two expressions given for B_2 the first one, which involves a drift index, is better suited for design office practice. If we assume drifts as large as these values, we are probably being quite conservative. That is, the real structures may very well not drift this much.

11-5 MODIFICATION OR C_m FACTORS

In Sections 11-3 and 11-4 the subject of moment magnification due to lateral deflections was introduced and the factors B_1 and B_2 were presented with which the moment increases could be estimated. In the expression for B_1 a term C_m called the modification factor was included. The purpose of this factor is to modify or reduce the value of B_1 in certain cases where it otherwise would be too large.

If a column is subject to moments at its ends which bend it in single curvature (( )) the lateral deflection situation is appreciably worse than if the same column had been bent in double curvature (( )) with the same moments. For the latter case there is an inflection point at middepth if the end moments are equal and the lateral deflection there is zero. Furthermore the lateral deflection is not very large at any point in the member and secondary bending is not large. However, if C_m were not included in the equation the value of B_1 would be the same for both situations.

To account for cases such as the one just described where the moment is reduced due to end support conditions the LRFD Specification includes C_m in the expression for B_1. Modification factors are based on the rotational restraint at the member ends and on the moment gradients in the members. The LRFD Specification (H1-2a) includes two categories of C_m as described in the next few paragraphs.

In Category 1 the members are prevented from joint translation or sidesway and they are not subject to transverse loading between their ends. For such members the modification factor is to be determined as follows.

$$C_m = 0.6 - 0.4 \frac{M_1}{M_2}$$

In this expression M_1/M_2 is the ratio of the smaller moment to the larger moment at the ends of the unbraced length in the plane of bending under consideration. The ratio is negative if the moments cause the member to bend in single curvature and positive if they bend the members in reverse or double curvature. As previously described a member in single curvature has larger lateral deflections than a member bent in reverse curvature. With larger lateral deflections the moments due to the axial loads will be larger.

Category 2 applies to members which are subjected to transverse loading between the joints and which are braced against joint translation or sidesway in the plane of loading. The compression chord of a truss with a purlin load between its joints is a typical example of this category. The LRFD Specification states that the value of C_m may be taken as follows.

a. For members with restrained ends $C_m = 0.85$.
b. For members with unrestrained ends $C_m = 1.0$.

Instead of these values C_m (Category 2) may be determined for various end conditions and loads by the values given in Table 11-1, which is a reproduction of Table C-H1.1 of the Commentary on the LRFD Specification. In the expressions given in the table f_a is the computed axial stress in the member while F'_e is the Euler buckling stress divided by a factor of safety of 23/12. Thus

$$F'_e = \frac{12\pi^2 E}{23(kL_b/r_b)^2}$$

TABLE 11-1

Case	ψ	C_m
(a)	0	1.0
(b)	-0.4	$1 - 0.4\dfrac{f_a}{F'_e}$
(c)	-0.4	$1 - 0.4\dfrac{f_a}{F'_e}$
(d)	-0.2	$1 - 0.2\dfrac{f_a}{F'_e}$
(e)	-0.3	$1 - 0.3\dfrac{f_a}{F'_e}$
(f)	-0.2	$1 - 0.2\dfrac{f_a}{F'_e}$

In this expression L_b and r_b are the actual unbraced length and the corresponding radius of gyration in the plane of bending.

11-6 INTERACTION EQUATIONS FOR AXIAL COMPRESSION LOADS AND BENDING

The same interaction equations are used for members subject to axial compression and bending as were used for members subject to axial tension and bending. However, some of the terms involved in the equations are defined somewhat differently. For instance, P_u and P_n refer to compressive forces rather than tensile ones, ϕ_c is 0.85 for axial compression, and ϕ_b is 0.9 for bending.

To analyze a particular beam column or a member subject to both bending and axial compression we need to make both a first-order and a second-order analysis to obtain the bending moments. The first-order moment is usually obtained by making an elastic analysis and consists of the moments M_{nt} (these are the moments in beam columns caused by gravity loads) and the moments M_{lt}

(these are the moments in beam columns due to the lateral loads, that is, due to lateral translation).

Theoretically if both the loads and frame are symmetrical, M_{lt} will be zero. Similarly if the frame is braced, M_{lt} will be zero. For practical purposes you can have lateral deflections in taller buildings with symmetrical dimensions and loads. If the buildings are below 20 stories, however, we can probably neglect these deflections. If above 20 stories we should consider them.

Examples 11-3 and 11-4 illustrate the application of the interaction formulas to two beam columns.

EXAMPLE 11-3

A 12-ft-long W12×170 consisting of A36 steel is used in a braced symmetrical frame to support the following factored forces caused by symmetrical gravity loads: $P_u = 600$ k, $M_{ux} = 200$ ft-k, and $M_{uy} = 90$ ft-k. Is the member satisfactory if it is bent in reverse curvature with equal end moments about both axes and if it has no intermediate loads?

Solution.   Using a W12×170 ($A = 50.0$ in.^{2}, $r_x = 5.74$ in., $r_y = 3.22$ in., $Z_x = 275$ in.^{3}, $Z_y = 126$ in.^{3}, $L_p = 13.4$ ft)
For a braced frame $K = 1.0$ and thus

$$\frac{K_x L_x}{r_x} = \frac{(1.0)(12 \times 12)}{5.74} = 25.09$$

$$\frac{K_y L_y}{r_y} = \frac{(1.0)(12 \times 12)}{3.22} = 44.72$$

$$\phi_c F_{cr} = 27.55 \text{ ksi (from LRFD Table 3-36)}$$

$$\phi_c P_n = (27.55)(50.0) = 1377.5 \text{ k (or from}$$
$$\text{column tables where 1380 k is given)}$$

$$\frac{P_u}{\phi P_n} = \frac{600}{1377.5} > 0.2$$

$$\therefore \text{ Must use LRFD Formula H1-1a}$$

For a braced frame M_{lt} (the moment due to lateral loads) = 0

$$C_{mx} = C_{my} = 0.6 - 0.4 \frac{M_1}{M_2} = 0.6 - 0.4(+1.0) = 0.2$$

From Table 9 page 6-131 in the LRFD Manual values of P_{ex} and P_{ey} are determined as follows:

$$\frac{P_{ex}}{A_g} = 454.84 \text{ ksi} \quad \text{for} \quad \frac{K_x L_x}{r_x} = 25.09$$

$$P_{ex} = (454.84)(50.0) = 22{,}742 \text{ k}$$

$$\frac{P_{ey}}{A_g} = 143.16 \text{ ksi} \quad \text{for} \quad \frac{K_y L_y}{r_y} = 44.72$$

$$P_{ey} = (143.16)(50.0) = 7158 \text{ k}$$

$$B_{1x} = \frac{C_{mx}}{1 - P_u/P_{ex}} = \frac{0.2}{1 - 600/22,742} = <1.0 \qquad \underline{\text{Use 1.0}}$$

$$B_{1y} = \frac{C_{my}}{1 - P_u/P_{ey}} = \frac{0.2}{1 - 600/7158} = <1.0 \qquad \underline{\text{Use 1.0}}$$

$$B_{2x} = B_{2y} = 0 \text{ since frame is braced}$$

$$M_{ux} = B_1 M_{ntx} + B_2 M_{ltx} = (1.0)(200) + 0 = 200 \text{ ft-k}$$

$$M_{uy} = B_1 M_{nty} + B_2 M_{lty} = (1.0)(90) + 0 = 90 \text{ ft-k}$$

$$\phi_b M_{nx} = \phi_b M_p \quad \text{since} \quad L_b < L_p$$

$$= \frac{(0.9)(36)(275)}{12} = 742.5 \text{ ft-k}$$

$$\phi_b M_{ny} = \frac{(0.9)(36)(126)}{12} = 340.2 \text{ ft-k}$$

Applying LRFD Formula H1-1a

$$\frac{P_u}{\phi P_n} + \frac{8}{9}\left(\frac{M_{ux}}{\phi_b M_{nx}} + \frac{M_{uy}}{\phi_b M_{ny}}\right) = \frac{600}{1377.5} + \frac{8}{9}\left(\frac{200}{742.5} + \frac{90}{340.2}\right)$$

$$= 0.911 < 1.0 \qquad \text{OK}$$

EXAMPLE 11-4

A W14×145 consisting of A36 steel is used as a column in a symmetrical unbraced frame to support a factored load $P_u = 800$ k due to gravity and wind and a factored bending moment due to wind about the x axis $= M_x = 290$ ft-k. Other data: $L_x = L_y = 14$ ft and $K = 1.0$; drift index $\Delta_{oh}/L = 0.0025$; $\Sigma P_u = 20,000$ k and $\Sigma H = 650$ k.

Solution. Using a W14×145 ($L_p = 16.6$ ft and $Z_x = 260$ in.^{3}), $\phi_c P_n = 1190$ k from column table, Part 2, LRFD Manual.

$$\frac{P_u}{\phi_c P_n} = \frac{800}{1190} > 0.2$$

$$\therefore \text{ Must use LRFD Formula H1-1a}$$

Noting that $M_{ntx} = M_{nty} = M_{lty} = 0$

$$B_2 = \frac{1}{1 - \Sigma P_u/\Sigma H)(\Delta_{oh}/L)} = \frac{1}{1 - \left(\dfrac{20,000}{650}\right)(0.0025)} = 1.08$$

$$M_{ux} = B_1 M_{nt} + B_2 M_{lt} = 0 + (1.08)(290) = 313.2 \text{ ft-k}$$

Since $L_b < L_p$

$$M_{nx} = \phi_b F_y Z_x = \frac{(0.9)(36)(260)}{12} = 702 \text{ ft-k}$$

$$\frac{P_u}{\phi P_n} + \frac{8}{9}\left(\frac{M_{ux}}{\phi_b M_{nx}} + \frac{M_{uy}}{\phi_b M_{ny}}\right)$$

$$= \frac{800}{1190} + \frac{8}{9}\left(\frac{313.2}{702} + 0\right)$$

$$= 1.07 > 1.00 \qquad\qquad\qquad\qquad\qquad \text{NG}$$

Section is unsatisfactory

When the young engineer begins to design actual structures, he or she is often distressed over the fact that real structures do not fall into the exact textbook situations learned back in school. As a result he or she must frequently make assumptions or interpolate between handbook values. For instance: Is a certain connection fixed or is it simple or is it really somewhere in between? The decision made may decidedly affect the values of moments, K values, C_m values, and so on.

The author likes to think that Example 11-5 which follows is a very practical problem in that these types of assumptions have to be made. In this case the author assumed that the end conditions of the member being checked are somewhere in between ⊿——⊿ and ⊿——⊧ . He then computed moments and C_m values for the two cases and averaged them together. It is thought that this procedure is within the spirit of the LRFD Specification.

EXAMPLE 11-5

For the truss shown in Fig. 11-4(a) a W8×31 is used as a continuous top chord member from joint L_0 to joint U_3. If the member consists of A36 steel, does it have sufficient strength to resist the factored loads shown in part (b) of Fig. 11-4? Part (b) shows the portion of the chord from L_0 to U_1 and the 16-k load represents the effect of a purlin. It is assumed that lateral support is provided for this member at its ends and center line.

Solution. Using a W8×31 ($A = 9.13$ in.^{2}, $Z_x = 30.4$ in.^{3}, $r_x = 3.47$ in., $r_y = 2.02$ in., and $L_p = 8.4$ ft)

For a frame braced against sidesway $K = 1.0$

$$\frac{K_x L_x}{r_x} = \frac{(1.0)(12 \times 13)}{3.47} = 44.96$$

$$\frac{K_y L_y}{r_y} = \frac{(1.0)(12 \times 6.5)}{2.02} = 38.61$$

$$\phi_c F_{cr} = 27.50 \text{ ksi (from LRFD Table 3-36)}$$

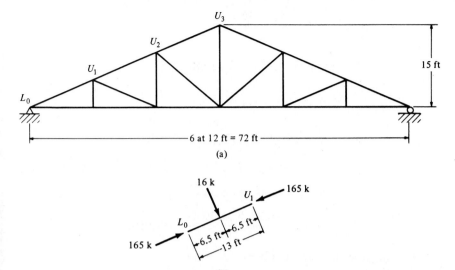

Figure 11-4

$$\phi_c P_n = (27.50)(9.13) = 251 \text{ k}$$

$$\frac{P_u}{\phi_c P_n} = \frac{165}{251} = \ >0.2$$

∴ Must use LRFD Formula H1-1a

For a braced frame M_{lt} (the moment due to lateral loads) = 0

$$f_a = \frac{165}{9.13} = 18.07 \text{ ksi}$$

$$F'_e = \frac{12\pi^2 E}{23(K_x L_x/r_x)^2} = \frac{(12)(\pi)^2(29 \times 10^3)}{(23)(44.96)^2} = 73.88 \text{ ksi}$$

From Table C-H1.1 in Manual

For $C_m = 1 - 0.2\left(\dfrac{18.07}{73.88}\right) = 0.951$

For $C_m = 1 - 0.3\left(\dfrac{18.07}{73.88}\right) = 0.927$

Average $C_m = 0.94$

For $M_u = \dfrac{(16)(13)}{4} = 52 \text{ ft-k}$

For $M_u = \dfrac{(3)(16)(13)}{16} = 39 \text{ ft-k}$

$$\text{Average } M_u = \frac{52 + 39}{2} = 45.5 \text{ ft-k}$$

$$\frac{P_{ex}}{A_g} = 141.6 \text{ ksi for } \frac{K_x L_x}{r_x} = 44.96$$

$$P_{ex} = (141.6)(9.13) = 1292.8 \text{ k}$$

$$B_{1x} = \frac{C_{mx}}{1 - P_u/P_{ex}} = \frac{0.94}{1 - 165/1292.8} = 1.15$$

$$M_u = B_1 M_{nt} + B_2 M_{lt} = (1.15)(45.5) + 0 = 52.33 \text{ ft-k}$$

$$\phi_b M_{nx} = \phi_b M_p \quad \text{since } L_b < L_p$$

$$= \frac{(0.9)(36 \times 30.4)}{12} = 82.1 \text{ ft-k}$$

Applying LRFD Formula H1.1a

$$\frac{165}{251} + \frac{8}{9}\left(\frac{49.14}{82.1} + 0\right) = 1.22 > 1.0 \qquad\qquad \text{NG}$$

11-7 DESIGN OF BEAM-COLUMNS

The design of beam-columns involves a trial-and-error procedure. A trial section is selected by some process and is then checked with the appropriate interaction formula. If the section does not satisfy the equation or if it's too much on the safe side (that is, if it's overdesigned) another section is selected and the interaction equation is applied again. Probably the first thought of the reader is "I sure hope that we can select a good section the first time and not have to go through all of that rigamarole more than once." We certainly can make a good initial estimate, and that's the topic of the remainder of this section.

A common method used for selecting sections to resist both moments and axial loads is the *equivalent axial load or effective axial load method*. In this method the axial load (P_u) and the bending moment (M_{ux} and/or M_{uy}) are replaced with a fictitious concentric load P_{ueff} equivalent to the actual design axial load plus the design moment.

It is assumed for this discussion that it is desired to select the most economical section to resist both a moment and an axial load. By a trial-and-error procedure it is possible eventually to select the lightest section. Somewhere, however, there is a fictitious axial load that will require the same section as the one required for the actual moment and the actual axial load. This fictitious load is called the equivalent axial load or the effective axial load P_{ueff}.

Equations are used to convert the bending moment into an estimated equivalent axial load P'_u which is added to the actual design axial load P_u. The total of $P_u + P'_u$ is the equivalent or effective axial load P_{ueff}, and it is used to enter the concentric column tables of the LRFD Manual for choice of a trial section. In the formula for P_{ueff} which follows m is a factor given in Table B on page 2-10 of the LRFD Manual and U is a factor provided in the column tables.

$$P_{ueff} = P_u + M_{ux}m + M_{uy}\,mU$$

To apply this expression a value of m is taken from the first approximation section of Table B and U is assumed equal to 2. *In applying the equation the moments* M_{ux} *and* M_{uy} *must be used in ft-k.* The equation is solved for P_{ueff} and a column is selected from the concentrically loaded column tables for that load. Then the equation for P_{ueff} is solved again using a revised value of m from the subsequent approximations part of Table B and the value of U is taken from the column tables for the column initially selected. Another shape is selected and the process is continued until m and U stabilize (that is, until the column size selected does not change).

Finally it is necessary to check the trial column size with the appropriate LRFD interaction equation (H1-1a) or (H1-1b). The equivalent axial load equation yields sections that are generally on the conservative side. For this reason the designer may select a section by the equivalent axial load method and then try the interaction equation on a section one or two sizes smaller. This procedure may very well provide significant savings in steel weight.

Examples 11-6 and 11-7 illustrate the design of beam-columns using the equivalent axial load procedure.

Limitations of the LRFD P_{ueff} Tables

The application of the equivalent axial load formula and Table B of Part 2 of the LRFD Manual results in economical beam-column designs unless the moment becomes quite large in comparison with the axial load. For such cases the members selected will be capable of supporting the loads and moments but may very well be rather uneconomical. The tables for concentrically loaded columns of Part 2 of the Manual are limited to the W14s, W12s, and shallower sections, but when the moment is large in proportion to the axial load there will often be a much deeper and appreciably lighter section such as a W27 or W30 which may be a good bit lighter and which will satisfy the appropriate interaction equations.

EXAMPLE 11-6

Select a 12-ft W section consisting of A36 steel to support the factored load $P_u = 500$ k and the factored moments $M_{ntx} = 200$ ft-k and $M_{nty} = 80$ ft-k. The member is to be used in a braced frame with $M_{ltx} = M_{lty} = 0$ and $K_x L_x = K_y L_y = 12$ ft and $C_m = 0.85$.

Solution

1st approximation ($m = 2.8$, $u = 2.0$)

$$P_{ueff} = 500 + (200)(2.8) + (80)(2.8)(2.0) = 1508 \text{ k}$$

2nd approximation. Try W12×190 ($m = 2.5$, $U = 1.33$)

$$P_{ueff} = 500 + (200)(2.5) + (80)(2.5)(1.33) = 1266 \text{ k}$$

3rd approximation. Try W12×170 ($m = 2.5$, $U = 1.35$)

$$P_{ueff} = 500 + (200)(2.5) + (80)(2.5)(1.35) = 1270 \text{ k}$$

Concentric load table still calls for W12×170

∴ We will assume a smaller section and
check it with the appropriate interaction equation.

Try W12×152 ($A = 44.7$ in.^{2}, $r_x = 5.66$ in., $r_y = 3.19$ in., $L_p = 13.3$ ft, $Z_x = 243$ in.^{3}, $Z_y = 111$ in.^{3})

$$\frac{K_x L_x}{r_x} = \frac{(12)(12)}{5.66} = 25.44$$

$$\frac{K_y L_y}{r_y} = \frac{(12)(12)}{3.19} = 45.14$$

$$\lambda_{cx} = \frac{KL}{r\pi} \sqrt{\frac{F_y}{E}} = \frac{25.44}{\pi} \sqrt{\frac{36}{29,000}} = 0.285$$

$$\lambda_{cy} = \frac{45.14}{\pi} \sqrt{\frac{36}{29,000}} = 0.506$$

$$P_{ex} = \frac{A_g F_y}{\lambda_{cx}^2} = \frac{(44.7)(36)}{(0.285)^2} = 19,812 \text{ k}$$

$$P_{ey} = \frac{(44.7)(36)}{(0.506)^2} = 6285 \text{ k}$$

$$B_{1x} = \frac{C_{mx}}{1 - P_u/P_{ex}} = \frac{0.85}{1 - 500/19,812} = <1.0 \qquad \underline{\text{Use } 1.0}$$

$$B_{1y} = \frac{C_{my}}{1 - P_u/P_{ey}} = \frac{0.85}{1 - 500/6285} = <1.0 \qquad \underline{\text{Use } 1.0}$$

$$M_{ux} = (1.0)(200) = 200 \text{ ft-k}$$

$$M_{uy} = (1.0)(80) = 80 \text{ ft-k}$$

Since $L_b < L_p$

$$\phi_b M_{nx} = \frac{(0.9)(36)(243)}{12} = 656.1 \text{ ft-k}$$

$$\phi_b M_{ny} = \frac{(0.9)(36)(111)}{12} = 299.7 \text{ ft-k}$$

$$\phi_c P_n \text{ from column tables} = 1230 \text{ k}$$

$$\frac{P_u}{\phi P_n} = \frac{500}{1230} > 0.2$$

$$\therefore \text{ Use LRFD Formula H1-1a}$$

$$\frac{500}{1230} + \frac{8}{9}\left(\frac{200}{656} + \frac{80}{299.7}\right) = 0.915 < 1.0 \qquad \text{OK}$$

$$\text{Use W12} \times 152$$

EXAMPLE 11-7

Select the lightest satisfactory W12 section (A36 steel) for a 12-ft column in a building frame for the situation to be described. It is to be unbraced in the plane of the frame but is to be braced at each story out of plane so that $K_y = 1.0$. In the plane of the frame K_x has been estimated to equal 1.50.

A first-order analysis has been made with the factored loads and the results are given in Fig. 11-5. It is assumed that the $1.2D + 0.5L + 1.3W$ load combination controls.

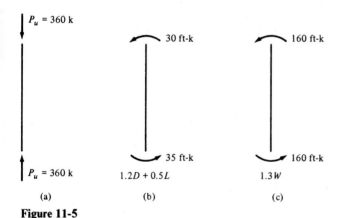

Figure 11-5

Solution

Trial column size

$$P_{ueff} = P_u + M_{ux}m + M_{uy}mU$$

$$= 360 + (35 + 160)(2.5) + 0 = 848 \text{ k}$$

A W12×106 is required by the equivalent axial load formula but is probably conservative.

$\therefore$ Try a W12×96 ($A = 28.2$ in.^{2}, $r_x = 5.44$ in., $r_y = 3.09$ in., $L_p = 12.9$ ft, $\phi_b M_p = 397$ ft-k).

$$\left(\frac{KL}{r}\right)_x = \frac{(1.50)(12 \times 12)}{5.44} = 39.71$$

$$\left(\frac{KL}{r}\right)_y = \frac{(1.0)(12 \times 12)}{3.09} = 46.60 \leftarrow$$

$$\phi_c F_{cr} = 27.29 \text{ ksi}$$

$$\phi_c P_n = (27.29)(28.2) = 769.6 \text{ k}$$

$$\frac{P_u}{\phi_c P_n} = \frac{360}{769.6} = 0.468 > 0.2$$

$$\therefore \text{ Must use Formula H1-1a}$$

Computing the magnification factors

$$\lambda_{cx} = \frac{KL}{r\pi}\sqrt{\frac{F_y}{E}} = \frac{39.71}{\pi}\sqrt{\frac{36}{29,000}} = 0.4454$$

$$\lambda_{cy} = \frac{46.60}{\pi}\sqrt{\frac{36}{29,000}} = 0.5226$$

$$P_{ex} = \frac{A_g F_y}{\lambda_{cx}^2} = \frac{(28.2)(36)}{(0.4454)^2} = 5117 \text{ k}$$

$$P_{ey} = \frac{A_g F_y}{\lambda_{cy}^2} = \frac{(28.2)(36)}{(0.5226)^2} = 3717 \text{ k}$$

$$C_m = 0.6 - 0.4\left(+\frac{30}{35}\right) = 0.257$$

$$B_1 = \frac{C_m}{1 - P_u/P_e} = \frac{0.257}{1 - 360/5117} = 0.276$$

$$<1.0 \therefore \text{ Use } B_1 = 1.0$$

Here the author assumes that ΣP_u and ΣP_e for all the columns in this line are each four times as large as the values are for this one column.

$$C_m = 1.0$$

$$B_2 = \frac{C_m}{1 - (\Sigma P_u/\Sigma P_e)} = \frac{1.0}{1 - 4 \times 360/4 \times 3717} = 1.107$$

Determining final moments

$$M_u \text{ at top of column} = (1.0)(30) + (1.107)(160) = 207.1 \text{ ft-k}$$
$$M_u \text{ at bottom of column} = (1.0)(35) + (1.107)(160) = 212.1 \text{ ft-k}$$

A W12×96 has an $L_p = 12.9$ ft $> L_b$ of 12 ft

$$\therefore \phi_b M_{nx} = 397 \text{ ft-k}$$

$$\frac{P_u}{\phi_c P_n} + \frac{8}{9}\left(\frac{M_{ux}}{\phi_b M_{nx}} + \frac{M_{uy}}{\phi_b M_{ny}}\right) = \frac{360}{769.6} + \frac{8}{9}\left(\frac{212.1}{397} + 0\right) = 0.943 < 1.0$$

$$\underline{\underline{\text{Use W12×96}}}$$

PROBLEMS

11-1. A W14×26 tension member with no holes consisting of A36 steel is subjected to a factored tensile load P_u = 35 k and a factored bending moment M_{ux} of 30 ft-k. Is the member satisfactory if L_b = 4.0 ft and if C_b = 1.0? (*Ans.* 0.345 < 1.00   OK)

11-2. A W12×58 tension member with no holes and F_y = 50 ksi is subjected to P_u = 80 k and M_{ux} = 42 ft-k. Is the member satisfactory if L_b = 8.0 ft and C_b = 1.0?

11-3. Repeat Prob. 11-1 if the member is also subject to a factored bending moment M_{uy} = 12 ft-k. (*Ans.* 1.148 > 1.00   NG)

11-4. Select the lightest W10 available to support a factored tensile load P_u = 45 k and a factored bending moment M_{uy} = 30 ft-k. Assume the member consists of A36 steel and has no holes.

11-5. Select the lightest W12 section available to support a factored tensile load P_u = 160 k using A36 steel to support a factored bending moment M_{ux} = 75 ft-k, L_b = 8.0 ft, and C_b = 1.0. The member is assumed to have no holes. (*Ans.* W12×35)

11-6. A W10×49 pin-connected beam-column, which is not subjected to sidesway is 15 ft long. A load P_u = 230 k is applied to the column at its upper end with an eccentricity of 2 in. so as to cause bending about the major axis of the section. Check the adequacy of the member if it consists of A36 steel and is used in a braced frame so that $M_{ltx} = M_{lty} = 0$. K_y = 1.0 while K_x = 1.45. Assume C_b = 1.0 and $C_{mx} = C_{my}$ = 0.85.

11-7. Sidesway is prevented for the beam-column shown in the accompanying illustration. If bending is about the major axis, is the member satisfactory if it consists of A36 steel? Assume $K_x = K_y$ = 1.0. (*Ans.* 0.887 < 1.00   OK)

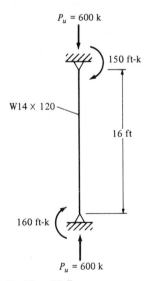

Problem 11-7

11-8. A pin-connected W10×49 consisting of A36 steel is used as a beam-column. If sidesway is possible in the x direction and a load P_u = 300 k is applied 1.50 in. off center at both ends so as to cause single curvature bending about the x axis, is the member satisfactory? Length = 12 ft. K_x = 1.4, K_y = 1.0. $M_{ltx} = M_{lty} = 0$.

11-9. The W10×30 truss member shown in the accompanying illustration consists of A36 steel and is fixed at its ends in the x and y directions. The purlin shown is assumed to provide pinned support in the y direction at the midpoint of the member. Is the member satisfactory to support the factored loads shown? (*Ans.* 0.776 < 1.00   OK)

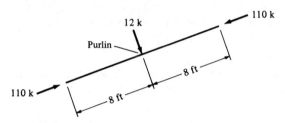

Problem 11-9

11-10. Repeat Prob. 11-9 if the ends of the member are assumed to be pinned and braced against lateral movement.

11-11. Select a W14 section consisting of A36 steel to support a factored load $P_u = 600$ k and the factored moments $M_{ntx} = 100$ ft-k and $M_{nty} = 40$ ft-k. The 14-ft member is to be used in a braced frame with M_{ltx} and M_{lty} 0. Assume $K_x = K_y = 1.0$ and $C_m = 0.85$. (*Ans.* W14×109)

11-12. Repeat Prob. 11-11 if a W14 section consisting of 50-ksi steel is to be used.

11-13. A W14×109 section (A36 steel) is used for a 14-ft column in a building frame. It is unbraced in the plane of the frame but is braced at each story out of plane so that $K_y = 1.00$. K_x has been estimated to equal 1.40. A first-order analysis has been made with the factored loads, and the results are shown in the accompanying illustration. Is the member satisfactory? (*Ans.* 0.734 < 1.0   OK)

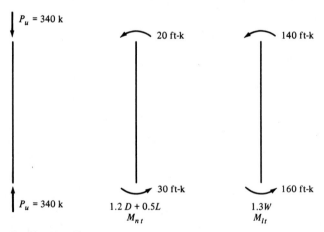

Problem 11-13

11-14. Repeat Prob. 11-13 if the moments bend the member in single curvature.

11-15. Select a W12 with $F_y = 50$ ksi for the load and moments of Prob. 11-13. (*Ans.* W12×72)

Chapter 12

Bolted Connections

12-1 INTRODUCTION

For quite a few decades riveting was the accepted method used for connecting the members of steel structures. For the last few decades, however, bolting and welding have been the methods used for making structural steel connections, and riveting is almost extinct. This chapter and the next are almost entirely devoted to bolted connections, although some brief remarks are presented at the end of Chapter 13 concerning rivets.

Bolting of steel structures is a very rapid field erection process which requires less skilled labor than does riveting or welding. This gives bolting a distinct economic advantage over the other connection methods in the United States where labor costs are so very high. Even though the purchase price of a high-strength bolt is several times that of a rivet the overall cost of bolted construction is cheaper than that for riveted construction because of reduced labor and equipment costs and the smaller number of bolts required to resist the same loads.

12-2 TYPES OF BOLTS

There are several types of bolts which can be used for connecting steel members. They are described in the following paragraphs.

Unfinished bolts are also called ordinary or common bolts. They are classified by the ASTM as A307 bolts and are made from carbon steels with stress-strain characteristics very similar to those of A36 steel. They are available in diameters from $\frac{5}{8}$ to $1\frac{1}{2}$ in. in $\frac{1}{8}$-in. increments.

A307 bolts generally have square heads and nuts to reduce costs, but hexagonal heads are sometimes used because they have a little more attractive appearance, are easier to turn and easier to hold with the wrenches, and they require less turning space. As they have relatively large tolerances in shank and thread dimensions, their design strengths are appreciably smaller than those for rivets or high-strength bolts. They are primarily used in light structures subjected to static loads and for secondary members (such as purlins, girts, bracing, platforms, small trusses, and so forth).

Designers are often guilty of specifying high-strength bolts for connections when common bolts would be satisfactory. *The strength and advantages of common bolts have usually been greatly underrated in the past.* The analysis and design of A307 bolted connections are handled exactly as are riveted connections in every way except that the allowable stresses are different.

High-strength bolts are made from medium carbon heat-treated steel and from alloy steel and have tensile strengths two or more times those of ordinary bolts. There are two basic types, the A325 bolts (made from a heat-treated medium carbon steel) and the higher strength A490 bolts (also heat-treated but made from an alloy steel). High-strength bolts are used for all types of structures, from small buildings to "skyscrapers" and monumental bridges. These bolts were developed to overcome the weaknesses of rivets—primarily insufficient tension in their shanks after cooling. The resulting rivet tensions may not be large enough to hold them in place during the application of severe impactive and vibrating loads. The result is that they may become loose and vibrate and may eventually have to be replaced. High-strength bolts may be tightened until they have very high tensile stresses so that the connected parts are clamped tightly together between the bolt and nut heads permitting loads to be transferred primarily by friction.

Sometimes high-strength bolts are made from A449 steel in sizes larger than the $1\frac{1}{2}$-in. maximum diameter A325 and A490 bolts. These larger bolts may also be used as high-strength anchor bolts and for threaded rods of many different diameters.

12-3 HISTORY OF HIGH-STRENGTH BOLTS

The joints obtained using high-strength bolts are superior to riveted joints in performance and economy and they are the leading field method of fastening structural steel members. C. Batho and E. H. Bateman first claimed in 1934 that

High-strength bolt. (Courtesy of Bethlehem Steel Corporation.)

high-strength bolts could satisfactorily be used for the assembly of steel structures[1] but it was not until 1947 that the Research Council on Riveted and Bolted Structural Joints of the Engineering Foundation was established. This group issued their first specifications in 1951, and high-strength bolts were adopted by both building and bridge engineers for both static and dynamic loadings with amazing speed. They not only quickly became the leading method of making field connections, but they also were found to have many applications for shop connections. The construction of the Mackinac Bridge in Michigan involved the use of more than one million high-strength bolts.

Connections that were formerly made with ordinary bolts and nuts were not too satisfactory when they were subjected to vibratory loads because the nuts frequently became loose. For many years this problem was dealt with by using some type of locknut, but the modern high-strength bolts furnish a far superior solution.

12-4 ADVANTAGES OF HIGH-STRENGTH BOLTS

Among the many advantages of high-strength bolts, partly explaining their great success, are the following.

1. Smaller crews are involved as compared with riveting. Two two-person bolting crews can easily turn out over twice as many bolts in a day as the number of rivets driven by the standard four-person riveting crew. The result is quicker steel erection.
2. In comparison with rivets, fewer bolts are needed to provide the same strength.
3. Good bolted joints can be made by people with a great deal less training and experience than is necessary to produce welded and riveted connections of equal quality. The proper installation of high-strength bolts can be learned in a matter of hours.
4. No erection bolts are required which may have to be later removed (depending on specifications) as in welded joints.
5. Though quite noisy, bolting is not nearly as bad as riveting.
6. Cheaper equipment is used to make bolted connections.
7. No fire hazard is present and no danger present from the tossing of hot rivets.
8. Tests on riveted joints and fully tensioned bolted joints under identical conditions definitely show that bolted joints have a higher fatigue strength. Their fatigue strength is also equal to or greater than that obtained with equivalent welded joints.
9. Where structures are to be later altered or disassembled changes in connections are quite simple because of the ease of bolt removal.

[1]C. Batho and E. H. Bateman, "Investigations on Bolts and Bolted Joints," H. M. Stationery Office (London, 1934).

12-5 SNUG-TIGHT AND FULLY TENSIONED BOLTS

Not all high-strength bolts have to be fully tensioned, says the LRFD. Such a process is expensive, as is inspection to see that they are fully tensioned. Thus the LRFD specification requires that those bolts which have to be fully tensioned must be clearly identified on drawings. These will be the bolts used in slip-critical connections and in connections subject to direct tension. Slip-critical connections will be required when the working loads cause a large number of stress changes possibly causing fatigue problems. In Section J1.9 of the LRFD Specification a detailed list of connections which must be made with fully tensioned bolts is presented. Included are connections for supports of running machinery or for live loads producing impact and stress reversal; column splices in all tier structures 200 ft or more in height; connections of all beams and girders to columns and other beams or girders on which the bracing of the columns is dependent for structures over 125 ft in height; and so on.

Other bolts need only be tightened to a snug-tight condition. *Snug-tight* is defined as the situation existing when all the plies of a connection are in firm contact with each other. It usually means the tightness obtained by the full effort of a man with a spud wrench or the tightness obtained after a few impacts of an impact wrench. Obviously there is some variation in the degrees of tightness obtained under these conditions. Snug-tight bolts must be clearly identified on both design and erection drawings.

Table 12-1, which is a reproduction of Table J3.1 in Part 6 of the LRFD Manual, presents the fastener tensions required for slip-resistant connections and for connections subject to direct tension. To be fully tensioned both A325 and

TABLE 12-1   FASTENER TENSION REQUIRED FOR SLIP-CRITICAL CONNECTIONS AND CONNECTIONS SUBJECT TO DIRECT TENSION

Bolt size (in.)	A325 Bolts	A490 Bolts
$\frac{1}{2}$	12	15
$\frac{5}{8}$	19	24
$\frac{3}{4}$	28	35
$\frac{7}{8}$	39	49
1	51	64
$1\frac{1}{8}$	56	80
$1\frac{1}{4}$	71	102
$1\frac{3}{8}$	85	121
$1\frac{1}{2}$	103	148

[a]Equal to 0.70 of minimum tensile strength of bolts, rounded off to nearest kip, as specified in ASTM specifications for A325 and A490 bolts with UNC threads.

A490 bolts are tightened to at least 70 percent of their specified minimum tensile strengths.

The quality-control provisions specified in the manufacture of the A325 and A490 bolts are more stringent than those for the A449 bolts. As a result, despite the method of tightening, the A449 bolts may not be used in slip-resistant connections.

Although many engineers felt that there would be some slippage as compared with rivets (because of the fact that the hot driven rivets more nearly filled the holes) the results show that there is less slippage in fully tensioned high-strength bolted joints than in riveted joints under similar conditions.

It is interesting to note that the nuts used for fully tensioned high-strength bolts need no special provisions for locking. Once these bolts are installed and sufficiently tightened to produce the tension required, there is almost no tendency for the nuts to come loose. There are, however, a few situations where they will work loose under heavy vibrating loads. What do we do then? Some steel erectors have replaced the offending bolts with longer ones with two fully tightened nuts. Others have welded the nuts onto the bolts. Apparently the results have been somewhat successful.

12-6 METHODS FOR FULLY TENSIONING HIGH-STRENGTH BOLTS

We have already commented on the tightening required for snug-tight bolts. For fully tensioned bolts several methods of tightening are available. These methods, including the turn-of-the-nut method, the calibrated wrench method, and the use of alternate design bolts and direct tension indicators, are permitted without preference by the LRFD Specification.

Turn-of-the-Nut Method   The bolts are brought to a snug-tight condition and then they are given from one-third to one full turn depending on their length and the slope of the surfaces under their heads and nuts. (The amount of turn given can easily be controlled by marking the snug-tight position with paint or crayon.)

Calibrated Wrench Method   With this method the bolts are tightened with an impact wrench that is adjusted to stall at that certain torque which is theoretically necessary to tension a bolt of that diameter and ASTM classification to the desired tension. It is necessary that wrenches be calibrated daily and that hardened washers be used. Particular care needs to be given to protecting the bolts from dirt and moisture at the job site.

Direct Tension Indicator   The direct tension indicator consists of a hardened washer which has protrusions on one face in the form of small arches. The arches will be flattened as a bolt is tightened. The amount of gap at any one time is a measure of the bolt tension. For fully tensioned bolts the gaps should measure about 0.015 in. or less.

Alternate Design Fasteners   In addition to the preceding methods there are some alternate design fasteners which can be tensioned quite satisfactorily. Bolts with splined ends which extend beyond the threaded portion of the bolts are one example. Special wrench chucks are used to tighten the nuts until the splined ends shear off.

A maximum bolt tension is not specified in any of the preceding tightening methods. This means that the load can be tightened to the highest load that will not break it and the bolt will still do the job. Should the bolt break, another one is put in with no damage done. It might be noted that the nuts are stronger than the bolt and the bolt will break before the nut strips.

For fatigue situations where members are subjected to constantly fluctuating loads the slip-resistant connection is very desirable. If the force to be carried is less than the frictional resistance and thus no forces are applied to the bolts how could we ever have a fatigue failure of the bolts? The slip-resistant connection is a serviceability limit state in that it is based on working loads. For a slip-resistant joint the working loads are not permitted to exceed the permissible frictional resistance.

Other situations where slip-resistant connections are desirable include joints where bolts are used in oversized holes, joints where bolts are used in slotted holes where the loads are applied parallel or nearly so to the slots, joints that

Torquing the nut for a high-strength bolt with an air-driven impact wrench. (Courtesy of Bethlehem Steel Corporation.)

are subjected to significant force reversals, and joints where bolts and welds resist shear together on a common *faying surface*. (The faying surface is the contact or shear area between the members.)

12-7 SLIP-RESISTANT CONNECTIONS AND BEARING-TYPE CONNECTIONS

When high-strength bolts are fully tensioned they clamp the parts being connected tightly together. The result is a considerable resistance to slipping on the faying surface. This resistance is equal to the clamping force times the coefficient of friction.

If the shearing load is less than the permissible frictional resistance the connection is referred to as a slip-resistant one. If the load exceeds the frictional resistance the members will slip on each other and will tend to shear off the bolts, and at the same time the connected parts will push or bear against the bolts as shown in Fig. 12-1.

The surfaces of joints including the area adjacent to washers need to be free of loose scale, dirt, burs, and other defects which might prevent the parts from solid seating. It is necessary for the surface of the parts to be connected to have slopes of not more than 1 to 20 with respect to the bolt heads and nuts unless beveled washers are used. For slip-resistant joints the faying surfaces must also be free from oil, paint, and lacquer.

If the faying surfaces are galvanized, the slip factor will be reduced to almost half of its value for clean mill scale surfaces. The slip factor, however, may be significantly improved if the surfaces are subjected to hand wire brushing or to "brush off" grit blasting. However, such treatments do not seem to provide increased slip resistance for sustained loadings where there seems to be a creep-like behavior.[2]

The 1983 AASHTO Specifications permit hot-dip galvanization if the coated surfaces are scored with wire brushes or sandblasted after galvanization and before steel erection.

The ASTM Specification permits the galvanization of the A325 bolts themselves but not the A490 bolts. There is a danger of embrittlement of this higher-strength steel during galvanization due to the possibility that hydrogen may be introduced into the steel in the pickling operation of the galvanization process.

If special faying surface conditions (such as blast-cleaned surfaces or blast-cleaned surfaces with special slip-resistant coatings applied) are used to increase the slip resistance the designer may increase the values used here to the ones given by the Research Council on Structural Joints in Part 6 of the LRFD Manual.

---

[2]J. W. Fisher and J. H. A. Struik, *Guide to Design Criteria for Bolted and Riveted Joints* (New York: John Wiley & Sons, 1974) pp. 205–206.

12-8 MIXED JOINTS

Bolts may on occasion be used in combination with welds and on other occasions with rivets (as where they are added to old riveted connections to enable them to carry increased loads). The LRFD Specification contains some specific rules for these situations.

Bolts in Combination with Welds   For new work neither A307 common bolts nor high-strength bolts designed for bearing-type connections may be considered to share the load with welds. (Before the connection's ultimate strength is reached the bolts will slip, with the result that the welds will carry a larger proportion of the load—the actual proportion being difficult to determine.) For such circumstances welds will have to be proportioned to resist the entire loads.

If high-strength bolts are designed for slip-critical conditions they may be allowed to share the load with welds. Should we be making welded alterations to a structure that was designed for slip resistance the welds need only be proportioned for the additional strength needed.

High-Strength Bolts in Combination with Rivets   High-strength bolts may be considered to share loads with rivets for new work or for alterations of existing connections that were designed as slip-critical. (The ductility of the rivets allows the capacity of both sets of fasteners to act together.)

12-9 SIZES OF HOLES FOR BOLTS AND RIVETS

In addition to the standard size bolt and rivet holes which are $\frac{1}{16}$ in. larger in diameter than the bolts or rivets there are three types of enlarged holes: oversized, short-slotted, and long-slotted. Oversized holes will on occasion be very useful in speeding up steel erection. In addition they give some latitude for adjustments in plumbing up frames during erection. Table 12-2, which is Table

TABLE 12-2   NOMINAL HOLE DIMENSIONS

Bolt dia.	Hole dimensions			
	Standard (dia.)	Oversize (dia.)	Short-slot (width × length)	Long-slot (width × length)
$\frac{1}{2}$	$\frac{8}{16}$	$\frac{5}{8}$	$\frac{9}{16} \times \frac{11}{16}$	$\frac{9}{16} \times 1\frac{1}{4}$
$\frac{5}{8}$	$\frac{11}{16}$	$\frac{13}{16}$	$\frac{11}{16} \times \frac{7}{8}$	$\frac{11}{16} \times 1\frac{9}{16}$
$\frac{3}{4}$	$\frac{13}{16}$	$\frac{15}{16}$	$\frac{13}{16} \times 1$	$\frac{13}{16} \times 1\frac{7}{8}$
$\frac{7}{8}$	$\frac{15}{16}$	$1\frac{1}{16}$	$\frac{15}{16} \times 1\frac{1}{8}$	$\frac{15}{16} \times 2\frac{3}{16}$
1	$1\frac{1}{16}$	$1\frac{1}{4}$	$1\frac{1}{16} \times 1\frac{5}{16}$	$1\frac{1}{16} \times 2\frac{1}{2}$
$\geq 1\frac{1}{8}$	$d + \frac{1}{16}$	$d + \frac{5}{16}$	$(d + \frac{1}{16}) \times (d + \frac{3}{8})$	$(d + \frac{1}{16}) \times (2.5 \times d)$

J3.5 of the LRFD Manual, provides the nominal dimensions of the various kinds of enlarged holes permitted for the different bolt sizes.

The situations where we may use the various types of enlarged holes are now described.

Oversized holes may be used in all plies of connections as long as the applied load does not exceed the permissible slip resistance. They may not be used in bearing-type connections. It is necessary for hardened washers to be used over oversized holes that are located in outer plies.

Short-slotted holes may be used regardless of the direction of the applied load if the permissible slip resistance is larger than the applied force. Should the load be applied in a direction approximately normal (between 80 and 100 degrees) to the slot these holes may be used in any or all plies of connections for bearing-type connections. It is necessary to use washers (hardened if high-strength bolts are being used) over short-slotted holes in an outer ply.

Long-slotted holes may be used in *only one* of the connected parts of slip-critical or bearing-type connections at any one faying surface. For slip-critical joints these holes may be used in any direction but for bearing-type connections the loads must be normal (between 80 and 100 degrees) to the axes of the slotted holes. If long-slotted holes are used in an outer ply they will need to be covered with plate washers or a continuous bar. For high-strength bolted connections the washers or bar do not have to be hardened but they must be made of structural-grade material and may not be less than $\frac{5}{16}$ in. thick.

12-10 LOAD TRANSFER AND TYPES OF JOINTS

The following paragraphs present a few of the elementary types of bolted or riveted joints subjected to axial forces (that is, the loads are assumed to pass through the centers of gravities of the groups of connectors). For each of these joint types some comments are made about the methods of load transfer. Eccentrically loaded connections are discussed in Chapter 13.

For this initial discussion the reader is referred to part (a) of Figure 12-1. It is assumed that the plates shown are connected with a group of snug-tight bolts. In other words, the bolts are not tightened sufficiently so as to significantly squeeze the plates together. If there is assumed to be little friction between the plates they will slip a little due to the applied loads shown. As a result the loads in the plates will tend to shear the connectors off on the plane between the plates and press or bear against the sides of the bolts as shown in the figure. These connectors are said to be in *single shear and bearing* (also called unenclosed bearing). They must have sufficient strength to satisfactorily resist these forces and the members forming the joint must be sufficiently strong to prevent the connectors from tearing through.

Should rivets be used instead of the snug-tight bolts the situation is somewhat different because hot-driven rivets cool and shrink and squeeze or clamp the connected pieces together with sizable forces which greatly increase the friction between the pieces. As a result a large portion of the loads being transferred

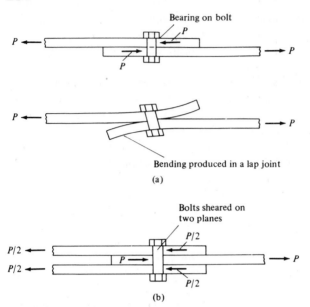

Figure 12-1   (a) Lap joint. (b) Butt joint.

between the members is transferred by friction. The clamping forces produced in riveted joints, however, are generally not considered to be dependable, and as a result specifications normally consider such connections to be snug tight with no frictional resistance. The same assumption is made for A307 common bolts which are not tightened so as to have large dependable tensions.

Fully tensioned high-strength bolts are in a different class altogether. Using the tightening methods previously described a very dependable tension is obtained in the bolts resulting in large clamping forces and large dependable amounts of frictional resistance to slipping. Unless the loads to be transferred are larger than the frictional resistance the entire forces are resisted by friction and the bolts are not really placed in shear or bearing. If the load exceeds the frictional resistance there will be slippage with the result that the bolts will be placed in shear and bearing.

The Lap Joint

The joint shown in part (a) of Figure 12-1 is referred to as a lap joint. This type of joint has a disadvantage in that the center of gravity of the force in one member is not in line with the center of gravity of the force in the other member. A couple is present which causes an undesirable bending in the connection as shown in the figure. For this reason the lap joint, which is desirably used only for minor connections, should be designed with at least two fasteners in each line parallel to the length of the member to minimize the possibility of a bending failure.

The Butt Joint

A butt joint is formed when three members are connected as shown in Figure 12-1 (b). If the slip resistance between the members is negligible the members will slip a little and tend to shear off the bolts simultaneously on the two planes of contact between the members. Again the members are bearing against the bolts and the bolts are said to be in *double shear and bearing* (also called enclosed bearing). The butt joint is more desirable than the lap joint for two main reasons. These are:

1. The members are arranged so that the total shearing force P is split into two parts, causing the force on each plane to be only about one-half of what it would be on a single plane if a lap joint were used. From a shear standpoint, therefore, the load-carrying ability of a group of bolts in double shear is theoretically twice as great as the same number of bolts in single shear.
2. A more symmetrical loading condition is provided. (In fact the butt joint does provide a symmetrical situation if the outside members are the same thickness and resist the same forces. The result is a reduction or elimination of the bending described for a lap joint.)

Double-Plane Connections

The double-plane connection is one in which the bolts are subjected to single shear and bearing but in which bending moment is prevented. This type of connection, which is shown for a hanger in Fig. 12-2 (a), subjects the bolts to single shear on two different planes.

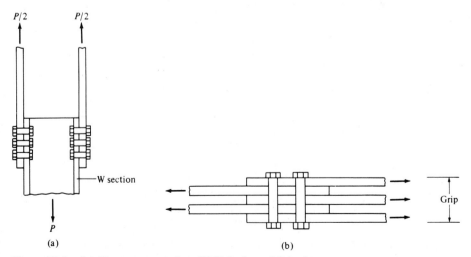

Figure 12-2 (a) Hanger connection. (b) Bolts in multiple shear.

Miscellaneous

Bolted connections generally consist of lap or butt joints or some combination of them but there are other cases. For instance there are occasionally joints in which more than three members are being connected and the bolts are in multiple shear as shown in Fig. 12-2 (b). Although the bolts in this connection are being sheared on more than two planes, the usual practice is to consider no more than double shear for strength calculations. It does not seem physically possible for shear failures to occur simultaneously on three or more planes. Several other types of bolted connections are discussed in this chapter and the next. These include bolts in tension, bolts in shear and tension, etc.

12-11 FAILURE OF BOLTED JOINTS

Figure 12-3 shows several ways in which failure of bolted joints can occur. To be able to satisfactorily design bolted joints it is necessary to understand these possibilities. These are described as follows.

1. The possibility of failure in a lap joint by shearing of the bolt on the plane between the members (single shear) is shown in part (a).
2. In part (b) the possibility of a tension failure of one of the plates through a bolt hole is shown.
3. A possible failure of the bolts and/or plates by bearing between the two is given in (c).

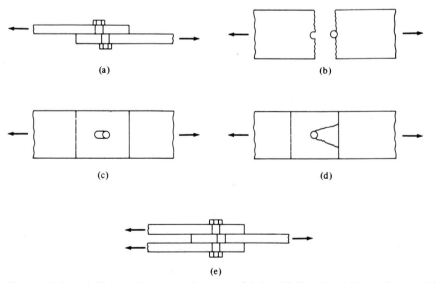

(a)   (b)

(c)   (d)

(e)

Figure 12-3   (a) Failure by single shearing of bolt. (b) Tension failure of plate. (c) Crushing failure of plate. (d) Shear failure of plate behind bolt. (e) Double shear failure of a butt joint.

4. Part (d) shows another possibility in the shearing out of part of the member.

5. A butt joint is shown in (e) with the possibility of a shear failure of the bolts along two planes (double shear).

12-12 SPACING AND EDGE DISTANCES OF BOLTS

Before minimum spacings and edge distances can be discussed it is necessary for a few terms to be explained. The following definitions are given for a group of bolts in a connection and are shown in Fig. 12-4.

Pitch is the center-to-center distance of bolts in a direction parallel to the axis of the member.

Gage is the center-to-center distance of bolt lines perpendicular to the axis of the member.

The *edge distance* is the distance from the center of a bolt to the adjacent edge of a member.

The *distance between bolts* is the shortest distance between fasteners on the same or different gage lines.

Minimum Spacings

Bolts should be placed a sufficient distance apart to permit efficient installation and to prevent tension failures of the members between fasteners. The LRFD Specification (J3.9) provides a minimum center-to-center distance for standard, oversized, or slotted fastener holes equal to not less than $2\frac{2}{3}$ diameters (with $3d$ being preferred). If we are measuring along a line of transmitted force, this distance may have to be increased. There the distance may not be less than $3d$ if the bearing strength R_n is determined by either of the expressions $2.4dtF_u$ or $2.0dtF_u$ (these values will be discussed in the next section of this chapter). Otherwise the minimum center-to-center distance of standard holes is to be determined by the

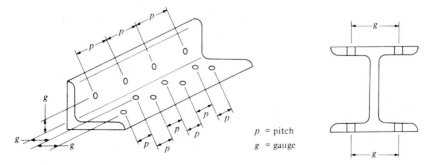

p = pitch

g = gauge

Figure 12-4

TABLE 12-3 VALUES OF SPACING INCREMENT C_1 FOR DETERMINING MINIMUM SPACINGS OF ENLARGED HOLES

Nominal dia. of fastener	Oversize holes	Perpendicular to line of force	Slotted holes	
			Parallel to line of force	
			Short slots	Long slots*
$\leq \frac{7}{8}$	$\frac{1}{8}$	0	$\frac{3}{16}$	$1\frac{1}{2}d - \frac{1}{16}$
1	$\frac{3}{16}$	0	$\frac{1}{4}$	$1\frac{7}{16}$
$\geq 1\frac{1}{8}$	$\frac{1}{4}$	0	$\frac{5}{16}$	$1\frac{1}{2}d - \frac{1}{16}$

*When length of slot is less than maximum allowed in Table 12.2, C_1 may be reduced by the difference between the maximum and actual slot lengths.

expression to follow in which P is the force transmitted by one fastener to the critical part, ϕ is 0.75, and t is the thickness of the critical connected part and d_h is the diameter of the standard size hole.

$$\text{Minimum distance center to center} = \frac{P}{\phi F_u t} + \frac{d_h}{2}$$

Should the holes be oversized or slotted the minimum center-to-center distance is as determined above plus the applicable increment C_1 from Table 12-3 (which is Table J3.6 of the LRFD Manual). The clear distance between these enlarged holes may never be less than one bolt diameter.

Minimum Edge Distances

Bolts should not be placed too near the edges of a member for two major reasons. The punching of holes too close to the edges may cause the steel opposite the hole to bulge out or even crack. The second reason applies to the ends of members where there is danger of the fastener tearing through the metal. The usual practice is to place the fastener a minimum distance from the edge of the plates equal to about 1.5 to 2.0 times the fastener diameter so the metal there will have a shearing strength at least equal to that of the fasteners. For more exact information it is necessary to refer to the specification being used. The LRFD Specification (J4.10) states that the distance from the center of a standard hole to the edge of a connected part may not be less than the applicable value given in Table 12-4 (which is Table J3.7 of the LRFD Manual) nor the value obtained from substituting, when applicable, into the formula to be described in the next paragraph.

In the direction of transmitted force the LRFD Specification states that the minimum edge distance may not be less than $1\frac{1}{2}d$ when the bearing strength R_n is determined by either of the expressions $2.4dtF_u$ or $2.0dtF_u$. Otherwise the minimum edge distance is to be determined by the formula to follow.

Minimum edge distance in the direction of the transmitted force

$$= \frac{P}{\phi F_u t} \qquad \text{with } \phi = 0.75$$

TABLE 12-4 MINIMUM EDGE DISTANCE FOR
STANDARD HOLES
(Center of Standard[a] Hole to Edge
of Connected Part)

Nominal rivet or bolt diameter (in.)	At sheared edges	At rolled edges of plates, shapes, or bars or gas cut edges[b]
$\frac{1}{2}$	$\frac{7}{8}$	$\frac{3}{4}$
$\frac{5}{8}$	$1\frac{1}{8}$	$\frac{7}{8}$
$\frac{3}{4}$	$1\frac{1}{4}$	1
$\frac{7}{8}$	$1\frac{1}{2}$[c]	$1\frac{1}{8}$
1	$1\frac{3}{4}$[c]	$1\frac{1}{4}$
$1\frac{1}{8}$	2	$1\frac{1}{2}$
$1\frac{1}{4}$	$2\frac{1}{4}$	$1\frac{5}{8}$
Over $1\frac{1}{4}$	$1\frac{3}{4}\times$ diameter	$1\frac{1}{4}\times$ diameter

[a]For oversized or slotted holes, see Table 12.5.
[b]All edge distances in this column may be reduced $\frac{1}{8}$ in. when the hole is at a point where stress does not exceed 25 percent of the maximum design strength in the element.
[c]These may be $1\frac{1}{4}$ in. at the ends of beam connection angles.

Should the holes be oversized or slotted the minimum edge distance may not be less than the value required for a standard hole plus an increment C_2 obtained from Table 12-5 (which is Table J3.8 of the LRFD Manual). Another minimum edge distance value is provided by the LRFD Specification for end connections bolted to beam webs and designed for beam shear reactions only.

Maximum Edge Distances

Many specifications give maximum distances bolts can be placed from the edge of a connection. The LRFD maximum (J3.11) is 12 times the plate thickness but not to exceed 6 in. If bolts are too far from the edges, openings may develop between the members being connected. Special pitch and edge distance limita-

TABLE 12-5 VALUES OF EDGE DISTANCE INCREMENT C_2
FOR ENLARGED HOLES
Maximum Edge Distances

Nominal diameter of fastener (in.)	Oversized holes	Short slots	Long slots*	Parallel to edge
$\leq\frac{7}{8}$	$\frac{1}{16}$	$\frac{1}{8}$		0
1	$\frac{1}{8}$	$\frac{1}{8}$	$\frac{3}{4}d$	
$\leq1\frac{1}{8}$	$\frac{1}{8}$	$\frac{3}{16}$		

(Slotted holes — Perpendicular to edge: Short slots, Long slots*)

*When length of slot is less than maximum allowable (see Table 12-2), C_2 may be reduced by one-half the difference between the maximum and actual slot lengths.

tions are given in Section E4 of the LRFD Specification for bolted joints in unpainted steel members which are exposed to atmospheric conditions. Maximum spacings for bolts may also be given for compression members so that buckling will not occur between bolts.

Holes cannot be punched very close to the web-flange junction of a beam or the junction of the legs of an angle. They can be drilled but this rather expensive practice should not be followed unless there is an unusual situation. Even if the holes are drilled in these locations there may be considerable difficulty in placing and tightening the bolts in the limited available space.

12-13 BEARING-TYPE CONNECTIONS—LOADS PASSING THROUGH CENTER OF GRAVITY OF CONNECTIONS

Shearing Strength   In bearing-type connections it is assumed that the loads to be transferred are larger than the frictional resistance caused by tightening the bolts with the result that the members slip a little on each other putting the bolts in shear and bearing. The design strength of a bolt in single shear equals ϕ times the nominal shearing strength of the bolt in ksi times its cross-sectional area. The LRFD ϕ values for shear are 0.65 for high-strength bolts and rivets and 0.60 for A307 common bolts.

The nominal shearing strengths of bolts and rivets are given in Table J3.2 of the LRFD Specification. For A325 bolts the values are 54 ksi if threads are not excluded from shear planes and 72 ksi if threads are excluded. (The values are 67.5 and 90 ksi for A490 bolts.) Should a bolt be in double shear its shearing strength is considered to be twice its single shear value.

The student may very well wonder what is done in design practice concerning threads excluded or not excluded from the shear planes. If normal bolt and member sizes are used, the threads will almost always be excluded from the shear plane. It is true that some extremely conservative individuals always assume the threads are not excluded from the shear plane.

Bearing Strength   The design strength of a bolt in bearing equals ϕ times the nominal bearing strength of the connected part in ksi times the diameter of the bolt times the thickness of the member that bears against the bolt. (For countersunk bolts and rivets $\frac{1}{2}$ of the depth of the countersink should be deducted, says the LRFD Specification J3.2) When the distance called L in the direction of the force from the center of a regular or oversized hole (or from the center of the end of a slotted hole) to the edge of a connected part is not less than $1\frac{1}{2}$ times the bolt diameter d and the distance center to center of the holes is not less than $3d$ and two or more bolts are used in the direction of the line of force the bearing strength is

$\phi R_n = \phi 2.4 dt F_u$ for standard or short-slotted holes $\phi = 0.75$

$\phi R_n = \phi 2.0 dt F_u$ for long-slotted holes perpendicular to the load $\phi = 0.75$

Should deformations around the hole not be a design consideration the two preceding expressions may be replaced with

$$\phi R_n = \phi 3.0 dt F_u \qquad \text{with } \phi = 0.75$$

Though the design bearing strength of a bolt with a small end distance is reduced *the strengths of the other bolts in the connection are not reduced.* The value of ϕR_n for a single bolt or for two or more bolts in the line *each with an end distance less than $1\frac{1}{2}d$* is to be determined with the following expression:

$$\phi R_n = \phi L t F_u \qquad \text{with } \phi = 0.75$$

Tests of bolted joints have shown that neither the bolts nor the metal in contact with the bolts actually fail in bearing. However, these tests have shown that the efficiency of the connected parts in tension and compression is affected by the magnitude of the bearing stress. Therefore, the nominal bearing strengths given by the LRFD Specification are values above which they feel the strength of the connected parts is impaired. In other words these apparently very high design bearing stresses are not really bearing stresses at all but rather indexes of the efficiencies of the connected parts. If bearing stresses larger than the values given are permitted the holes seem to elongate more than about $\frac{1}{4}$ in. and impair the strength of the connections.

Minimum Connection Strength   The LRFD Specification (Section J1.5) states that except for lacing, sag rods, and girts, connections must be provided with design strengths sufficient to support factored loads of at least 10 kips.

Example 12-1 illustrates the calculations involved in determining the strength of a bearing-type connection. Using a similar procedure the number of bolts required for a certain loading condition is calculated in Example 12-2.

The values given for the strength of bolts in this chapter, whether bearing-type or slip-critical, can be obtained from the tables entitled "Shear Design load in kips" and "Bearing Design load in kips" in Part 5 of the LRFD Manual.

EXAMPLE 12-1

Determine the design strength P_u of the bearing-type connection shown in Fig. 12-5. The steel is A36, the bolts $\frac{7}{8}$-in. A325, the holes are standard sizes, the threads are excluded from the shear plane, edge distances are $>1\frac{1}{2}d$, and the distance center to center of the holes is $>3d$.

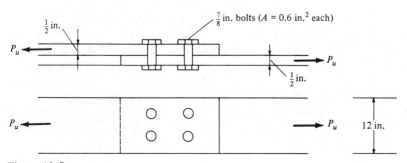

Figure 12-5

Solution. Design strength of plates:

$$A_g = (\tfrac{1}{2})(12) = 6.0 \text{ in.}^2$$

$$A_n = 6.00 - (2)(1.0)(\tfrac{1}{2}) = 5.0 \text{ in.}^2 = A_e$$

$$P_u = \phi_t F_y A_g = (0.9)(36)(6.0) = 194.4 \text{ k}$$

$$P_u = \phi_t F_u A_e = (0.75)(58)(5.0) = 217.5 \text{ k}$$

Bolts in single shear and bearing on $\tfrac{1}{2}$ in.:

$$P_u = \phi(0.6)(72)(4) = (0.65)(0.6)(72)(4) = 112.3 \text{ k} \leftarrow$$

$$P_u = \phi 2.4 dt F_u = (0.75)(2.4)(\tfrac{7}{8})(\tfrac{1}{2})(58)(4) = 182.7 \text{ k}$$

$$\underline{\text{Design } P_u = 112.3 \text{k}}$$

EXAMPLE 12-2

How many $\tfrac{3}{4}$-in. A325 bolts in standard-size holes with threads excluded from the shear plane are required for the bearing-type connection shown in Fig. 12-6? Use A36 steel and assume edge distances $>1\tfrac{1}{2}d$ and the distance center to center of holes to be $>3d$.

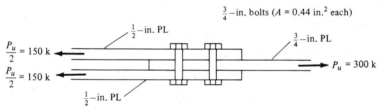

Figure 12-6

Solution. Bolts in double shear and bearing on $\tfrac{3}{4}$ in.:

Design shear strength per bolt $= (0.65)(2 \times 0.44)(72) = 41.2 \text{ k} \leftarrow$

Design bearing strength per bolt $= (0.75)(2.4)(\tfrac{3}{4})(\tfrac{3}{4})(58) = 58.7 \text{ k}$

$$\text{No. of bolts required} = \frac{300}{41.2} = 7+$$

$$\underline{\text{Use 8 or 9 bolts (depending on arrangement)}}$$

Where cover plates are bolted to the flanges of W sections the bolts must carry the longitudinal shear on the plane between the plates and the flanges. With reference to the cover-plated beam of Fig. 12-7 the unit longitudinal shearing stress to be resisted between a cover plate and the W flange can be determined with the expression $f_v = VQ/Ib$. The total shear force across the flange for a 1-in. length of the beam equals $(b)(1.0)(VQ/Ib) = VQ/I$.

The LRFD Specification (E4) provides a maximum permissible spacing for bolts used in the outside plates of built-up members. It equals the thinner outside plate thickness times $127/\sqrt{F_y}$ or 12 in., whichever is smaller.

The spacing of pairs of bolts in Fig. 12-7 can be determined by dividing the design strength of two bolts by the shear force to be taken per inch at a particular

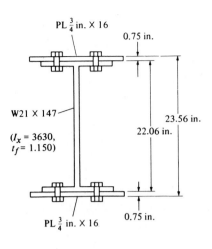

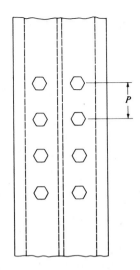

Figure 12-7

section. The theoretical spacings will vary as the external shear varies along the span. Example 12-3 illustrates the calculations involved in determining bolt spacing for a cover-plated beam.

EXAMPLE 12-3

At a certain section in the cover-plated beam of Fig. 12-7 the external factored shear V_u is 275 k. Determine the spacing required for $\frac{7}{8}$-in. A325 bolts used in a bearing-type connection. Assume the $1\frac{1}{2}d$ and $3d$ edge and center-to-center distance requirements are met and that the bolt threads are excluded from the shear plane. The steel is A36.

Solution

$$I_x = 3630 + (2)(16 \times \tfrac{3}{4})(11.405)^2 = 6752 \text{ in.}^4$$

Factored shear to be taken per inch $= \dfrac{V_u Q}{I}$

$$= \frac{(275)(16 \times \tfrac{3}{4} \times 11.405)}{6752} = 5.574 \text{ k/in}$$

Bolts in single shear and bearing on 0.75 in.

Design shear strength for 2 bolts $= (2)(0.65)(0.6)(72) = 56.16 \text{ k} \leftarrow$

Design bearing strength for 2 bolts $= (2)(0.75)(2.4)(\tfrac{7}{8})(\tfrac{3}{4})(58) = 137 \text{ k}$

$$p = \frac{56.16}{5.574} = 10.07 \text{ in.} \qquad \text{(Say 10 in. on center)}$$

$$\text{Maximum } p = \frac{3}{4}\frac{127}{\sqrt{36}} = 15.88 \text{ in. or 12 in.}$$

Place bolts 10 in. O.C.

The assumption has been made that the loads applied to a bearing-type connection are equally divided between the bolts. For this distribution to be correct the plates must be perfectly rigid and the bolts perfectly elastic, but actually the plates being connected are elastic too and have deformations which decidedly affect the bolt stresses. The effect of these deformations is to cause a very complex distribution of load in the elastic range.

Should the plates be assumed to be completely rigid and nondeforming, all bolts would be deformed equally and have equal stresses. This situation is shown in part (a) of Fig. 12-8. Actually the loads resisted by the bolts of a group are probably never equal (in the elastic range) when there are more than two bolts in a line. Should the plates be deformable, the plate stresses and thus the deformations will decrease from the ends of the connection to the middle as shown in part (b) of Fig. 12-8. The result is that the highest stressed elements of the top plate will be over the lowest stressed elements of the lower plate, and vice versa. The slip will be greatest at the end bolts and smallest at the middle bolts. The bolts at the ends will then have stresses much greater than those in the inside bolts.

The greater the spacing of bolts in a connection the greater will be the variation in bolt stresses due to plate deformation; therefore, the use of compact joints is very desirable as they will tend to reduce the variation in bolt stresses. It might be interesting to consider a theoretical (although not practical) method of roughly equalizing bolt stresses. The theory would involve the reduction of the thickness of the plate toward its end in proportion to the reduced stresses by stepping. This procedure, which is shown in Fig. 12-8 (c), would tend to equalize the

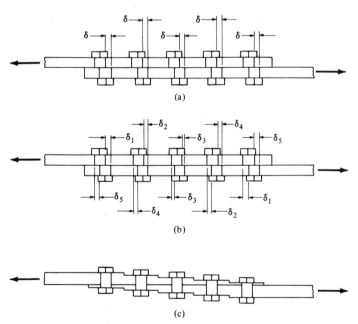

Figure 12-8 (a) Assuming nondeforming plates. (b) Assuming deformable plates. (c) Stepped joint (impractical).

Three high-strength bolted structures in Constitution Plaza Complex, Hartford, Conn., using approximately 195,000 bolts. (Courtesy of Bethlehem Steel Corporation.)

deformations of the plate and thus the bolt stresses. A similar procedure would be to scarf the overlapping plates.

The calculation of the theoretically correct elastic stresses in a bolted group based on plate deformations is a tedious problem and is rarely handled in the design office. On the other hand the analysis of a bolted joint based on the plastic theory is a very simple problem. In this theory the end bolts are assumed to be stressed to their yield point. Should the total load on the connection be increased, the end bolts will deform without resisting additional load, the next bolts in the line will have their stresses increased until they too are at the yield point, and so forth. Plastic analysis seems to justify to a certain extent the assumption of rigid plates and equal bolt stresses which is usually made in design practice. This assumption is used in the example problems of this chapter.

When there are only a few bolts in a line the plastic theory of equal stresses seems to be borne out very well but when there are a large number of bolts in a line the situation changes. Tests have clearly shown that the end bolts will fail before the full redistribution takes place.[3]

For load carrying bolted joints it is common for specifications to require a minimum of two or three fasteners. The feeling is that a single connector may fail to live up to its specified strength because of improper installation, material weakness, etc., but if several fasteners are used the effects of one bad fastener in the group will be overcome.

[3]*Trans. ASCE* 105 (1940), p. 1193.

12-14 SLIP-CRITICAL CONNECTIONS—LOADS PASSING THROUGH CENTER OF GRAVITY OF CONNECTIONS

The situations where slip-critical connections are desirable were previously described in Section 12-5. They are particularly useful for fatigue-type situations where members are subjected to constantly fluctuating loads. *Slip-critical connections must be checked at both service loads and at factored load conditions.* (1) *The design slipping resistances must be equal to or greater than the computed service load slipping forces.* (2) *The design strengths as bearing-type connections must be equal to or greater than the factored forces.*

If bolts are tightened to their required tensions for slip-critical connections (see Table 12-1) there is very little chance of their bearing against the plates which they are connecting. In fact tests show that there is very little chance of slip occurring unless there is a calculated shear of at least 50 percent of the total bolt tension. This means as we have said all along that slip-critical bolts are not stressed in shear; however, the LRFD Specification provides allowable shear strengths (they are really permissible friction values on the faying surfaces) so that the designer can handle the connections in just about the same manner he or she uses for bearing-type connections. The LRFD assumes the bolts are in "shear" and no bearing and the nominal shear strengths for high-strength bolts are given in Table 12-6 (which is Table J3.4 of the Manual). $\phi = 1.0$ except its 0.85 for long slotted holes with load parallel to slot.

The preceding discussion concerning slip-critical joints does not present the whole story because during erection the joints may be assembled with bolts, and as the members are erected their weights will often push the bolts against the side of the holes before they are tightened and put them in some bearing and shear.

Example 12-4 presents the design of a slip-critical connection for a lap joint. First the number of bolts required for the service load limit state of no slippage is determined. Then the number of bolts required for the factored load limit state is computed assuming the slip resistance is overcome and the bolts subjected to bearing and shear.

TABLE 12-6 NOMINAL SLIP-CRITICAL SHEAR STRENGTH, KSI, OF HIGH-STRENGTH BOLTS[a]

Type of bolt	Nominal shear strength		
	Standard size holes	Oversized and short-slotted holes	Long-slotted holes[b]
A325	17	15	12
A490	21	18	15

[a]Class A (slip coefficient 0.33). Clean mill scale and blast cleaned surfaces with class A coatings. For design strengths with other coatings see RCSC "Load and Resistance Factor Design Specification for Structural Joints Using ASTM A325 or A490 Bolts."

[b]Tabulated values are for the case of load application transverse to the slot. When the load is parallel to the slot multiply tabulated values by 0.85.

Bridge over Allegheny River at Kittaning, PA. (Courtesy American Bridge Company.)

EXAMPLE 12-4

It is desired to design a slip-critical connection for the plates shown in Fig. 12-9 to resist the axial service loads $P_D = 30$ k and $P_L = 50$ k using 1-in. A325 high-strength bolts with threads excluded from the shear plane and with standard-size holes, $L > 1\frac{1}{2}d$ and distance center to center of bolts $>3d$.

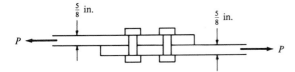

Figure 12-9

Solution

Slip-critical design (service loads)

Bolts in "single shear" and no bearing

"ss" strength of one bolt $= (\phi)(0.785)(17) = (1.0)(0.785)(17) = 13.35$ k

$$\text{No. of bolts required} = \frac{80}{13.35} = 5.99 \qquad \underline{\text{say } 6}$$

Design as bearing-type connection (factored loads)

$$P_u = (1.2)(30) + (1.6)(50) = 116 \text{ k}$$

Bolts in single shear and bearing on $\frac{5}{8}$ in.

$$\text{ss strength of one bolt} = (0.65)(0.785)(72) = 36.8 \text{ k} \leftarrow$$

$$\text{Bearing strength of one bolt} = (0.75)(2.4)(1.0)(\tfrac{5}{8})(58) = 65.25 \text{ k}$$

$$\text{No. of bolts required} = \frac{116}{36.8} = 3^+$$

Use 6 bolts

EXAMPLE 12-5

The connection shown in Fig. 12-10 is made with $\frac{7}{8}$-in. A325 bearing-type bolts in standard-size holes with the threads excluded from the shear planes. The beam and gusset plates consist of A36 steel. Check the following items: (a) the tensile strengths of the W section and the gusset plates, (b) the strength of the bolts in double shear and bearing, and (c) the block shear strength of the W section crosshatched areas shown in part (b) the figure.

(a)   Tensile design strength of W section

$$P_u = \phi_t F_y A_g = (0.9)(36)(11.2) = 362.9 \text{ k} > 330 \text{ k} \qquad \text{OK}$$

$$A_n = 11.2 - (4)(1)(0.515) = 9.14 \text{ in.}^2$$

$$U = 0.85 \text{ since } b_f < \tfrac{2}{3} d$$

$$A_e = (0.85)(9.14) = 7.77 \text{ in.}^2$$

$$P_u = \phi_t F_u A_e = (0.75)(58)(7.77) = 338 \text{ k} > 330 \text{ k} \qquad \text{OK}$$

Tensile design strength of gusset plates

$$P_u = \phi_t F_y A_g = (0.9)(36)(\tfrac{1}{2} \times 12)(2) = 388.8 \text{ k} > 330 \text{ k} \qquad \text{OK}$$

$$A_n \text{ of 2 plates} = [(\tfrac{1}{2})(12) - (2)(1)(\tfrac{1}{2})]2 = 10 \text{ in.}^2 \leftarrow$$

$$0.85 A_g = (0.85)(\tfrac{1}{2})(12)(2) = 10.2 \text{ in.}^2$$

$$P_u = \phi_t F_u A_n = (0.75)(58)(10) = 435 \text{ k} > 330 \text{ k} \qquad \text{OK}$$

(b)   Bolts in ss and bearing on $\frac{1}{2}$ in.

$$\text{ss design strength of bolts} = (0.65)(0.6)(72)(12) = 337 \text{ k} > 330 \text{ k} \qquad \text{OK}$$

Bearing design strength of bolts

$$= (0.75)(2.4)(\tfrac{7}{8})(\tfrac{1}{2})(58)(12) = 548.1 \text{ k} > 330 \text{ k} \qquad \text{OK}$$

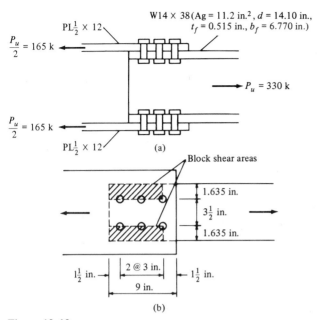

Figure 12-10

(c)   Block shearing strength for the W section

Tension fracture and shear yielding

$$P_{bs} = \phi[F_u A_{nt} + 0.6F_y A_{vg}]$$

$$= 0.75[(58)(\tfrac{1}{2})(1.635 - \tfrac{1}{2} \times 1) + (0.6)(36)(7\tfrac{1}{2})(\tfrac{1}{2})]4$$

$$= 341.8 \text{ k} > 330 \text{ k} \qquad\qquad\qquad\qquad\qquad \text{OK}$$

Shear fracture and tension yielding

$$P_{bs} = \phi[F_y A_{tg} + 0.6F_u A_{ns}] \text{ noting there are}$$
$$2\tfrac{1}{2} \text{ holes on the net area shear plane}$$
shown in part (b) of Fig. 12-10.

$$= 0.75[(36)(1.635)(\tfrac{1}{2}) + (0.6)(58)(7.5 - 2.5 \times 1)(\tfrac{1}{2})]4$$

$$= 349 \text{ k} > 330 \text{ k} \qquad\qquad\qquad\qquad\qquad \text{OK}$$

The calculations for riveted connections and for connections made with A307 common bolts are made almost exactly as are the ones for high-strength bolts in bearing-type connections. The only differences are that the shear strength values for these connections are much smaller and the shear ϕ factor for A307 bolts is 0.60 instead of the 0.65 used for rivets and high-strength bolts. The LRFD Specification does not permit the design of slip-critical joints using rivets or common bolts. Rivet and common bolt examples are presented in Chapter 13.

PROBLEMS

For each of the problems listed the following information is to be used *unless otherwise indicated*: (a) LRFD Specification; (b) standard-size holes; (c) threads of bolts excluded from shear plane; (d) members have clean mill scale surfaces (class A); (e) edge distances are $>1\frac{1}{2}d$ and distances center to center of holes are $>3d$.

12-1. The plates used in the lap joint shown in the accompanying illustration are each $\frac{5}{8} \times 12$ in. and consist of A36 steel. Determine the design load P_u which can be resisted if $\frac{3}{4}$-in. A325 bolts are used in a bearing-type connection. (*Ans.* 185.3 k)

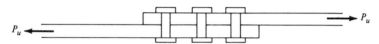

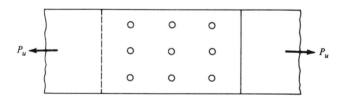

Problem 12-1

12-2. Rework Prob. 12-1 if 1-in. A490 bolts are used in a bearing-type connection.

12-3. If $\frac{7}{8}$-in. A325 bolts are used in the bearing-type connection shown in the accompanying illustration determine its design tensile capacity. Steel is A36. (*Ans.* 340.2 k)

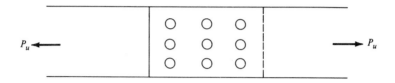

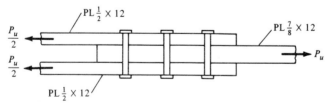

Problem 12-3

12-4. The truss member shown in the accompanying illustration consists of two C12 × 25s (A36 steel) connected to a 1-in. gusset plate. How many $\frac{7}{8}$-in. A325 bolts are required to develop the full design tensile capacity of the member if it is used as a bearing-type connection?

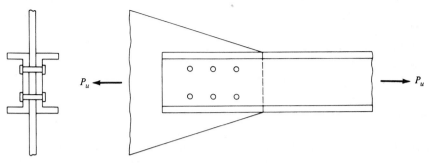

Problem 12-4

12-5. Rework Prob. 12-4 if 1-in. A490 bolts and A572 steel ($F_y = 50$ ksi, $F_u = 65$ ksi) are used. (*Ans.* 5.93, say 6 bolts)

12-6. For the connection shown in the accompanying illustration P_u equals 750 k. Determine the number of 1-in. A325 bolts required for a bearing-type connection. A36 steel.

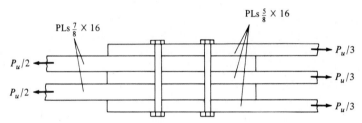

Problem 12-6

12-7. Rework Prob. 12-6 using $\frac{7}{8}$-in. A490 bolts. (*Ans.* 10.68, say 12 bolts)

12-8. Determine the number of bolts required for a slip-critical connection for the plates shown in the accompanying illustration. The axial service loads are $P_D = 50$ k and $P_L = 90$ k; the bolts are $\frac{7}{8}$-in. A325 high-strength bolts and A36 steel are to be used.

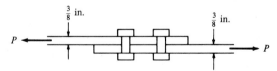

Problem 12-8

12-9. Repeat Prob. 12-8 if the threads are not excluded from the shear plane. (*Ans.* 13.73, say 14 or 15 bolts)

12-10. How many $\frac{3}{4}$-in. A325 high-strength bolts are required for the slip-critical connection shown in the accompanying illustration? $P_D = 120$ k and $P_L = 80$ k. A36 steel.

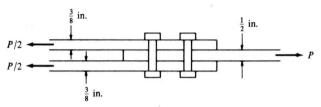

Problem 12-10

12-11. Repeat Prob. 12-6 if $P_D = 130$ k and $P_L = 160$ k and if 1-in. A325 bolts are to be used in a slip-critical connection. A36 steel. (*Ans.* 10.87, say 12 bolts)

12-12. For the beam shown in the accompanying illustration what is the required spacing of $\frac{7}{8}$-in. A325 bolts in a bearing-type connection at a section where the external shear V_u is 300 k? A36 steel.

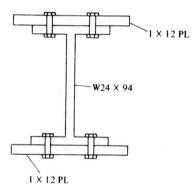

Problem 12-12

12-13. The cover-plated section shown in the accompanying illustration is used to support a uniform load $w_D = 12$ klf (includes beam weight effect and $w_L = 15$ klf for an 18-ft simple span). If $\frac{7}{8}$-in. A325 bolts are used in a slip-critical connection work out a spacing diagram for the entire span. A36 steel. (*Ans.* Use $6\frac{1}{2}$-in. spacing as required by LRFD Specification E.4)

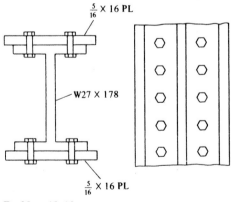

Problem 12-13

12-14. For the section shown in the accompanying illustration determine the required spacing of $\frac{3}{4}$-in. A490 bolts for a slip-critical connection if the member consists of A572 grade 60 steel ($F_u = 75$ ksi). $V_D = 100$ k and $V_L = 140$ k.

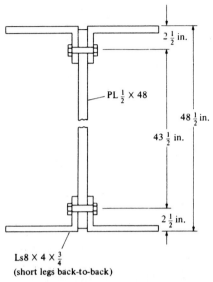

$2\frac{1}{2}$ in.

PL $\frac{1}{2}$ × 48

$48\frac{1}{2}$ in.

$43\frac{1}{2}$ in.

$2\frac{1}{2}$ in.

Ls8 × 4 × $\frac{3}{4}$
(short legs back-to-back)

Problem 12-14

12-15. For an external shear V_u of 900 k determine the spacing required for $\frac{3}{4}$-in. A325 web bolts in a bearing-type connection for the built-up section shown in the accompanying illustration. A36 steel. (*Ans.* 5.69 in., say $5\frac{1}{2}$ in.)

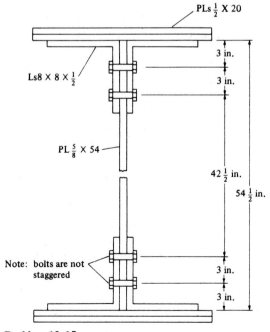

PLs $\frac{1}{2}$ × 20

3 in.

Ls8 × 8 × $\frac{1}{2}$

3 in.

PL $\frac{5}{8}$ × 54

$42\frac{1}{2}$ in.

$54\frac{1}{2}$ in.

Note: bolts are not
staggered

3 in.

3 in.

Problem 12-15

Chapter 13

Bolted Connections Continued and Historical Notes on Rivets

13-1 BOLTS SUBJECTED TO ECCENTRIC SHEAR

Eccentrically loaded bolt groups are subjected to shears and bending moments. The student may feel such situations are rare but the truth is that they are much more common than he or she suspects. For instance, in a truss it is desirable to have the center of gravity of a member lined up exactly with the center of gravity of the bolts at its end connections. This feat is not quite as easy to accomplish as it may seem and connections are often subjected to moments.

Eccentricity is quite obvious in Fig. 13-1 (a) where a beam is connected to a column with a plate. In part (b) of the figure another beam is connected to a column with a pair of web angles. It is obvious that this connection must resist some moment because the center of gravity of the load from the beam does not coincide with the reaction from the column.

The LRFD Specification presents values for computing the design strengths of individual bolts or rivets but does not specify a method for computing the forces on these fasteners when they are eccentrically loaded. As a result the method of analysis to be used is left up to the designer.

Three general approaches for the analysis of eccentrically loaded connections have been developed through the years. The first of the methods is the very conservative *elastic method* in which friction or slip resistance between the connected parts is neglected. In addition these connected parts are assumed to be perfectly rigid. This type of analysis has been commonly used since at least 1870.[1,2]

---

[1]W. McGuire, *Steel Structures* (Englewood Cliffs, N.J.: Prentice-Hall, 1968), p. 813.

[2]C. Reilly, "Studies of Iron Girder Bridges," *Proc. Inst. Civil Engrs.* 29 (London, 1870).

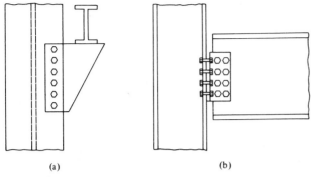

(a)   (b)

Figure 13-1

Tests have shown that the elastic method usually provides very conservative results. As a consequence various *reduced* or *effective eccentricity methods* have been proposed.[3] The analysis is handled just as it is in the elastic method except that smaller eccentricities and thus smaller moments are used in the calculations.

The third method, called the *ultimate strength method,* provides the most realistic values as compared with test results but is extremely tedious to apply, at least with hand-held calculators. The tables in the LRFD Manual for eccentrically loaded connections are based on the ultimate strength method and enable us to solve most of these types of problems quite easily as long as the bolt and rivet patterns are symmetrical. The remainder of this section is devoted to these three analysis methods.

Elastic Analysis For this discussion the bolts of Fig. 13-2 (a) are assumed to be subjected to a load P which has an eccentricity of e from the c.g. (center of gravity) of the bolt group. To consider the force situation in the bolts an upward and downward force, each equal to P, is assumed to act at the c.g. of the bolt group. This situation, shown in part (b) of the figure, in no way changes the bolt forces. The force in a particular bolt will, therefore, equal P divided by the number of bolts in the group as seen in part (c), plus the force due to the moment caused by the couple shown in part (d) of the figure.

The magnitude of the forces in the bolts due to the moment Pe will now be considered. The distances of each bolt from the c.g. of the group are represented by the values d_1, d_2, etc., in Fig. 13-3. The moment produced by the couple is assumed to cause the plate to rotate about the c.g. of the bolt connection with the amount of rotation or strain at a particular bolt being proportional to its distance from the c.g. (For this derivation the gusset plates are again assumed to be perfectly rigid and the bolts are assumed to be perfectly elastic.) Rotation is greatest at the bolt which is the greatest distance from the c.g. as will be the stress since stress is proportional to strain in the elastic range.

[3]T.R. Higgins, "New Formulas for Fasteners Loaded Off Center," *Engr. News Record* (May 21, 1964).

Bridge over New River Gorge in Fayette County, West VA near Charleston. (Courtesy American Bridge Company.)

The rotation is assumed to produce forces of r_1, r_2, r_3, and r_4 respectively on the bolts in the figure. The moment transferred to the bolts must be balanced by resisting moments of the bolts as follows.

$$M_{c.g.} = Pe = r_1 d_1 + r_2 d_2 + r_3 d_3 + r_4 d_4 \tag{1}$$

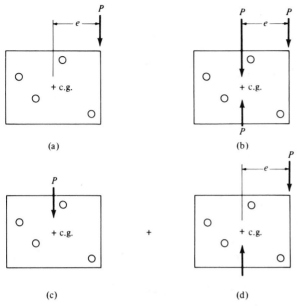

(a)

(b)

(c)

+

(d)

Figure 13-2

As the force caused on each bolt is assumed to be directly proportional to the distance from the c.g. the following expression can be written.

$$\frac{r_1}{d_1} = \frac{r_2}{d_2} = \frac{r_3}{d_3} = \frac{r_4}{d_4}$$

and writing each r value in terms of r_1 and d_1

$$r_1 = \frac{r_1 d_1}{d_1} \quad r_2 = \frac{r_1 d_2}{d_1} \quad r_3 = \frac{r_1 d_3}{d_1} \quad r_4 = \frac{r_1 d_4}{d_1}$$

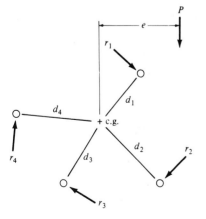

Figure 13-3

Substituting these values into equation (1) and simplifying,

$$M = \frac{r_1 d_1^2}{d_1} + \frac{r_1 d_2^2}{d_1} + \frac{r_1 d_3^2}{d_1} + \frac{r_1 d_4^2}{d_1}$$

$$= \frac{r_1}{d_1}(d_1^2 + d_2^2 + d_3^2 + d_4^2)$$

Therefore,

$$M = \frac{r_1 \Sigma d^2}{d_1}.$$

The force on each bolt can now be written as follows.

$$r_1 = \frac{M d_1}{\Sigma d^2} \quad r_2 = \frac{d_2}{d_1} r_1 = \frac{M d_2}{\Sigma d^2} \quad r_3 = \frac{M d_3}{\Sigma d^2} \quad r_4 = \frac{M d_4}{\Sigma d^2}$$

Each value of r is perpendicular to the line drawn from the c.g. to the particular bolt. It is usually more convenient to break these down into vertical and horizontal components. In accomplishing this purpose reference is made to Fig. 13-4.

The horizontal and vertical components of the distance d_1 are represented by h and v respectively and the horizontal and vertical components of force r_1, are represented by H and V respectively in this figure. It is now possible to write the following ratio from which H can be obtained.

$$\frac{r_1}{d_1} = \frac{H}{v}$$

$$H = \frac{r_1 v}{d_1} = \left(\frac{M d_1}{\Sigma d^2}\right)\left(\frac{v}{d_1}\right)$$

Therefore,

$$H = \frac{M v}{\Sigma d^2}$$

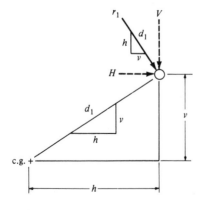

Figure 13-4

By a similar procedure V is found to equal

$$V = \frac{Mh}{\Sigma d^2}$$

EXAMPLE 13-1

Determine the force in the most stressed bolt of the group shown in Fig. 13-5 using the elastic analysis method.

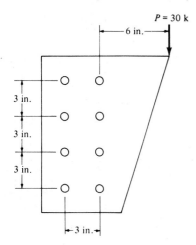

Figure 13-5

Solution

A sketch of each bolt and the forces applied to it by the direct load and the clockwise moment are shown in Fig. 13.6. From this sketch the student will see that the upper right-hand bolt and the lower right-hand bolt are the most stressed and have equal stresses.

$$e = 6 + 1.5 = 7.5 \text{ in.}$$

$$M = Pe = (30)(7.5) = 225 \text{ in.-k}$$

$$\Sigma d^2 = \Sigma h^2 + \Sigma v^2$$

$$\Sigma d^2 = (8)(1.5)^2 + (4)(1.5^2 + 4.5^2) = 108$$

$$H = \frac{Mv}{\Sigma d^2} = \frac{(225)(4.5)}{108} = 9.38 \text{ k}$$

$$V = \frac{Mh}{\Sigma d^2} = \frac{(225)(1.5)}{108} = 3.13 \text{ k}$$

$$\frac{P}{8} = \frac{30}{8} = 3.75 \text{ k}$$

$$\downarrow 3.13 \text{ k}$$
$$\downarrow 3.75 \text{ k}$$
$$O \leftarrow 9.38 \text{ k}$$

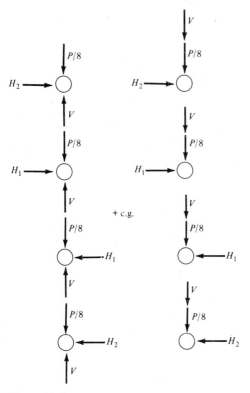

Figure 13-6

Force in the lower right-hand bolt (equal to force in the upper right-hand bolt) is determined as follows:

$$r = \sqrt{(6.88)^2 + (9.38)^2} = 11.63 \text{ k}$$

Should the eccentric load be inclined it can be broken down into vertical and horizontal components and the moment of each about the c.g. of the bolt group determined. Various design formulas can be developed which will enable the engineer to directly design eccentric connections but in all probability the process of assuming a certain number and arrangement of bolts, checking stresses, and redesigning is probably just as satisfactory.

Reduced Eccentricity Method   The elastic analysis method just described appreciably overestimates the moment forces applied to the connectors. As a result quite a few proposals have been made through the years which make use of an effective eccentricity, thus in effect taking into account the slip resistance on the faying or contact surfaces. One set of reduced eccentricity values which were fairly common at one time follow:

1. With one gage line of fasteners and where n is the number of fasteners in the line,

$$e_{\text{effective}} = e_{\text{actual}} - \frac{1 + 2n}{4}$$

2. With two or more gage lines of fasteners symmetrically placed and where n is the number of fasteners in each line,

$$e_{\text{effective}} = e_{\text{actual}} - \frac{1 + n}{2}$$

The reduced eccentricity values for two fastener arrangements are shown in Fig. 13-7.

To analyze a particular connection with the reduced eccentricity method, the value of $e_{\text{effective}}$ is computed as described above and used to compute the eccentric moment. Then the elastic procedure is used for the remainder of the calculations.

Ultimate Strength Method   The elastic and reduced eccentricity methods for analyzing eccentrically loaded fastener groups are both based upon the assumption that the behavior of the fasteners is elastic. A much more realistic method of analysis is the ultimate strength method, which is described in the next few paragraphs. The values given in the tables of Part 5 of the LRFD Manual for eccentrically loaded fastener groups were computed using this method.

If one of the outermost bolts or rivets in an eccentrically loaded connection begins to slip or yield the connection will not fail. Instead the magnitude of the eccentric load may be increased, the inner bolts will resist more load, and failure will not occur until all of the bolts slip or yield.

The eccentric load tends to cause both a relative rotation and translation of the connected material. In effect this is equivalent to pure rotation of the connec-

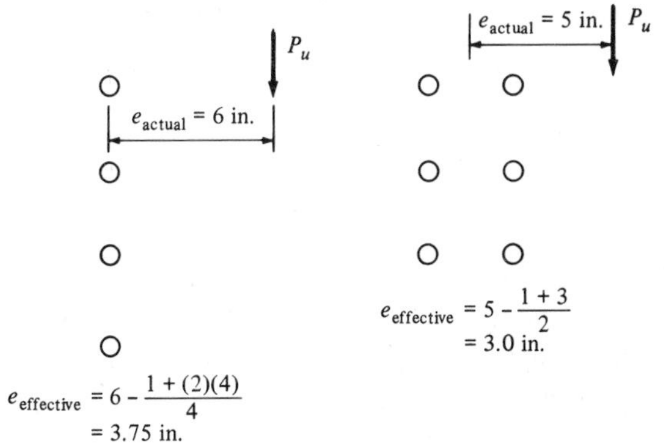

$$e_{\text{effective}} = 6 - \frac{1 + (2)(4)}{4}$$
$$= 3.75 \text{ in.}$$

$$e_{\text{effective}} = 5 - \frac{1 + 3}{2}$$
$$= 3.0 \text{ in.}$$

Figure 13-7

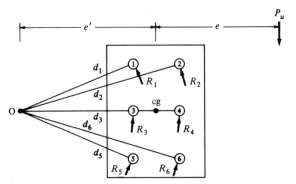

Figure 13-8

tion about a single point called the *instantaneous center of rotation*. An eccentrically loaded bolted connection is shown in Fig. 13-8, and the instantaneous center is represented by point 0. It is located a distance e' from the center of gravity of the bolt group.

The deformations of these bolts are assumed to vary in proportion to their distances from the instantaneous center. The ultimate shear force which one of them can resist is not equal to the pure shear force which a bolt can resist. Rather it is dependent upon the load-deformation relationship in the bolt. Studies by Crawford and Kulak[4] have shown that this force may be closely estimated with the following expression:

$$R = R_{ult}(1 - e^{-10\Delta})^{0.55}$$

In this formula R_{ult} is the ultimate shear load for a single fastener equaling 74 k for an A325 bolt, e is the base of the natural logarithm (2.718), and Δ is equal to the total deformation of a bolt experimentally determined as equal to 0.34 in. The coefficients 10.0 and 0.55 were also experimentally obtained. Figure 13-9 illustrates this load-deformation relationship.

This expression clearly shows that the ultimate shear load taken by a particular bolt in an eccentrically loaded connection is affected by its deformation. Thus, the load applied to a particular bolt is dependent upon its position in the connection with respect to the instantaneous center of rotation.

The resisting forces of the bolts of the connection of Fig. 13-8 are represented with the letters R_1, R_2, R_3, and so on. Each of these forces is assumed to act in a direction perpendicular to a line drawn from point 0 to the center of the bolt in question. For this symmetrical connection the instantaneous center of rotation will fall somewhere on a horizontal line through the center of gravity of the bolt group. This is the case because the sum of the horizontal components of the R forces must be zero as also must be the sum of the moments of the horizontal components about point 0. The position of point 0 on the horizontal line may be found by a tedious trial-and-error procedure to be described here.

[4] S.F. Crawford and G.L. Kulak, "Eccentrically Loaded Bolt Connections," *Journal of Structural Division*, ASCE 97, ST3 (March 1971), pp. 765–783.

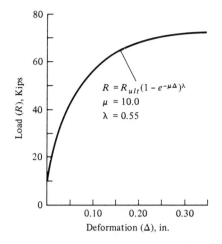

Figure 13-9 Ultimate shear force R in a single bolt at any given deformation.

With reference to Fig. 13-8 the moment of the eccentric load about point 0 must be equal to the summation of the moments of the R resisting forces about the same point. If we knew the location of the instantaneous center we could compute R values for the bolts with the Crawford-Kulak formula and determine P_u as follows:

$$P_u(e' + e) = \Sigma Rd$$

$$P_u = \frac{\Sigma Rd}{e' + e}$$

The location of the instantaneous center is not known, however. Its position is estimated, the R values determined, and P_u calculated as described. It will be noted that P_u must be equal to the summation of the vertical components of the R resisting forces (ΣR_v). If the value is computed and equals the P_u computed by the formula above we have the correct location for the instantaneous center. If not we try another location, and so on.

In Example 13-2 the author demonstrates the very tedious trial- and error calculations necessary to locate the instantaneous center of rotation for a symmetrical connection consisting of four bolts. In addition the design strength P_u of the connection is determined.

To solve such a problem it is very convenient to set the calculations up in a table similar to the one used in the solution to follow. In the table shown the h and v values given are the horizontal and vertical components of the d distances from point 0 to the center of gravities of the bolts. The bolt which is located at the greatest distance from point 0 is assumed to have a Δ value of 0.34 in. The Δ values for the other bolts are assumed to be proportionate to their distances from point 0. The Δ values so determined are used in the R formula.

A set of tables entitled "Eccentric Loads on Fastener Groups" is presented in Part 5 of the LRFD Manual. The values in these tables were determined by the procedure described here. A large percentage of the practical cases that the designer will encounter are included in the tables. Should some other situation not covered be faced, the designer may very well decide to use the more conservative elastic procedure previously described.

EXAMPLE 13-2

The bearing-type 7/8-in. A325 bolts of the connection of Fig. 13.10 have a design shear strength $R_u = (0.65)\,(0.60)\,(72) = 28.1$ k. Locate the instantaneous center of rotation of the connection using the trial-and-error procedure and determine the value of P_u.

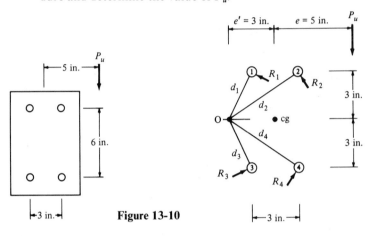

Figure 13-10   **Figure 13-11**

Solution.   By trial and error: Try a value of $e' = 3$ in., reference being made to Fig. 13-11.

Bolt No.	h(in.)	v(in.)	d(in.)	Δ(in.)	R(kips)	R_v(kips)	Rd(in.-kips)
1	1.5	3	3.3541	0.211	26.17	11.70	87.78
2	4.5	3	5.4083	0.34	27.58	22.95	149.16
3	1.5	3	3.3541	0.211	26.17	11.70	87.78
4	4.5	3	5.4083	0.34	27.58	22.95	149.16
						$\Sigma = 69.30$	$\Sigma = 473.88$

$$P_u = \frac{\Sigma Rd}{e' + e} = \frac{473.88}{3 + 5} = 59.24 \text{ k not} = 69.30 \text{ k} \qquad \text{NG}$$

After several trials assume $e = 2.40$ in.

Bolt No.	h(in.)	v(in.)	d(in.)	Δ(in.)	R(kips)	R_v(kips)	Rd(in.-kips)
1	0.90	3	3.1321	0.216	26.27	7.55	82.28
2	3.90	3	4.9204	0.34	27.58	21.86	135.70
3	0.90	3	3.1321	0.216	26.27	7.55	82.28
4	3.90	3	4.9204	0.34	27.58	21.86	135.70
						$\Sigma = 58.82$	$\Sigma = 435.96$

$$P_u = \frac{435.96}{2.4 + 5} = 58.91 \text{ k almost} = 58.82 \text{ k} \qquad \text{OK}$$

$$\underline{\underline{P_u = 58.9 \text{ k}}}$$

Although the development of this method of analysis was actually based on bearing-type connections where slip may occur, both theory and load tests have shown that it may conservatively be applied to slip-critical connections.[5]

The ultimate-strength method may be expanded to include inclined loads and unsymmetrical bolt arrangements but the trial-and-error calculations are extraordinarily long for such situations.

Examples 13-3 and 13-4 which follow provide illustrations of the use of the ultimate strength tables of the LRFD Manual for both analysis and design.

EXAMPLE 13-3

Repeat Example 13-2 using the ultimate strength tables entitled "Eccentric Loads on Fastener Groups" in Part 5 of the LRFD Manual.

Solution

$$P_u = C \times \phi r$$

$$C = 2.10 \text{ from Table XI}$$

$$P_u = (2.10)(28.1) = \underline{\underline{59 \text{ k}}} \qquad \text{OK}$$

EXAMPLE 13-4

Determine the number of 7/8-in. A325 bolts in standard size holes required for the connection shown in Fig. 13-12. Use A36 steel and assume the connection is to be a bearing type with threads excluded from the shear plane. Further assume that the bolts are in single shear and bearing on 1/2 in. Use the ultimate strength analysis method presented in the LRFD Manual.

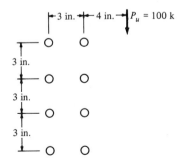

Figure 13-12

Solution. Using Table XI page 5-66 of LRFD Manual

$$x_0 = 5.5 \text{ in.}$$

[5]G.L. Kulak, "Eccentrically Loaded Slip-Resistant Connections," *Engineering Journal,* AISC, 12, No. 2 (second quarter, 1975), pp. 52–55.

Bolts in single shear and bearing on $\frac{1}{2}$ in.

$$\phi r_v = \text{shear design strength per fastener}$$

$$= (0.65)(0.6)(72) = 28.1 \text{ k} \leftarrow$$

$$\phi r_v = \text{bearing design strength per fastener}$$

$$= (0.75)(2.4)(\tfrac{7}{8})(\tfrac{1}{2})(58) = 45.7 \text{ k}$$

$$C = \frac{P_u}{\phi r_v} = \frac{100}{28.1} = 3.56$$

By interpolation in table we need 4 bolts in each line.

Note: If the situation faced by the designer does not fit the ultimate strength tables given in the LRFD Manual it is recommended that he or she use the conservative elastic procedure to handle the problem whether it's analysis or design.

13-2 BOLTS SUBJECTED TO SHEAR AND TENSION

The bolts used for a good many structural steel connections are subjected to a combination of shear and tension. One obvious case is shown in Fig. 13-13 where a diagonal brace is attached to a column. The 111.8-k component shown in the

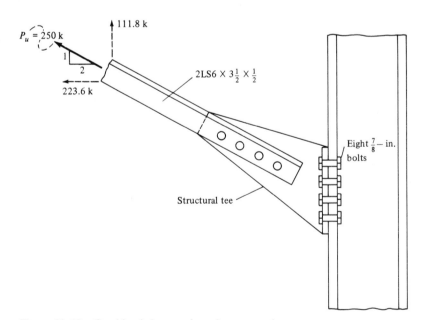

Figure 13-13   Combined shear and tension connection.

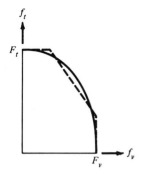

Figure 13-14 Bolts in a bearing-type connection subject to combined shear and tension.

figure is trying to shear the bolts off at the face of the column while the 223.6 k component is tending to pull off their heads.

Tests on bearing-type bolts subject to combined shear and tension show that their ultimate strengths can be represented with an elliptical interaction curve as shown in Fig. 13-14. Particularly note the values F_t and F_v. F_t is the limiting tensile stress if there is no shear and F_v is the limiting shearing stress if there is no externally applied tension.

The three straight dashed lines shown in Fig. 13-14 very closely represent the test result interaction curve. Equations for these lines are given in Table 13.1 (Table J3.3 of the LRFD Specification). In these expressions f_v and f_t are respectively the computed shear and tension stresses in the bolts due to the factored loads. The maximum values given in the table (as the 68 ksi for A325 bolts with threads not excluded from the shear plane) equal ϕ (which is 0.75) times the nominal strength of the bolts if they are subjected to external tensile loads only.

Example 13-5 illustrates the calculations involved in checking a high-strength bearing-type bolted connection for shear and tension. (There is another stress situation which may affect this connection. It is called *prying action*, but as it is not discussed until section 13-4 it is not considered in this example.)

Table 13.1 TENSION STRESS LIMIT (F_t), ksi, FOR FASTENERS IN BEARING-TYPE CONNECTIONS

Description of fasteners	Threads included in the shear plane	Threads excluded from the shear plane
A307 bolts	$39 - 1.8f_v \leq 30$	
A325 bolts	$85 - 1.8f_v \leq 68$	$85 - 1.4f_v \leq 68$
A490 bolts	$106 - 1.8f_v \leq 84$	$106 - 1.4f_v \leq 84$
Threaded parts A449 bolts over 1½-in. diameter	$0.73F_u - 1.8f_v \leq 0.56F_u$	$0.73F_u - 1.4f_v \leq 0.56F_u$
A502 grade 1 rivets	$44 - 1.3f_v \leq 34$	
A502 grade 2 rivets	$59 - 1.3f_v \leq 45$	

EXAMPLE 13-5

The tension member shown in Fig. 13-13, is connected to the column shown with eight 7/8-in. A325 high-strength bolts in a bearing-type connection with the threads excluded from the shear plane and standard-size holes. Is this a sufficient number of bolts to resist the applied load according to the LRFD Specification?

Solution

$$\text{Shearing stress } f_v = \frac{111.8}{(8)(0.6)} = 23.29 \text{ ksi}$$

$$\text{Tension stress } f_t = \frac{223.6}{(8)(0.6)} = 46.58 \text{ ksi}$$

$$\text{Limiting tension stress } F_t = 85 - (1.4)(23.29) = 52.39 \text{ ksi}$$

$$> 46.58 \text{ ksi} \qquad\qquad \text{OK}$$

Connection is satisfactory

When an axial tension force is applied to a slip-critical connection, the clamping force will be reduced and the design shear strength must be decreased in some proportion to the loss in clamping or prestress. This is accomplished in the LRFD Specification (Section J3.5) by requiring that the nominal slip-critical shear strengths given in LRFD Table J3.4 be multiplied by the reduction factor $(1 - T/T_b)$. T is the service tensile force applied to one bolt and T_b is the minimum pretension load for one bolt in a slip-critical connection as given in Table 12.1 (LRFD Table J3.1). For such a situation ϕ is equal to 1.0 unless we have long-slotted holes with the load applied in the direction of the slot. In the latter case ϕ is to be taken as 0.85, says the LRFD Specification.

For bolts in slip-critical connections with standard holes the nominal shear strengths at service load conditions are to be determined as follows:

$$\text{For A325 bolts } \leq \left(1 - \frac{T}{T_b}\right)(17)$$

$$\text{For A490 bolts } \leq \left(1 - \frac{T}{T_b}\right)(21)$$

EXAMPLE 13-6

Rework Example 13-5 assuming that a slip-critical connection is to be used. The service loads are $P_D = 75$ k and $P_L = 90$ k.

Solution

$$\text{Service load} \quad P = 75 + 90 = 165 \text{ k}$$

$$P_v = \left(\frac{1}{\sqrt{5}}\right)(165) = 73.8 \text{ k}$$

$$P_h = \left(\frac{2}{\sqrt{5}}\right)(165) = 147.6 \text{ k}$$

$$f_v = \frac{73.8}{(8)(0.6)} = 15.38 \text{ ksi}$$

$$f_n = \frac{147.6}{(8)(0.6)} = 30.75 \text{ ksi}$$

$$\text{Limiting shear strength} = \left(1 - \frac{T}{T_b}\right)(17)$$

$$= \left(1 - \frac{147.6/8}{39}\right)(17) = 8.95 \text{ ksi} < 15.38 \text{ ksi} \qquad \text{NG}$$

Connection is unsatisfactory if slip-critical

13-3 TENSION LOADS ON BOLTED JOINTS

Bolted and riveted connections subjected to pure tensile loads have been avoided as much as possible in the past by designers. The use of tensile connections was probably "forced on" them more often for wind-bracing systems in tall buildings than for any other situation. There are some other locations where they have been used, however, such as hanger connections for bridges, flange connections for piping systems, etc. Figure 13-15 shows a hanger-type connection with an applied tensile load.

Hot-driven rivets and fully tensioned high-strength bolts are not free to shorten, with the result that large tensile forces are produced in them during their installation. These initial tensions are actually close to the yield points. There has always been considerable reluctance among designers to apply tensile loads to connectors of this type for fear that the external loads might easily increase their already present tensile stresses and cause them to fail. The truth of the matter,

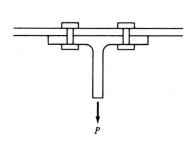

Hanger Connection

Figure 13-15 Hanger connection.

however, is that when external tensile loads are applied to connections of this type the connectors will probably experience little if any change in stress.

Hot-driven rivets which have cooled and shrunk or fully tensioned high-strength bolts actually prestress the joints in which they are used against tensile loads. (To follow this discussion the student may like to think of a prestressed concrete beam which has external compressive loads applied at each end.) The tensile stresses in the connectors squeeze together the members being connected. If a tensile load is applied to this connection at the contact surface, it cannot exert any additional load on the bolts or rivets until the members are pulled apart and additional strains put on the bolts or rivets. The members cannot be pulled apart until a load is applied which is larger than the total tension in the connectors of the connection. This statement means that the joint is prestressed against tensile forces by the amount of stress initially put in the shanks of the connectors.

Another way of saying this is that if a tensile load P is applied at the contact surface it tends to reduce the thickness of the plates somewhat but at the same time the contact pressure between the plates will be correspondingly reduced and the plates will tend to expand by the same amount. The theoretical result then is no change in plate thickness and no change in connector tension. This situation continues until P equals the connector tension. At this time an increase in P will result in separation of the plates and thereafter the tension in the connector will equal P.

Should the load be applied to the outer surfaces there will be some immediate strain increase in the connector. This increase will be accompanied by an expansion of the plates even though the load does not exceed the prestress, but the increase will be very slight because the load will go to the plate and connectors roughly in proportion to their stiffness. As the plate is stiffer it will receive most of the load. An expression can be developed for the elongation of the bolt based on the bolt area and the assumed contact area between the plates. Depending on the contact area assumed, it will be found that unless P is greater than the bolt tension its stress increase will be in the range of 10 percent. Should the load exceed the prestress the bolt stress will rise appreciably.

The preceding rather lengthy discussion is approximate but should explain why an ordinary tensile load applied to a riveted or bolted joint will not change the stress situation very much.

The LRFD tensile design strength of bolts, rivets, and threaded parts is given by the expression to follow which is independent of any initial tightening force.

$$P_u = \phi F_t A_g$$

When fasteners are loaded in tension there is usually some bending due to the deformation of the connected parts. As a result the value of ϕ in this expression is a rather small 0.75. Table J3.2 of the LRFD Specification gives values of F_t for the different kinds of connectors with the values of rivets and threaded parts being quite conservative.

In this expression A_g is the nominal body area of a rivet or the unthreaded portion of a bolt or its threaded part not including upset rods. An upset rod has its

Figure 13-16 A round upset rod.

ends made larger than the regular rod and the threads are placed in the enlarged section so that the area at the root of the thread is larger than that of the regular rod. An upset rod is shown in Fig. 13-16. The use of upset rods is not usually economical and should be avoided unless a large order is being made.

If an upset rod is used the nominal tensile strength of the threaded portion is set equal to $0.75 F_u$ times the cross-sectional area at its major thread diameter. This value must be larger than F_y times the nominal body area of the rod before upsetting.

Example 13-7 illustrates the determination of the strength of a tension connection.

EXAMPLE 13-7

Determine the design tensile strength of the bolts of the hanger connection of Fig. 13-15 if eight $\frac{7}{8}$-in. A490 high-strength bolts are used.

Solution

$$P_u = (0.75)(8)(0.6)(112.5) = \underline{\underline{405 \text{ k}}}$$

The load applied to a tensile connection shall be the sum of the factored external loads plus any tension forces that result from prying action as described in the next section.

13-4 PRYING ACTION

A further consideration that should often be given to tensile connections is the possibility of prying action. A tensile connection is shown in Fig. 13-17(a) which is subjected to prying action as illustrated in part (b) of the same figure. Should the flanges of the connection be quite thick and stiff or have stiffener plates as shown in Fig. 13-17(c), the prying action will probably be negligible but this is not the case if they are thin and flexible and have no stiffeners.

It is usually desirable to limit the number of rows of rivets or bolts in a tensile connection because a large percentage of the load is carried by the inner rows of multirow connections even at ultimate load. The tensile connection shown in Fig. 13-18 illustrates this point as the prying action will throw a large part of the load to the inner connectors, particularly if the plates are thin and flexible. For connections subjected to pure tensile loads estimates should be made of possible prying action and its magnitude.

The additional force in the bolts resulting from prying action should be added to the tensile force resulting directly from the applied forces. The actual determination of prying forces is quite complex, and research on the subject is

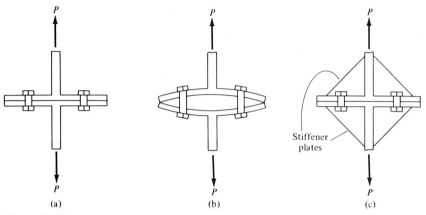

Figure 13-17

still being conducted. Several empirical formulas have been developed which approximate test results. Among these are the LRFD expressions included in this section. The reader should realize that we don't know very much about prying action and our formulas keep changing almost yearly.

Only fully tensioned bolts should be used for connections for which the applied loads subject the bolts to axial tension. This is true whether or not the connections are classified as slip-critical and whether or not they are subject to fatigue loads and whether or not there is prying action. If snug-tight bolts are used for any of these situations the tensile loads will immediately start increasing bolt tensions.

Hanger and other tension connections should be so designed as to prevent significant deformations. The most important item in such designs is the use of rigid flanges. Rigidity is more important than bending resistance. To achieve this goal the distance b shown in Fig. 13-19 should be made as small as possible with a minimum value equal to the space required to use a wrench for tightening the bolts. Information concerning wrench clearance dimensions is presented in a

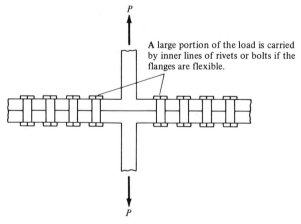

Figure 13-18

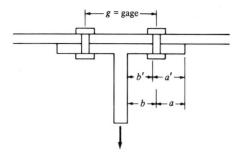

Figure 13-19

table entitled "Threaded Fasteners Assembling Clearances" on page 5-164 of the LRFD Manual.

On pages 5-119 through 5-125 of the LRFD Manual a detailed procedure is presented for designing hanger connections and computing prying forces. This method is for factored forces only. If it is desired to consider service loads for the purposes of investigating fatigue, deflections, or drift limitations the reader may refer to the service load formulas for prying which were presented in the eighth edition of the *AISC Manual of Steel Construction.*

For reasons of space limitations only one numerical example is presented here. Three detailed examples for designing hangers and computing prying action are included in the LRFD Manual.

To determine the prying force in a hanger connection the LRFD Manual presents a long set of empirical equations. Reference should be made to Fig. 13-19 or 13-20 for several of the letters used in these expressions.

B = design tensile strength of each bolt

T = tension force applied to each bolt *not including prying action.* (This force is really fictitious unless the tensile load exceeds the prestress due to tensioning the bolts.)

b = $g/2 - t_w/2$ where g is the gage. It must be sufficient for wrench clearance as provided in the LRFD Manual

a = distance from bolt center line to the edge of tee flange or angle leg but not more than 1.25 b

$b' = b - d/2$ where d = bolt diameter

$a' = a + d/2$

ρ = length of connection tributary to each bolt

d' = width of bolt hole parallel to tee stem

δ = ratio of the net area at the bolt line to the gross area at the face of the stem = $1 - d'/\rho$

p = b'/a

$\alpha = 1/\delta \, [T/B/(t_f/t_c)^2 - 1]$

t_c = flange thickness required to develop B in bolts with no prying action =

$$\sqrt{\frac{4.44\,Bb'}{\rho F_y}}\,.$$

Q = factored prying force per bolt = $B\delta\alpha p(t/t_c)^2$

B_c = factored load per bolt including prying action = $T + Q$

EXAMPLE 13-8

A 10-in.-long WT8×22.5 (t_f = 0.565 in., t_w = 0.345 in., and b_f = 7.035 in.) is connected to a W36×150 as shown in Fig. 13-20 with four 7/8-in. A325 high-strength bolts. If A36 steel is used are the bolts satisfactory? Include the effect of prying action.

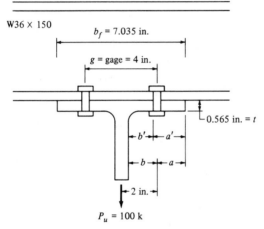

Figure 13-20

Solution

$$B = (0.75)(0.6)(90) = 40.5 \text{ k}$$

$$T = \frac{100}{4} = 25 \text{ k}$$

Determining the values of b, a, b', ρ, d', δ, and p for use in subsequent expressions,

$$b = 2.0 - \frac{t_w}{2} = 2 - \frac{0.345}{2} = 1.827 \text{ in.} > 1\tfrac{3}{8} \text{ in.}$$

which is required for wrench clearance (page 5-164 LRFD Manual)

$$a = \frac{b_f}{2} - 2.0 = \frac{7.035}{2} - 2.0 = 1.517 \text{ in.}$$

Note that $1.25b = (1.25)(1.827) = 2.284$ in. > 1.517 in. $\therefore$ Use $a = 1.517$ in., says LRFD.

$$b' = b - \frac{d}{2} = 1.827 - \frac{0.875}{2} = 1.389 \text{ in.}$$

$$\rho = \frac{10}{2} = 5 \text{ in.}$$

$$d' = \frac{15}{16} \text{ in.} = 0.937 \text{ in.}$$

$$\delta = 1 - \frac{d'}{\rho} = 1 - \frac{0.937}{5} = 0.813 \text{ in.}$$

$$p = \frac{b'}{a} = \frac{1.389}{1.517} = 0.916 \text{ in.}$$

Computing α, t_c, Q, and B_c,

$$t_c = \sqrt{\frac{(4.44)(40.5)(1.389)}{(5)(36)}} = 1.178$$

$$\alpha = \frac{1}{0.813}\left[\frac{25/40.5}{(0.565/1.178)^2} - 1\right] = 2.070$$

$$Q = (40.5)(0.813)(2.070)(0.916)\left(\frac{0.565}{1.178}\right)^2 = 14.36 \text{ k}$$

$$B_c = 25 + 14.36 = 39.36 \text{ k} < 40.5 \text{ k}$$

Connection is satisfactory

13-5 HISTORICAL NOTES ON RIVETS

For many years rivets were the accepted method for connecting the members of steel structures. Today, however, they no longer provide the most economical connections. They are still occasionally used for fasteners but their use has declined to such a degree that most steel fabricators in the United States have discontinued riveting altogether. It is, however, desirable for the designer to be familiar with rivets even though he or she will seldom if ever design riveted structures. He or she may have to analyze an existing riveted structure for new loads or for an expansion of the structure. The purpose of these sections is to present only a very brief introduction to the analysis and design of rivets.

The rivets used in construction work were usually made of a soft grade of steel which would not become brittle when heated and hammered with a riveting gun to form the head. The usual rivet consisted of a cylindrical shank of steel with a rounded head on one end. It was heated in the field to a cherry-red color (ap-

Handling hot rivets, Chicago, Ill. (Courtesy of IR Construction Products Co.)

proximately 1800°F), inserted in the hole, and a head formed on the other end probably with a portable rivet gun operated by compressed air. The rivet gun, which had a depression in its head to give the rivet head the proper shape, applied a rapid succession of blows to the rivet.

For riveting done in the shop the rivets were probably heated to a light cherry-red color and driven with a pressure-type riveter. This type of riveter, usually called a "bull" riveter, squeezed the rivet with a pressure of perhaps as high as 50 to 80 tons (445 to 712 kN) and drove the rivet with one stroke. Because of this great pressure the rivet in its soft state was forced to fill the hole very satisfactorily. This type of riveting was much to be preferred over that done with the pneumatic hammer but no greater nominal strengths were allowed by riveting specifications. The bull riveters were built for much faster operation than were the portable hand riveters, but the latter riveters were needed for places that are not easily accessible (i.e., field erection).

As the rivet cooled it shrunk, or contracted and squeezed together the parts being connected. The squeezing effect actually caused considerable transfer of stress between the parts being connected to take place by friction. The amount of friction was not dependable, however, and the specifications did not permit its inclusion in the strength of a connection. Rivets shrink diametrically as well as lengthwise and actually become somewhat smaller than the holes which they are assumed to fill. (Permissible strengths for rivets are actually given in terms of the nominal cross-sectional areas of the rivets before driving.)

Some shop rivets were driven cold with tremendous pressures. Obviously the cold-driving process worked better for the smaller size rivets probably $\frac{3}{4}$ in. in diameter or less although larger ones have been successfully used. Cold-driven rivets fill the holes better, eliminate the cost of heating, and are stronger because the steel is cold worked. There is, however, a reduction of clamping force since the rivets do not shrink after driving.

13-6 TYPES OF RIVETS

The sizes of rivets used in ordinary construction work were $\frac{3}{4}$ in. and $\frac{7}{8}$ in. in diameter but they could be obtained in standard sizes from $\frac{1}{2}$ in. to $1\frac{1}{2}$ in. in $\frac{1}{8}$-in. increments. (The smaller sizes were used for small roof trusses, signs, small towers, etc., while the larger sizes were used for very large bridges or towers and very tall buildings.) The use of more than one or two sizes of rivets or bolts on a single job is usually undesirable because it is expensive and inconvenient to punch different-size holes in a member in the shop, and the installation of different-size rivets or bolts in the field may be confusing. Some cases arise where it is absolutely necessary to have different sizes, as where smaller rivets or bolts are needed for keeping the proper edge distance in certain sections, but these situations should be avoided if possible.

Rivet heads were usually round in shape, called button heads; but if clearance requirements dictated the head was flattened or even countersunk and chipped flush. These situations are shown in Fig. 13-21.

The countersunk and chipped-flush rivets do not have sufficient bearing areas to develop full strength and should be discounted 50 percent in design (LRFD Specification J3.2). A rivet with a flattened head was to be preferred to a countersunk rivet but if a smooth surface was required the countersunk and chipped-flush rivet were necessary. This latter type of rivet was appreciably more expensive than the button head type in addition to being weaker, and they were not used unless absolutely necessary.

There are three ASTM classifications for rivets for structural steel applications as described in the following paragraphs.

ASTM Specification A502, Grade 1

These rivets were used for most structural work. They had a low carbon content of about 0.80 percent, were weaker than the ordinary structural carbon steel, and had a higher ductility. The fact that these rivets were easier to drive than the higher-strength rivets was the main reason that when rivets were used they probably were A502, grade 1 regardless of the strength of the steel used in the structural members.

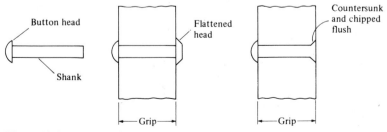

Figure 13-21 Types of rivets.

ASTM Specification A502, Grade 2

These carbon-manganese rivets have higher strengths than the grade 1 rivets and were developed for the higher-strength steels. Their higher strength permits the designer to use fewer rivets in a connection and thus smaller gusset plates.

ASTM Specification A502, Grade 3

These rivets have the same nominal strengths as the grade 2 rivets but they have much higher resistance to atmospheric corrosion equal to approximately four times that of carbon steel without copper.

13-7 STRENGTH OF RIVETED CONNECTIONS—RIVETS IN SHEAR

The factors determining the strength of a rivet are its grade, its diameter and the thickness and arrangement of the pieces being connected. The actual distribution of stress around a rivet hole is difficult to determine, if it can be determined at all; and to simplify the calculations it is assumed to vary uniformly over a rectangular area equal to the diameter of the rivet times the thickness of the plate.

The strength of a rivet in bearing equals the design bearing unit stress of the rivet times the diameter of the shank of the rivet times the thickness of the member that bears on the rivet. The strength of a rivet in single shear is the nominal shearing strength times the cross-sectional area of the shank of the rivet. Should a rivet be in double shear its shearing strength is considered to be twice its single-shear value.

The nominal tension and shearing strengths for rivets and A307 bolts are given in LRFD Table J3.2. These values are for static loads only and are repeated in Table 13.2. Notice that the nominal shear strength of the A307 bolts is not affected if the bolt threads are in the shear plane.

Examples 13-9 and 13-10 illustrate the calculations necessary to determine the design strengths of existing connections or to design riveted connections. Little comment is made here concerning A307 bolts. The reason is that all the calculations for these fasteners are made exactly as they are for rivets except that the shearing strengths given by the LRFD Specification are different. Only one brief example with these common bolts (13-11) is included.

TABLE 13.2 LRFD NOMINAL TENSILE AND SHEARING STRENGTHS FOR RIVETS AND A307 BOLTS

Fastener type	Tensile strength (ksi)	Shearing strength in bearing-type connections (ksi)
A502, grade 1, hot-driven rivets	45.0, $\phi = 0.75$	36.0, $\phi = 0.65$
A502, grade 2 or 3, hot-driven rivets	60.0, $\phi = 0.75$	48.0, $\phi = 0.65$
A307 bolts	45.0, $\phi = 0.75$	27.0, $\phi = 0.60$

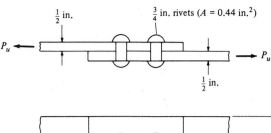

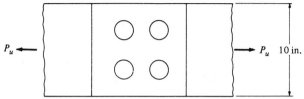

Figure 13-22

EXAMPLE 13-9

Determine the design strength P_u of the bearing-type connection shown in Fig. 13-22. A36 steel and A502, grade 1 rivets are used in the connection and it is assumed that standard-size holes are used and that edge distances and center-to-center distances are $>1\frac{1}{2}d$ and $3d$ respectively.

Solution. Design tensile force applied to plates

$$A_g = (\tfrac{1}{2})(10) = 5.00 \text{ in.}^2$$
$$A_n = [(\tfrac{1}{2})(10) - (2)(\tfrac{7}{8})(\tfrac{1}{2})] = 4.125 \text{ in}^2 = A_e$$
$$P_u = \phi_t F_y A_g = (0.90)(36)(5.00) = 162 \text{ k}$$
$$P_u = \phi_t F_u A_e = (0.75)(58)(4.125) = 179.4 \text{ k}$$

Rivets in single shear and bearing on $\frac{1}{2}$ in.

Design shearing strength of rivets = $(0.65)(0.44)(36)(4) = 41.2 \text{ k} \leftarrow$

Design bearing strength of rivets = $(0.75)(2.4)(\tfrac{3}{4})(\tfrac{1}{2})(58)(4) = 156.6 \text{ k}$

$$\underline{P_u = 41.2 \text{ k}}$$

EXAMPLE 13-10

How many $\frac{7}{8}$-in. A502, grade 1 rivets are required for the connection shown in Fig. 13-23 if the plates are A36 and if standard-size holes are used and the edge distances and center-to-center distances are $>1\frac{1}{2}d$ and $3d$ respectively?

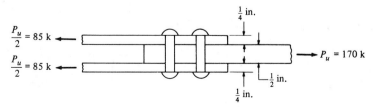

Figure 13-23

Solution. Bolts in double shear and bearing on $\frac{1}{2}$ in.:

$$\text{Design shear strength of 1 rivet} = (0.75)(2 \times 0.6)(36) = 28.1 \text{ k} \leftarrow$$

$$\text{Design bearing strength of 1 rivet} = (0.75)(2.4)(\tfrac{7}{8})(\tfrac{1}{2})(58) = 45.7 \text{ k}$$

$$\text{No. of rivets required} = \frac{170}{28.1} = 6.05$$

$$\underline{\text{Use six } \tfrac{7}{8}\text{-in. rivets}}$$

EXAMPLE 13-11

Repeat Example 13-10 using $\frac{7}{8}$-in. A307 bolts.

Solution. Bolts in double shear and bearing on $\frac{1}{2}$ in.:

$$\text{Design shear strength of 1 bolt} = (0.60)(2 \times 0.6)(27.0) = 19.44 \text{ k} \leftarrow$$

$$\text{Design bearing strength of 1 bolt} = (0.75)(2.4)(\tfrac{7}{8})(\tfrac{1}{2})(58) = 45.7 \text{ k}$$

$$\text{No. of bolts required} = \frac{170}{19.44} = 8.74$$

$$\underline{\text{Use 9 or 10 } \tfrac{7}{8}\text{-in. A307 bolts}}$$

PROBLEMS

For each of the problems listed the following information is to be used unless otherwise indicated: (a) A36 steel; (b) standard-size holes; (c) edge distances and center-to-center distances of bolts $\geq 1\frac{1}{2} \, d$ and $3 \, d$ respectively; (d) threads of bolts excluded from shear plane.

13-1 to 13-7. Determine the resultant load on the most stressed bolt in the eccentrically loaded connections shown using the elastic method.

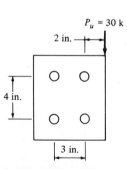

Problem 13-1   (*Ans.* 16.16 k)

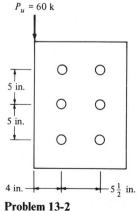

Problem 13-2

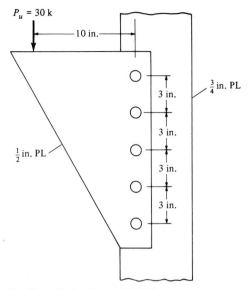

Problem 13-3 (*Ans.* 20.88 k)

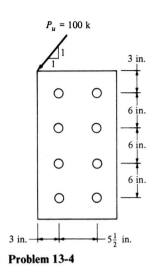

Problem 13-4

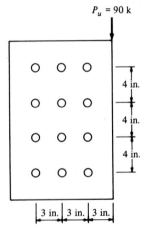

Problem 13-5 (*Ans.* 16.39 k)

Problem 13-6

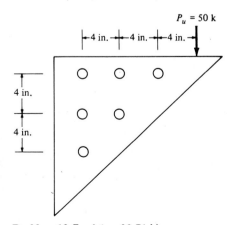

Problem 13-7 (*Ans.* 33.71 k)

13-8. Repeat Prob. 13-2 using the reduced eccentricity method.

13-9. Using the elastic method determine the design strength of the bearing-type connection shown. The bolts are $\frac{7}{8}$-in. A325 and are in single shear and bearing on $\frac{1}{2}$ in. (*Ans.* 93.2 k)

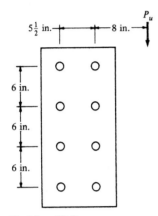

Problem 13-9

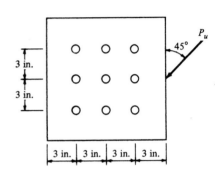

Problem 13-10

13-10. Using the elastic method determine the design strength P_u for the slip-critical connection shown. The 3/4-in. A325 bolts are in "double shear."

13-11. Repeat Prob. 13-9 using the ultimate strength tables entitled "Eccentric Loads on Fastener Groups" in Part 5 of the LRFD Manual. (*Ans.* 112 k)

13-12. Repeat Prob. 13-10 using the ultimate strength tables entited "Eccentric Loads on Fastener Groups" in Part 5 of the LRFD Manual.

13-13. Is the bearing-type connection shown in the accompanying illustration sufficient to resist the 130-k load which passes through the center of gravity of the bolt group? (*Ans,* OK, $F_t = 54.10$ ksi $> f_t$ of 29.43 ksi)

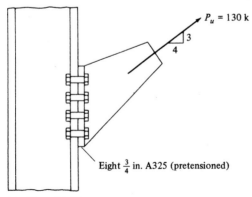

Problem 13-13

13-14. Repeat Prob. 13-13 if snug-tight bolts are used and if $P_D = 55$ k and $P_L = 40$ k.

13-15. If the load shown in the accompanying bearing-type illustration passes through the center of gravity of the bolt group, how large can it be? (*Ans.* 419.3 k)

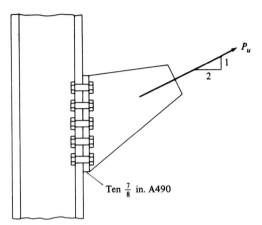

Ten $\frac{7}{8}$ in. A490

Problem 13-15

13-16. Repeat Prob. 13-15 if bolts are A325.

13-17. Determine the number of $\frac{3}{4}$-in. A325 bolts required in the angles and in the flange of the W shape shown in the accompanying illustration if a bearing-type (snug-tight) connection is used. (*Ans.* 4 in angles and 8 in W flange)

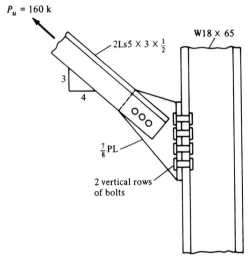

$P_u = 160$ k

2Ls5 × 3 × $\frac{1}{2}$

W18 × 65

$\frac{7}{8}$ PL

2 vertical rows of bolts

Problem 13-17

13-18. Are the bolts shown in this 16 in. long hanger satisfactory to resist direct tension and prying? There are eight $\frac{7}{8}$-in. A325 bolts and they are spaced 4 in. on center longitudinally.

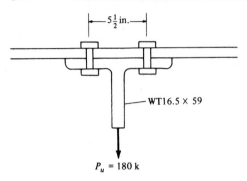

Problem 13-18

13-19. Determine the design strength P_u of the connection shown if $\frac{7}{8}$-in. A502 grade 1 rivets are used. (*Ans.* 168.8 k)

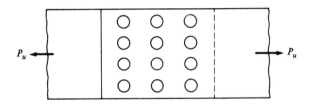

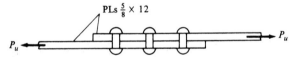

Problem 13-19

13-20. The truss tension member shown consists of a single-angle $5 \times 3 \times \frac{5}{16}$ angle and is connected to a $\frac{1}{2}$-in. gusset plate with five $\frac{7}{8}$-in. A502, grade 1 rivets. Determine P_u.

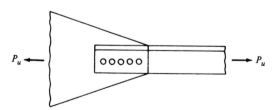

Problem 13-20

13-21. How many $\frac{3}{4}$-in. A502, grade 1 rivets are needed to carry the load shown in the accompanying illustration? (*Ans.* 12.62, say 14 or 15)

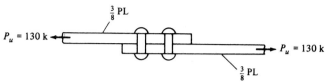

Problem 13-21

13-22. Repeat Prob. 13-21 if A307 bolts are used.

13-23. How many A502, grade 1 rivets with 1-in. diameters need to be used for the butt joint shown? (*Ans.* 4.35, say 5 or 6)

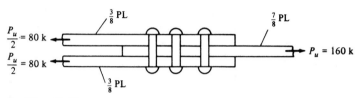

Problem 13-23

13-24. How many $\frac{7}{8}$-in. A307 bolts are required for the connection shown in the accompanying illustrations? Factored loads are shown.

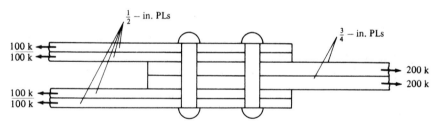

Problem 13-24

13-25. For the connection shown in the accompanying illustration $P_u = 650$ k, determine the number of $\frac{7}{8}$-in. A502, grade 2 rivets required. (*Ans.* 17.33, say 18)

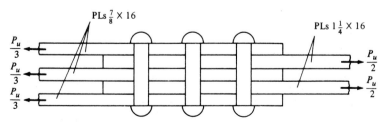

Problem 13-25

13-26. Determine the design strength P_u for the connection shown if 1-in. A502, grade 2 rivets are used.

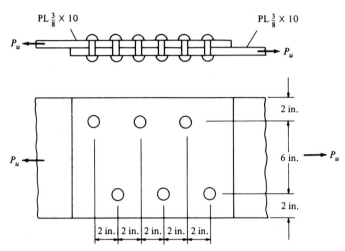

Problem 13-26

13-27. For the beam shown in the accompanying illustration what is the required spacing of $\frac{3}{4}$-in. A 307 bolts if $V_u = 160$ k? (*Ans.* 3.30 in., say $3\frac{1}{4}$ in.)

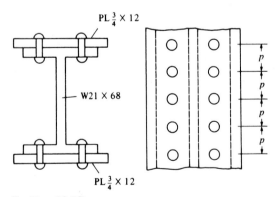

Problem 13-27

13-28. Is the connection shown sufficient to resist the 90-k load which passes through the center of gravity of the rivet group?

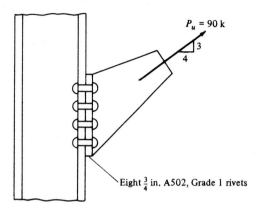

$P_u = 90$ k

3
4

Eight $\frac{3}{4}$ in. A502, Grade 1 rivets

Problem 13-28

13-29. If the load shown in the accompanying illustration passes through the center of gravity of the rivet group, how large can it be? (*Ans.* 179 k)

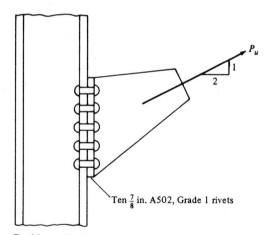

P_u

1
2

Ten $\frac{7}{8}$ in. A502, Grade 1 rivets

Problem 13-29

Chapter 14

Welded Connections

14-1 GENERAL

Welding is a process in which metallic parts are connected by heating their surfaces to a plastic or fluid state and allowing the parts to flow together and join (with or without the addition of other molten metal). It is impossible to determine when welding originated but it was several thousand years ago. Metalworking, including welding, was quite an art in ancient Greece at least three thousand years ago, but welding had undoubtedly been performed for many centuries before those days. Ancient welding was probably a forging process in which the metals were heated to a certain temperature (not to the melting stage) and hammered together.

Although modern welding has been available for a good many years, it has only come into its own in the last few decades for the building and bridge phases of structural engineering. The adoption of structural welding was quite slow for several decades because many engineers thought that welding had two great disadvantages—(1) that welds had reduced fatigue strength as compared with riveted and bolted connections and (2) that it was impossible to ensure a high quality of welding without unreasonably extensive and costly inspection.

These negative feelings persisted for many years, although tests seemed to indicate that neither reason was valid. Regardless of the validity of these fears they were widely held and undoubtedly slowed down the use of welding, particularly for highway bridges and to an even greater extent for railroad bridges. Today most engineers agree that welded joints have considerable fatigue strength. They will also admit that the rules governing the qualification of welders, the better techniques applied, and the excellent workmanship requirements of the AWS (American Welding Society) specifications make the inspection of

Fabrication of plate girders for Connecticut expressway bridge. (Courtesy of the Lincoln Electric Company.)

welding a much less difficult problem. Consequently welding is today permitted for almost all structural work other than for some bridges.

On the subject of fear of welding it is interesting to consider welded ships. Ships are subjected to severe impactive loadings which are difficult to predict and yet naval architects use all welded ships with great success. A similar discussion can be made for airplanes and aeronautical engineers. The slowest adoption of structural welding was for railroad bridges. These bridges are undoubtedly subjected to heavier live loads than highway bridges, larger vibrations, and more stress reversals; but are their stress situations as serious and as difficult to predict as those for ships and planes?

14-2 ADVANTAGES OF WELDING

Today it is possible to make use of the many advantages which welding offers, since the fatigue and inspection fears have been largely eliminated. Several of the many welding advantages are discussed in the following paragraphs.

1. To most persons the first advantage is in the area of economy, because the use of welding permits large savings in pounds of steel used. Welded structures allow the elimination of a large percentage of the gusset and splice plates necessary for riveted or bolted structures as well as the elimination of rivet or bolt heads. In some bridge trusses it may be possible to save up to 15 percent or more of the steel weight by using welding. Welding also requires appreciably less labor than does riveting because one welder can replace the standard four-person riveting crew.

2. Welding has a much wider range of application than riveting or bolting. Consider a steel pipe column and the difficulties of connecting it to other steel members by bolting. A bolted connection may be virtually impossible but a welded connection will present no difficulties whatsoever. The student can visualize many other similar situations where welding has a decided advantage.

3. Welded structures are more rigid because the members are often welded directly to each other. The connections for bolted structures are often made through connection angles or plates which deform due to load transfer, making the entire structure more flexible. On the other hand, greater rigidity can be a disadvantage where simple end connections with little moment resistance are desired. For such cases designers must be careful as to the type of joint they specify.

4. The process of fusing pieces together gives the most truly continuous structures. It results in one-piece construction and because welded joints are as strong as or stronger than the base metal no restrictions have to be placed on the joints. This continuity advantage has permitted the erection of countless slender and graceful steel statically indeterminate frames throughout the world. Some of the more outspoken proponents of welding have referred to riveted and bolted structures, with their heavy plates and large number of rivets or bolts, as looking like tanks or armored cars when compared with the clean, smooth lines of welded structures. For a graphic illustration of this advantage the student should compare the moment resisting connections of Fig. 15-4.

5. It is easier to make changes in design and to correct errors during erection (and at less expense) if welding is used. A closely related advantage has certainly been illustrated in military engagements during the past few wars by the quick welding repairs made to military equipment under battle conditions.

6. Another item that is often important is the relative silence of welding. Imagine the importance of this fact when working near hospitals or schools or when making additions to existing buildings. Anyone with close to normal hearing who has attempted to work in an office within several hundred feet of a bolted or riveted job will go along with this advantage.

7. Fewer pieces are used and as a result time is saved in detailing, fabrication, and field erection.

14-3 TYPES OF WELDING

Although both gas and arc welding are available almost all structural welding is arc welding. Sir Humphry Davy discovered in 1801 how to create an electric arc by bringing close together two terminals of an electric circuit of relatively high voltage. Although he is generally given credit for the development of modern welding, a good many years elapsed after his discovery before welding was actually performed with the electric arc. (His work was of the greatest importance to the modern structural world but it is interesting to know that many people say his greatest discovery was not the electric arc but rather a laboratory assistant whose name was Michael Faraday.) Several Europeans formed welds of one type or another in the 1880s with the electric arc, while in the United States the first patent for arc welding was given to Charles Coffin of Detroit in 1889.[1]

The figures to follow in this chapter show the necessity of supplying additional metal to the joints being welded to give satisfactory connections. In electric-arc welding the metallic rod, which is used as the electrode, melts off into the joint as it is being made. When gas welding is used it is necessary to introduce a metal rod known as a *filler* or *welding rod*.

In gas welding a mixture of oxygen and some suitable type of gas is burned at the tip of a torch or blowpipe held in the welder's hand or by an automatic machine. The gas used in structural welding is probably acetylene and the process is called oxyacteylene welding. The flame produced can be used for flame cutting of metals as well as for welding. Gas welding is rather easy to learn and the equipment used is rather inexpensive. It is, however, a rather slow process as compared with other means of welding, and is normally used for repair and maintenance work and not for the fabrication and erection of large steel structures.

In arc welding an electric arc is formed between the pieces being welded and an electrode held in the operator's hand with some type of holder or by an automatic machine. The arc is a continuous spark which upon contact brings the electrode and the pieces being welded to the melting point. The resistance of the air or gas between the electrode and the pieces being welded changes the electrical energy into heat. A temperature of somewhere between 6000 and 10,000°F is produced in the arc. As the end of the electrode melts small droplets or globules of the molten metal are formed and are actually forced by the arc across to the pieces being connected penetrating the molten metal to become a part of the weld. The amount of penetration can actually be controlled by the amount of current consumed. Since the molten droplets of the electrodes are actually propelled to the weld, arc welding can be successfully used for overhead work.

A pool of molten steel can hold a fairly large amount of gases in solution and if not protected from the surrounding air will chemically combine with oxygen and nitrogen. After cooling the welds will be relatively porous due to the little pockets formed by the gases. Such welds are relatively brittle and have much less resistance to corrosion. A weld can be shielded by using an electrode coated with certain mineral compounds. The electric arc causes the coating to melt and creates an inert gas or vapor around the area being welded. The vapor acts as a shield around the molten metal and keeps it from coming freely in contact with the surrounding air. It also deposits a slag in the molten metal, which has less density than the base metal and comes to the surface to protect the weld from the air while the weld cools. After cooling, the slag can easily be removed by peening and wire brushing (such removal being absolutely necessary before painting or application of another weld layer). A picture showing the elements of the shielded arc welding process is shown in Fig. 14-1. This figure is taken from the *Procedure Handbook of Arc Welding Design and Practice* published by the Lincoln Electric Company. Shielded metal arc welding is frequently abbreviated here with the letters SMAW.

[1]Lincoln Electric Company, *Procedure Handbook of Arc Welding Design and Practice,* 11th ed., 1957, Part I.

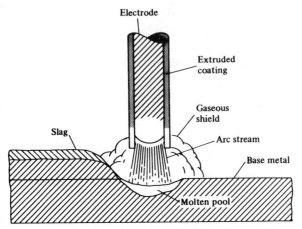

Figure 14-1   Elements of the shielded metal arc welding process (SMAW).

The type of welding electrode used is very important as it decidedly affects the weld properties such as strength, ductility, and corrosion resistance. Quite a number of different types of electrodes are manufactured, the type to be used for a certain job being dependent upon the type of metal being welded, the amount of material which needs to be added, the position of the work, etc. The electrodes fall into two general classes—the *lightly coated electrodes* and the *heavily coated electrodes*.

The heavily coated electrodes are normally used in structural welding because the melting of their coatings produces very satisfactory vapor shields around the work as well as slag in the weld. The resulting welds are stronger, more resistant to corrosion, and more ductile than are those produced with lightly coated electrodes. When the lightly coated electrodes are used, no attempt is made to prevent oxidation and no slag is formed. The electrodes are lightly coated with some arc stabilizing chemical such as lime.

Another type of welding is the submerged (or hidden) arc welding. In this method (frequently labeled SAW here) the arc is covered with a mound of granular fusible material and thus hidden from view. A bare metal electrode is fed from a reel and melted and deposited as filler material. SAW welds are quickly and efficiently made and are of high quality exhibiting high impact strength and corrosion resistance and good ductility. Furthermore they provide deeper penetration with the result that the area effective in resisting loads is larger. A large percentage of the welding done for bridge structures is SAW. If a single electrode is used, the size of the weld obtained with a single pass is limited. Multiple electrodes may be used, however, with no limit on weld size. Positions for SAW welds should be flat or horizontal.

14-4   WELDING INSPECTION

Three steps must be taken to ensure good welding for a particular job. These are: (1) establishing good welding procedures, (2) use of prequalified welders, and (3) employment of competent inspectors in shop and field.

Lincoln ML-3 Squirtwelder mounted on a self-propelled trackless trailer deposits this ¼-in. web-to-flange weld at 28 in./min. (Courtesy of the Lincoln Electric Company.)

When the procedures established by the AWS and AISC for good welding are followed and when welders are used who have previously been required to prove their ability, good results will probably be obtained. However, to make absolutely sure, well-qualified inspectors are needed.

Good welding procedure involves the selection of proper electrodes, current, and voltage; the properties of base metal and filler; and the position of welding, to name only a few factors. The usual practice for large jobs is to employ welders who have certificates showing their qualifications. In addition, it is not a bad practice to have each person make an identifying mark on each weld so that those persons frequently doing poor work can be located. This practice probably improves the general quality of work performed.

Visual Inspection

Another factor that will cause welders to perform better work is just the presence of an inspector who they feel knows good welding when he sees it. A good inspector should have done welding and spent much time observing the work of good welders. From this experience he or she should be able to know if a welder is obtaining satisfactory fusion and penetration. He or she should be able to recognize goods welds as to their shape, size, and general appearance. For instance, the metal in a good weld should approximate its original color after it has cooled. If it has been overheated it may have a rusty and reddish-looking color. An inspector can use various scales and gages to check the sizes and shapes of welds.

Visual inspection by a good person will probably give a good indication of the quality of welds but is not a perfect source of information as to the subsurface condition of the weld. There are several methods for determining the internal soundness of a weld. These include the use of penetrating dyes and magnetic particles, ultrasonic testing, and radiographic procedures. These methods can be used to detect internal defects such as porosity, weld penetration, and presence of slag.

Liquid Penetrants

Various types of dyes can be spread over weld surfaces and they will penetrate into surface cracks of the weld. After the dye has penetrated into the crack the excess surface material is wiped off and a powdery developer is used to draw the dye out of the cracks. The outlines of the cracks can then be seen with the eye. Several variations of this method are used to improve the visibility of the defects such as the use of fluorescent dyes. After the dye is drawn from the cracks the cracks are made to stand out brightly by examination under black light.[2]

Magnetic Particles

In this method the weld being inspected is magnetized electrically. Cracks that are at or near the surface of the weld cause north and south poles to form on each side of the cracks. Dry iron powdered filings or a liquid suspension of particles are placed on the weld. The patterns of these particles formed when many of them cling to the cracks show the locations of cracks and indicate their size and shape. A disadvantage of this method is that if multilayer welds are used the method has to be applied to each layer.

Ultrasonic Testing

In recent years the steel industry has applied ultrasonics to the manufacture of steel. Although the equipment is expensive the method is quite useful in welding inspection as well. Sound waves are sent through the material being tested and are reflected from the opposite side of the material. These reflections are shown on a cathode ray tube. Defects in the weld will affect the time of the sound transmission. The operator can read the picture on the tube and then locate flaws and learn how severe they are.

Radiographic Procedures

The more expensive radiographic methods can be used to check occasional welds in important structures. From these tests it is possible to make good estimates of

[2]James Hughes, "It's Superinspector," *Steelways* 25, no. 4 (New York: American Iron and Steel Institute, September/October, 1969), pp. 19–21.

the percentage of bad welds in a structure. The use of portable x-ray machines where access is not a problem and the use of radium or radioactive cobalt for making pictures are excellent but expensive methods of testing welds. These methods are satisfactory for butt welds (such as for the welding of important stainless steel piping at chemical and nuclear projects) but are not satisfactory for fillet welds because the pictures are difficult to interpret. A further disadvantage of these methods is the radioactive danger. Careful procedures have to be used to protect the technicians as well as nearby workers. On the average construction job this danger probably requires night inspection of welds when only a few workers are near the inspection area. (Normally a very large job would be required before the use of the extremely expensive radioactive materials could be justified.)

A properly welded connection can always be made much stronger (perhaps $1\frac{1}{2}$ or 2 times) than the plates being connected. As a result, the actual strength is much higher than is required by the specifications. The reasons for this extra strength are as follows: the electrode wire is made from premium steel, the metal is melted electrically (as is done in the manufacture of high-quality steels), and the cooling rate is quite rapid. As a result of these facts it is probably a rare occasion for a welder to make a weld of less strength than required by the design.

14-5 CLASSIFICATION OF WELDS

There are three separate classifications of welds described in the following paragraphs. These classifications are based upon the types of welds made, positions of welds, and types of joints.

Type of Weld

The two main types of welds are the *fillet welds* and the *groove welds*. In addition there are plug and slot welds which are not as common in structural work. These four types of welds are shown in Fig. 14-2.

The fillet welds will be shown to be weaker than groove welds; however, most structural connections (about 80 percent) are made with fillet welds. Any person who has had experience in steel structures will understand why fillet welds

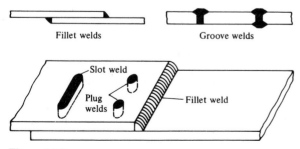

Fillet welds   Groove welds

Slot weld

Plug welds   Fillet weld

Figure 14-2

are more common than are groove welds. Groove welds (which are welds made in grooves between the members to be joined) are used when the members to be connected are lined up in the same plane. To use them in every situation would mean that the members would have to fit almost perfectly and unfortunately the average steel structure does not fit together in that manner. Many students have seen steel workers pulling and ramming steel members to get them in position. When members are allowed to lap over each other larger tolerances are allowable in erection and fillet welds are the welds used. Nevertheless groove welds are quite common for many connections such as column splices, butting of beam flanges to columns, etc., and they comprise about 15 percent of structural welding.

A plug weld is a circular weld passing through one member to another and joining the two together. A slot weld is a weld formed in a slot or elongated hole which joins one member to the other member through the slot. The slot may be partly or fully filled with weld material. These two expensive types of welds may occasionally be used when members lap over each other and the desired length of fillet welds cannot be obtained. They may also be used to stitch together parts of a member as the fastening of cover plates to a built-up member.

A plug or slot weld is not generally considered to be suitable for transferring tensile forces perpendicular to the faying surface. The reason is that there is not usually much penetration of the weld into the member behind the plug or slot—and yet resistance to tension is provided primarily by penetration.

Structural designers accept plug and slot welds as being satisfactory for stitching the different parts of a member together but many designers are not happy to use these welds for the transmission of shear forces. The penetration of the welds from the slots or plugs into the other members is questionable and in addition there can be critical voids down in the welds which cannot be detected with the usual inspection procedures.

Position

Welds are referred to as being flat, horizontal, vertical, and overhead. They are listed in the preceding sentence in order of their economy with the flat welds being the most economical and the overhead welds being the most expensive. A fairly good welder can do a very satisfactory job with a flat weld but it takes the very best welder to do a good job with an overhead weld. Although the flat welds can often be made with an automatic machine, much structural welding is done by hand. It has previously been indicated that the assistance of gravity is not necessary for the forming of good welds but it does speed up the process. The globules of the molten electrodes can be forced into the overhead welds against gravity and good welds will result but they are slow and expensive to make so it is desirable to avoid them whenever possible. These types of welds are shown in Fig. 14-3.

Type of Joint

Welds can be further classified according to the type of joint used as being: butt, lap, tee, edge, corner, etc. These joint types are shown in Fig. 14-4.

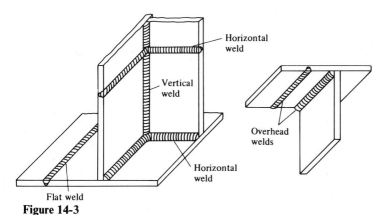

Flat weld
Figure 14-3

14-6 WELDING SYMBOLS

Figure 14-5 presents the method of making welding symbols, developed by the American Welding Society. With this excellent shorthand system a great deal of information can be presented in a small space on engineering plans and drawings with a few lines and numbers. These symbols remove the necessity of drawing in the welds and making long descriptive notes. It is certainly desirable for steel designers and draftsmen to use this standard system. Should most of the welds on a drawing be of the same size, a note to that effect can be given and the symbols omitted, except for the off-size welds.

The purpose of this section is not to show every possible type of symbol but rather to give a general idea of the appearance of welding symbols and the information which they can show. The reader can refer to the detailed information published by the AWS and reprinted in many handbooks (including the LRFD Manual). At first glance the information presented in Fig. 14-5 is probably quite confusing. For this reason a few very common symbols for fillet welds are presented in Fig. 14-6, together with an explanation of each.

14-7 GROOVE WELDS

When complete penetration groove welds are subjected to axial tension or axial compression the weld stress is assumed to equal the load divided by the net area

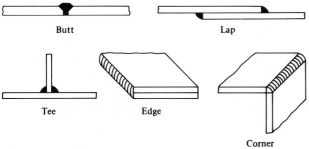

Figure 14-4

WELDED JOINTS
Standard symbols

BASIC WELD SYMBOLS

BACK	FILLET	PLUG OR SLOT	Groove or Butt						
			SQUARE	V	BEVEL	U	J	FLARE V	FLARE BEVEL

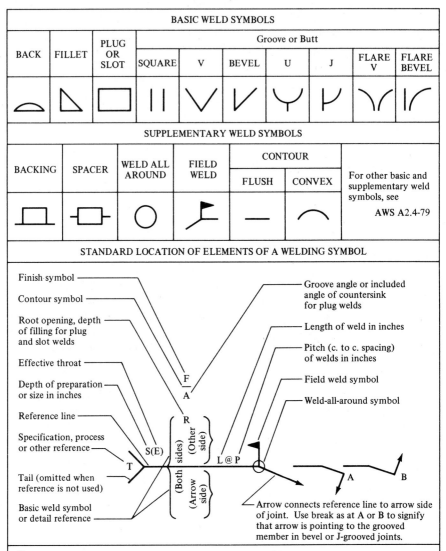

SUPPLEMENTARY WELD SYMBOLS

BACKING	SPACER	WELD ALL AROUND	FIELD WELD	CONTOUR		For other basic and supplementary weld symbols, see AWS A2.4-79
				FLUSH	CONVEX	

STANDARD LOCATION OF ELEMENTS OF A WELDING SYMBOL

Finish symbol

Contour symbol

Root opening, depth of filling for plug and slot welds

Effective throat

Depth of preparation or size in inches

Reference line

Specification, process or other reference

Tail (omitted when reference is not used)

Basic weld symbol or detail reference

Groove angle or included angle of countersink for plug welds

Length of weld in inches

Pitch (c. to c. spacing) of welds in inches

Field weld symbol

Weld-all-around symbol

R (Other side)

S(E) (Both sides) (Arrow side)

T

L @ P

F / A

A   B

Arrow connects reference line to arrow side of joint. Use break as at A or B to signify that arrow is pointing to the grooved member in bevel or J-grooved joints.

Note:

Size, weld symbol, length of weld and spacing must read in that order from left to right along the reference line. Neither orientation of reference line nor location of the arrow alters this rule.

The perpendicular leg of △, ⌐, ⌐, ⌐ weld symbols must be at left.

Arrow and Other Side welds are of the same size unless otherwise shown. Dimensions of fillet welds must be shown on both the Arrow Side and the Other Side Symbol.

The point of the field weld symbol must point toward the tail.

Symbols apply between abrupt changes in direction of welding unless governed by the "all around" symbol or otherwise dimensioned.

These symbols do not explicitly provide for the case that frequently occurs in structural work, where duplicate material (such as stiffeners) occurs on the far side of a web or gusset plate. The fabricating industry has adopted this convention: that when the billing of the detail material discloses the existence of a member on the far side as well as on the near side, the welding shown for the near side shall be duplicated on the far side.

Figure 14-5

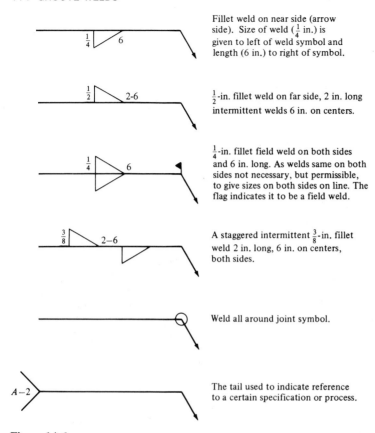

Fillet weld on near side (arrow side). Size of weld ($\frac{1}{4}$ in.) is given to left of weld symbol and length (6 in.) to right of symbol.

$\frac{1}{2}$-in. fillet weld on far side, 2 in. long intermittent welds 6 in. on centers.

$\frac{1}{4}$-in. fillet field weld on both sides and 6 in. long. As welds same on both sides not necessary, but permissible, to give sizes on both sides on line. The flag indicates it to be a field weld.

A staggered intermittent $\frac{3}{8}$-in. fillet weld 2 in. long, 6 in. on centers, both sides.

Weld all around joint symbol.

The tail used to indicate reference to a certain specification or process.

Figure 14-6

of the weld. Three types of groove welds are shown in Fig. 14-7. The square groove joint, shown in part (a) of the figure, is used to connect relatively thin material up to roughly $\frac{5}{16}$ in. (8 mm) thickness. As the material becomes thicker it is necessary to use the single-vee groove welds and the double-vee groove welds illustrated in parts (b) and (c) respectively of Fig. 14-7. For these two welds the members are bevelled before welding to permit full penetration of the weld.

The groove welds shown in Fig. 14-7 are said to have *reinforcement*. Rein-

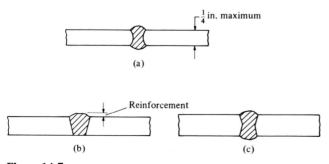

Figure 14-7

forcement is added weld metal that causes the throat dimension to be greater than the thickness of the welded material. Because of reinforcement groove welds may be referred to as 100, 125, 150 percent, etc., welds according to the amount of extra thickness at the weld. There are two major reasons for having reinforcement. These are: (1) reinforcement gives a little extra strength because the extra metal takes care of pits and other irregularities and (2) the welder can easily make the weld a little thicker than the welded material. It would be a difficult if not impossible task to make a perfectly smooth weld with no places that were thinner or thicker than the material welded.

Reinforcement undoubtedly makes groove welds stronger and better when they are to be subjected to static loads. When the connection is to be subjected to repeated and vibrating loads, however, reinforcement is not as satisfactory because stress concentrations seem to develop in the reinforcement and contribute to earlier failure. For such cases as these a common practice is to provide reinforcement and grind it off flush with the material being connected.

In Fig. 14-8 some of the edge preparations that may be necessary for groove welds are shown. In part (a) a bevel with a feathered edge is shown. When feathered edges are used there is a problem with burn-through. This may be lessened if a *land* is used such as shown in part (b) of the figure or a backup strip as shown in part (c). The backup strip is often a $\frac{1}{4}$-in. copper plate. Weld metal does not stick to copper and copper also has a very high conductivity which is useful in carrying away excess heat and reducing distortion. Sometimes steel backup strips are used but they will become a part of the weld and are left in place. A land should not be used together with a backup strip because there is a high possibility that a gas pocket might be formed, preventing full penetration. When double bevels are used as shown in part (d) of the figure, spacers are sometimes provided to prevent burn-through. The spacers are removed after one side is welded.

From the standpoints of strength, resistance to impact and stress repetition, and amount of filler metal required, groove welds are much to be preferred to fillet welds. From other standpoints, however, they are not so attractive and the vast majority of structural welding is fillet welding. Groove welds have higher residual stresses, and the preparations (such as the scarfing and veeing) of the edges of members for groove welds are expensive, but the major disadvantages

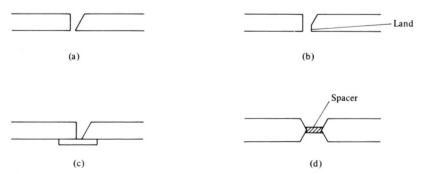

Figure 14-8 Edge preparation for groove welds. (a) Bevel with feathered edge. (b) Bevel with a land. (c) Bevel with a backup plate. (d) Double bevel with a spacer.

probably lie in the problems involved in getting the pieces to fit together in the field. The advantages of fillet welds in this respect were described in Section 14-5. For these reasons field butt joints are not used very often, except on small jobs and where members may be fabricated a little long and cut in the field to the lengths necessary for precise fitting.

Sometimes connections are designed where the groove welds do not extend for the full thicknesses of the parts being connected. These welds are referred to as *partial-penetration groove welds.* Special design requirements are given in specifications for these welds.

14-8 FILLET WELDS

Tests have shown that fillet welds are stronger in tension and compression than they are in shear, so the controlling fillet weld stresses given by the various specifications are shearing stresses. When practical, it is desirable to try to arrange welded connections so they will be subjected to shearing stresses only and not to a combination of shear and tension or shear and compression.

Fillet welds when tested to failure seem to fail by shear at angles of about 45° through the throat. Their strength is therefore assumed to equal the allowable shearing stress times the theoretical throat area of the weld. The theoretical throats of several fillet welds are shown in Fig. 14-9. The throat area equals the theoretical throat distance times the length of the weld. In this figure the root of the weld is the point where the faces of the original metal pieces intersect, and the theoretical throat of the weld is the shortest distance from the root of the weld to its diagrammatic face.

For the 45° or equal leg fillet the throat dimension is 0.707 times the leg of the weld, but it has a different value for fillet welds with unequal legs. The desirable fillet weld has a flat or slightly convex surface, although the convexity of the weld does not add to its calculated strength. At first glance the concave surface would appear to give the ideal fillet weld shape because stresses could apparently

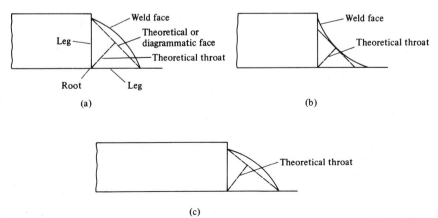

Figure 14-9   (a) Convex surface. (b) Concave surface. (c) Unequal leg fillet weld.

flow smoothly and evenly around the corner with little stress concentration. Years of experience, however, have shown that single-pass fillet welds of a concave shape have a greater tendency to crack upon cooling and this factor has proved to be of greater importance than the smoother stress distribution of convex types.

When a concave weld shrinks the surface is placed in tension tending to cause cracks, whereas when the surface of a convex weld shrinks it does not place the outer surface in tension. Rather as the face shortens it is placed in compression.

Another item of importance pertaining to the shape of fillet welds is the angle of the weld with the pieces being welded. The desirable value of this angle is in the vicinity of 45°. For 45° fillet welds the leg sizes are equal and such welds are referred to by the leg sizes (as a $\frac{1}{4}$-in. fillet weld). Should the leg sizes be different (not a 45° weld) both leg sizes are given in describing the weld as a $\frac{3}{8}$ by $\frac{1}{2}$-in. fillet weld).

The automatic submerged arc welding method (SAW) provides a deeper penetration than does the usual shielded arc welding process. As a result the LRFD permits the designer to use a larger throat area for welds made by this process. In Section J2.2a the LRFD Specification states that the effective throat thickness for SAW fillet welds with legs $\frac{3}{8}$ in. or less may equal the leg sizes. For larger leg sizes the effective throat thicknesses are considered to equal the theoretical throat thicknesses plus 0.11 in.

14-9 STRENGTH OF WELDS

For this discussion reference is made to Fig. 14-10. As previously indicated the stress in a weld is considered to equal the load P divided by the effective throat area of the weld. This method of determining the strength of fillet welds is used regardless of the direction of load. Tests have shown that transverse fillets are without question about a third stronger than are longitudinal fillets, but this fact is not recognized by most specifications in order to simplify design calculations. One reason why transverse fillet welds are stronger is because they are more uniformly stressed for their entire length, while the longitudinal fillets are stressed unevenly due to varying deformations along the length of the weld. Another reason for their greater strength is given by tests which show failure occurs at an angle other than 45°, giving them a larger effective throat area.

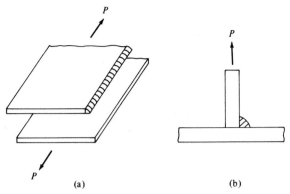

Figure 14-10   (a) Longitudinal fillet weld. (b) Transverse fillet weld.

14-10 LRFD REQUIREMENTS

When welds are made the electrode material should have properties of the base metal. If the properties are comparable the weld metal is referred to as the "matching" base metal. Table 4.1.1 of AWS D1.1 gives details concerning "matching" weld metals.

Table 14-1 (which is Table J2.3 of the LRFD Specification) provides nominal strengths for various types of welds including fillet welds, plug and slot welds, and complete-penetration and partial-penetration groove welds.

The design strength of a particular weld is taken as the lower value of ϕF_W (where F_W is the nominal strength of the weld) and ϕF_{BM} (where F_{BM} is the nominal strength of the base material).

For fillet welds the nominal strength for stress on the effective area of the weld is $0.60F_{EXX}$ (where F_{EXX} is the classification strength of the weld metal) and ϕ is 0.75. If we have tension or compression parallel to the axis of the weld the nominal strength of the base metal is F_y and ϕ is 0.90. The design shear strength of members being connected is taken as $\phi F_n A_{ns}$ where ϕ is 0.75, F_n is $0.6F_u$, and A_{ns} is the net area subject to shear.

The filler metal electrodes for shielded arc welding are listed as E60XX, E70XX, etc. In this classification the letter E represents an electrode while the first set of digits (as 60 or 70) indicates the minimum tensile strength of the weld in kips per square inch. The remaining digits designate the position, current, and other necessary information which are to be employed with a certain electrode.

In addition to the nominal stresses given in Table 14-1 there are several other LRFD provisions applying to welding. Among the more important are the following:

1. The minimum length of a fillet weld may not be less than four times the nominal leg size of the weld. Should its length actually be less than this value the weld size considered effective must be reduced to one-quarter of the weld length.

TABLE 14-1   DESIGN STRENGTH OF WELDS

Types of weld and stress[a]	Material	Resistance factor ϕ	Nominal strength F_{BM} or F_w	Required weld strength level[b,c]
Complete-penetration groove weld				
Tension normal to effective area	Base	0.90	F_y	"Matching" weld must be used.
Compression normal to effective area Tension or compression parallel to axis of weld	Base	0.90	F_y	Weld metal with a strength level equal to or less than "matching" may be used.
Shear on effective area	Base Weld electrode	0.90 0.80	$0.60\,F_y$ $0.60\,F_{EXX}$	
Partial-penetration groove welds				
Compression normal to effective area Tension or compression parallel to axis of weld[d]	Base	0.90	F_y	Weld metal with a strength level equal to or less than "matching" weld metal may be used.
Shear parallel to axis of weld	Base[e] Weld electrode	0.75	$0.60\,F_{EXX}$	
Tension normal to effective area	Base Weld electrode	0.90 0.80	F_y $0.60\,F_{EXX}$	
Fillet welds				
Stress on effective area	Base[e] Weld electrode	0.75	$0.60\,F_{EXX}$	Weld metal with a strength level equal to or less than "matching" weld metal may be used.
Tension or compression parallel to axis of weld[d]	Base	0.90	F_y	
Plug or slot welds				
Shear parallel to faying surfaces (on effective area)	Base[e] Weld electrode	0.75	$0.60\,F_{EXX}$	Weld metal with a strength level equal to or less than "matching" weld metal may be used.

[a]For definition of effective area, see Sec. J2 of the LRFD specification.

[b]For "matching" weld metal, see Table 4.1.1, AWS D1.1.

[c]Weld metal one strength level stronger than "matching" weld metal will be permitted.

[d]Fillet welds and partial-penetration groove welds joining component elements of built-up members, such as flange-to-web connections, may be designed without regard to the tensile or compressive stress in these elements parallel to the axis of the welds.

[e]The design of connected material is governed by Sec. J4 of the LRFD specification.

The all welded 56-story Toronto Dominion Bank Tower. (Courtesy of the Lincoln Electric Company.)

2. The maximum size of a fillet weld along edges of material less than $\frac{1}{4}$ in. thick equals the material thickness. For thicker material it may not be larger than the material thickness less $\frac{1}{16}$ in., unless the weld is specially built out to give a full throat thickness. For a plate with a thickness of $\frac{1}{4}$ in. or more it is desirable to keep the weld back at least $\frac{1}{16}$ in. from the edge so that the inspector can clearly see the edge of the plate and thus accurately determine the dimensions of the weld throat.

3. The minimum permissible size fillet welds of the LRFD Specification are given in Table 14-2 (Table J2.5 of the LRFD Specification). They vary from $\frac{1}{8}$ in. for $\frac{1}{4}$ in. or thinner material up to $\frac{5}{16}$ in. for material over $\frac{3}{4}$ in. in thickness.

TABLE 14-2   MINIMUM SIZE OF FILLET WELDS

Material thickness of thicker part joined (in.)	Minimum size of fillet weld[a] (in.)
To $\frac{1}{4}$ inclusive	$\frac{1}{8}$
Over $\frac{1}{4}$ to $\frac{1}{2}$	$\frac{3}{16}$
Over $\frac{1}{2}$ to $\frac{3}{4}$	$\frac{1}{4}$
Over $\frac{3}{4}$	$\frac{5}{16}$

[a]Leg dimension of fillet welds.

The smallest practical weld size is about $\frac{1}{8}$ in. and the most economical size is probably about $\frac{1}{4}$ or $\frac{5}{16}$ in. The $\frac{5}{16}$-in. weld is about the largest size that can be made in one pass with the shielded arc welded process (SMAW) and $\frac{1}{2}$ in. with the submerged arc process (SAW).

These minimum sizes were not developed on the basis of strength consider-ations but rather upon the fact that thick materials have a quenching or rapid cooling effect on small welds. If this happens the result is often a loss in weld ductility. In addition the thicker material tends to restrain the weld material from shrinking as it cools with the result that weld cracking can be a problem.

4. When practical, end returns (sometimes called boxing) should be made for fillet welds as shown in Fig. 14-11. The length of returns should not be less than twice the nominal size of the weld. When they are not used it is considered good practice by some designers to subtract two times the weld size from the effective weld length. End returns are useful in reducing the high stress concen-trations which occur at the ends of welds, particularly for connections where there is considerable vibration and eccentricity of load. The LRFD Specification (Section J2.2a), however, says that the effective length of fillet welds includes the length of the end returns used.

5. When longitudinal fillet welds are used for the connection of plates or bars, their length may not be less than the perpendicular distance between them because of shear lag, discussed in Chapter 3. Furthermore, the distance between fillet welds may not be greater than 8 in. for end connections unless the member is designed on the basis of its effective area in accordance with LRFD specifica-tion B3.

6. For lap joints the minimum amount of lap permitted is equal to 5 times the thickness of the thinner part joined but may not be less than 1 in. The purpose

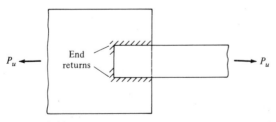

Figure 14-11

of this minimum lap is to keep the joint from rotating excessively [see Fig. 12.1(a)] when the connected parts are loaded.

14-11 DESIGN OF SIMPLE FILLET WELDS

Examples 14-1 through 14-3 illustrate the calculations necessary to determine the strength of various fillet welded connections while Example 14-4 presents the design of such a connection. In these and other problems weld lengths are selected no closer than to the nearest $\frac{1}{4}$ in., because closer work cannot be expected in shop or field.

EXAMPLE 14-1

Determine the design strength of a 1-in. length of $\frac{5}{16}$-in. fillet weld using (a) the shielded metal arc process (SMAW) and (b) the submerged arc process (SAW). Use E70 electrodes with a minimum tensile strength of 70 ksi = F_{EXX}.

Solution

(a) Shielded metal arc process

Effective throat thickness of weld = $(0.707)(\frac{5}{16}) = 0.221$ in.

$$\text{Design strength} = \phi F_w = (0.75)(0.60 \times 70)(0.221)(1.0)$$

$$= 6.96 \text{ k/in.}$$

(b) Submerged arc process

From LRFD Section J2.2a the effective throat thickness of weld = $\frac{5}{16}$ in.

$$\text{Design strength} = (0.75)(0.60 \times 70)(\tfrac{5}{16})(1.0)$$

$$= 9.84 \text{ k/in.}$$

Fillet welds may not be designed with a stress that is greater than the design stress on the adjacent members being connected. If the external force applied to the member (tensile or compressive) is parallel to the axis of the weld metal the design strength may not exceed the axial design strength of the member.

Examples 14-2 and 14-3 illustrate the calculations necessary to determine the design strength of plates connected with SMAW and SAW fillet welds. In each of these examples the shearing strength per inch of the welds controls and is multiplied by the total length of the welds to give the total capacity of the connections.

EXAMPLE 14-2

What is the design strength of the connection shown in Fig. 14-12 if A36 steel and E70 electrodes are used? The $\frac{7}{16}$-in. fillet welds shown were made by the SMAW process.

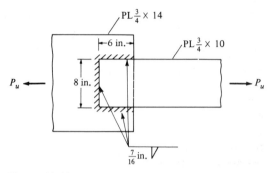

Figure 14-12

Solution

Effective throat thickness $= (0.707)(\frac{7}{16}) = 0.309$ in.

Capacity of weld/in. $= \phi F_w = (0.75)(0.60 \times 70)(0.309)(1.0)$

$= 9.73$ k/in. ←

Total capacity of weld $= (9.73)(20) = 194.65$ k ←

Design strength of plate $= \phi F_y A_g = (0.90)(36)(\frac{3}{4} \times 10)$

$= 243$ k

Design capacity $= 194.6$ k

EXAMPLE 14-3

Repeat Example 14-2 if SAW welds are used.

Solution

Effective throat thickness $= (0.707)(\frac{7}{16}) + 0.11 = 0.419$ in.

Capacity of weld/in. $= \phi F_w = (0.75)(0.60 \times 70)(0.419)(1.0)$

$= 13.20$ k/in.

Total capacity of weld $= (13.20)(20) = 264$ k

Design strength of plate $= \phi F_y A_g = (0.90)(36)(\frac{3}{4} \times 10)$

$= 243$ k ←

Design capacity $= 243$ k

EXAMPLE 14-4

Using A36 steel and E70 electrodes design SMAW fillet welds to resist a full-capacity load on the $\frac{3}{8} \times 6$ in. member shown in Fig. 14-13.

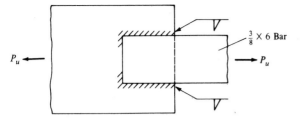

Figure 14-13

Solution

$P_u = \phi_t F_y A_g = (0.90)(36)(\frac{3}{8} \times 6) = 72.9$ k

Maximum weld size $= \frac{3}{8} - \frac{1}{16} = \frac{5}{16}$ in. (LRFD Section J2.2b)

Minimum weld size $= \frac{3}{16}$ in. (Table 14.2)

Use $\frac{5}{16}$-in. weld

Effective throat thickness of weld $= (0.707)(\frac{5}{16}) = 0.221$ in.

Capacity of weld per in. $= \phi F_w = (0.75)(0.60 \times 70)(0.221)(1.0) = 6.96$ k/in.

Length required $= \dfrac{72.9}{6.96} = 10.47$ in.

Use end returns not less than $2 \times \frac{5}{16} = \frac{10}{16}$ in. (Say 1 in.)

Leaving $10.47 - 2.0 = 8.47$ in. or $4\frac{1}{2}$ in. each side

However, must use $6 - 1 = 5$-in.-long welds each side as required by LRFD Specification J2.2b.

On some occasions the lengths available for the usual longitudinal fillet welds are not sufficient for the load to be resisted. For the situation shown in Fig. 14-14 it may be possible to develop sufficient strength by welding along the back of the channel at the edge of the plate if sufficient space is available. The dashed lines shown in this figure show this weld.

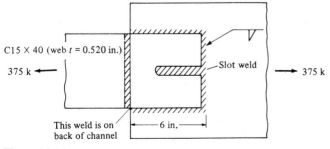

Figure 14-14

Another possibility is the use of slot welds as illustrated in Example 14-5. There are several LRFD requirements pertaining to slot welds which need to be mentioned here. This specification (J2.3) says that the width of a slot may not be less than the member thickness $+\frac{5}{16}$ in. (rounded off to the next greater odd $\frac{1}{16}$ in. since structural punches are made in odd 16th's diameters) nor may it be greater than $2\frac{1}{4}$ times the weld thickness. For members up to $\frac{5}{8}$ in. thickness the weld thickness must equal the plate thickness, and for members greater than $\frac{5}{8}$ in thickness the weld thickness may not be less than one-half the member thickness nor $\frac{5}{8}$ in. The maximum length permitted for slot welds is 10 times the weld thickness. The limitations given in specifications for the maximum sizes of plug or slot welds are caused by the detrimental shrinkage which occurs around these types of welds when they exceed certain sizes. Should holes or slots larger than those specified be used it is desirable to use fillet welds around the borders of the holes or slots rather than using a slot or plug weld. Slot and plug welds are normally used in conjunction with fillet welds in lap joints. Sometimes plug welds are used to fill in the holes temporarily used for erection bolts for beam and column connections. They may or may not be included in the calculated strength of these joints.

The strength of a plug or slot weld is equal to its design stress ϕF_W times its nominal area in the shearing plane. This area is equal to the area of contact at the base of the plug or slot. The length of a slot weld can then be determined from the following.

$$L = \frac{\text{load}}{(\text{width})(\text{design stress})}$$

Example 14-5 illustrates the design of the welds necessary to connect a channel to a plate. The calculations show that the ordinary side and end fillet welds do not provide sufficient strength in this case because of the limited space available. It is decided to use a slot weld to provide the remaining load resistance needed.

EXAMPLE 14-5

Design SMAW fillet welds to connect a C15 × 40 to the plate shown in Fig. 14-14. The load to be resisted is 375 k and E70 electrodes are to be used. As shown in the figure the channel may lap over the plate by only 6 in. due to space limitations. Space is not available for welding on back of channel.

Solution.   Because of limited space use

$$\text{Max weld size} = \text{web } t - \tfrac{1}{16} = \tfrac{1}{2} - \tfrac{1}{16} = \tfrac{7}{16} \text{ in.}$$

$$\text{Effective throat thickness} = (0.707)(\tfrac{7}{16}) = 0.309 \text{ in.}$$

$$\text{Capacity of weld/in.} = \phi F_w = (0.75)(0.60 \times 70)(0.309)$$

$$= 9.73 \text{ k/in.}$$

$$\text{Length required} = \frac{375}{9.73} = 38.54 \text{ in.} > 27 \text{ in. available}$$

Therefore use a slot weld

$$\text{Min width of slot} = 0.520 + \tfrac{5}{16} = \tfrac{13}{16} \text{ in.}$$

$$\text{Max width} = 2\tfrac{1}{4} \times \text{weld thickness}$$

$$= (2\tfrac{1}{4})(\text{web } t \text{ of channel}) = (2\tfrac{1}{4})(\tfrac{1}{2})$$

$$= 1\tfrac{1}{8} \text{ say } \tfrac{17}{16} \text{ (to odd } \tfrac{1}{16})$$

Use $\tfrac{15}{16}$ in.

$$\text{Capacity of } \tfrac{7}{16}\text{-in. fillet weld} = (9.73)(6 + 6 + 15 - \tfrac{15}{16}) = 253.6 \text{ k}$$

$$\text{Load to be resisted by slot weld} = 375 - 253.6 = 121.4 \text{ k}$$

$$\text{Length required for slot weld} = \frac{121.4}{(\tfrac{15}{16})(0.75)(0.60 \times 70)} = 4.11 \text{ in.}$$

Say $4\tfrac{1}{2}$ in.

$$\text{Max length permitted by LRFD} = (10)(\tfrac{1}{2})$$

$$= 5.00 \text{ in.} > 4\tfrac{1}{2} \text{ in.} \qquad \text{OK}$$

$$\text{Use } \tfrac{15}{16} \times 4\tfrac{1}{2} \text{ in. slot weld}$$

Alternate Solution. Should space have been available on the back of the channel next to the plate a $\tfrac{7}{16}$-in. fillet weld would carry $(15)(9.73) = 145.9$ k > 121.4 k   OK

14-12 DESIGN OF FILLET WELDS FOR TRUSS MEMBERS

Should the members of a welded truss consist of single angles, double angles, or similar shapes and be subjected to static axial loads only, the LRFD Specification (J1.6) permits the connections to be designed by the same procedures described in the preceding section. The designers can select the weld size, calculate the total length of the weld required, and place the welds around the member ends as they see fit. (It would not be logical, of course, to place the weld all on one side of a member such as for the angle of Fig. 14.15 because of the rotation possibility.) Example 14.6 illustrates the simple calculations involved in designing the welds for the ends of a truss member.

EXAMPLE 14-6

Using A36 steel and E70 electrodes design side and end fillet SMAW welds for the full capacity of a $6 \times 4 \times \tfrac{1}{2}$-in. angle tension member with the long leg connected. Assume static load.

Solution

Tensile capacity of angle $= \phi_t F_y A_g = (0.90)(36)(4.75) = 153.9$ k

Maximum weld size $= \frac{1}{2} - \frac{1}{16} = \frac{7}{16}$ in.

Minimum weld size $= \frac{3}{16}$ in. (from Table 14-2).

Use $\frac{5}{16}$-in. weld as maximum size which can be made in one pass

Effective throat thickness of weld $= (0.707)(\frac{5}{16}) = 0.221$ in.

Design strength of weld/in. $= \phi F_w = (0.75)(0.60 \times 70)(0.221)(1.0)$

$$= 6.96 \text{ k/in.}$$

$$\text{Length required} = \frac{153.9}{6.96} = 22.11 \text{ in.} \quad \underline{\text{Say 23 in.}}$$

Place welds as shown in Fig. 14-15

The student should carefully note that the centroid of the welds and the centroid of the statically loaded angle do not coincide in the connection selected in Example 14-6 and shown in Fig. 14-15. Should a welded connection be subjected to varying stresses (such as those occurring in a bridge member), it is considered desirable to place the welds so that their centroid will coincide with the centroid of the member (or the resulting torsion must be accounted for in design). If the member being connected is symmetrical the welds will be placed symmetrically, but if the member is not symmetrical the welds will not be symmetrical.

The force in an angle, such as the one shown in Fig. 14-16, is assumed to act along its center of gravity. If the center of gravity of weld resistance is to coincide

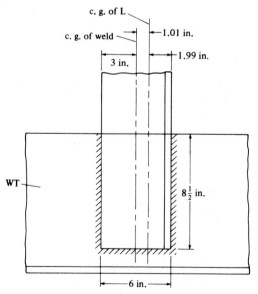

Figure 14-15

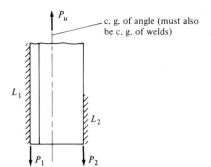

Figure 14-16

with the angle force the weld must be asymmetrically placed, or in this figure L_1 must be longer than L_2. (When angles are connected by rivets or bolts there is usually an appreciable amount of eccentricity, but in a welded joint eccentricity can be fairly well eliminated.) The information necessary to handle this type of weld design can be easily expressed in equation form but only the theory behind the equations is presented here.

For the angle shown in Fig. 14-16 the force acting along line L_2 (designated here as P_2) can be determined by taking moments about L_1. The member force and the weld resistance are to coincide and the moments of the two about any point must be zero. If moments are taken about L_1 the force P_1 (which acts along line L_1) will be eliminated from the equation and P_2 can be determined. In a similar manner P_1 can be determined by taking moments along L_2 or by $\Sigma V = 0$. Example 14-7 illustrates the design of fillet welds of this type. A similar problem is handled in Example 14-8 except an end fillet weld is included, thus permitting a shorter connection. The center of gravity and resistance of the end weld are known and can be easily included in the moment equations.

There are other possible solutions for the design of the welds for the angle considered in these two examples. Although the $\frac{7}{16}$-in. weld is the largest one permitted at the edges of the $\frac{1}{2}$-in. angle and at its end, a larger weld could be used on the other side next to the outstanding leg. From a practical point of view, however, the welds should be the same size because different size welds slow the welder down owing to the need to change electrodes to make different sizes.

It is also to be remembered from our fatigue discussion in Chapter 4 that should the estimated number of loading cycles during the structure's estimated life exceed 20,000 it will be necessary to study the stress range of the connection at service loads. This study may very well result in larger connections as required by Appendix K of the LRFD Specification.

EXAMPLE 14-7

Use A36 steel, E70 electrodes, and the SMAW process to design side fillet welds for the full capacity of the $5 \times 3 \times \frac{1}{2}$-in. angle tension member shown in Fig. 14-17. Assume the member is subjected to repeated stress variations, making any connection eccentricity undesirable. Check block shear strength of member.

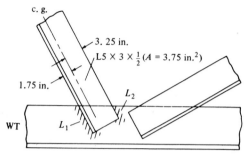

Figure 14-17

Solution. Tensile capacity of angle

$$P_u = \phi_t F_y A_g = (0.9)(36)(3.75) = 121.5 \text{ k} \leftarrow$$

or

$$P_u = \phi_t F_u A_g \text{ assuming } U = 0.87$$
$$= (0.75)(58)(0.87)(3.75) = 141.9 \text{ k}$$

Maximum weld size $= \frac{1}{2} - \frac{1}{16} = \frac{7}{16}$ in.

Use $\frac{5}{16}$-in. weld

Effective throat thickness of weld $= (0.707)(\frac{5}{16}) = 0.221$ in.

Capacity of weld/in. $= \phi F_w = (0.75)(0.60 \times 70)(0.221)(1.0) = 6.96 \text{ k/in.}$

Total weld length required $= \dfrac{121.5}{6.96} = 17.46$ in.

Taking moments about L_1 (see Fig. 4-17) to determine force P_2

$$(121.5)(1.75) - 5.00 \, P_2 = 0$$
$$P_2 = 42.53 \text{ k}$$
$$P_1 = P - P_1 = 121.5 - 42.53 = 78.97 \text{ k}$$
$$L_1 = \dfrac{78.97}{6.96} = 11.35 \text{ in.} \qquad \text{say } 11\tfrac{1}{2} \text{ in.}$$
$$L_2 = \dfrac{42.53}{6.96} = 6.11 \text{ in.} \qquad \text{say } 6\tfrac{1}{2} \text{ in.}$$

Use end returns $2 \times \frac{5}{16} = \frac{10}{16}$ (say 1 in.). The end return lengths may be subtracted from the side weld lengths making them $10\frac{1}{2}$ in. and $5\frac{1}{2}$ in., respectively.

Check block shearing strength assuming these dimensions

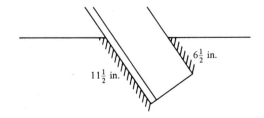

$$P_{bs} = \phi[F_u A_{nt} + 0.6 F_y A_{vg}]$$
$$= 0.75[(58)(5 \times \tfrac{1}{2}) + (0.6)(36)(18 \times \tfrac{1}{2})]$$
$$= 254.6 \text{ k} > 121.5 \text{ k} \qquad\qquad\qquad\qquad \text{OK}$$

$$P_{bs} = \phi[F_y A_{tg} + 0.6 F_u A_{ns}]$$
$$= 0.75[(36)(5 \times \tfrac{1}{2}) + (0.6)(58)(18 \times \tfrac{1}{2})]$$
$$= 302.4 \text{ k} > 121.5 \text{ k} \qquad\qquad\qquad\qquad \text{OK}$$

EXAMPLE 14-8

Rework Example 14.7 using fillet welds along the sides and end of the angle.

Solution. Assuming $\tfrac{5}{16}$-in. weld (capacity $= 6.96$ k/in. from Example 14-7)

Design strength of end weld $= (5.0)(6.96) = 34.8$ k

Taking moments about L_1 to determine force P_2,

$$(121.5)(1.75) - (2.50)(34.8) - 5.00\,P_2 = 0$$

$$P_2 = 25.12 \text{ k}$$

$$P_1 = 121.5 - 34.8 - 25.12 = 61.58 \text{ k}$$

$$L_1 = \frac{61.58}{6.96} = 8.85 \text{ in.} \qquad \underline{\text{say 9 in.}}$$

$$L_2 = \frac{25.12}{6.96} = 3.61 \text{ in.} \qquad \underline{\text{say 4 in.}}$$

It is rather convenient for design purposes to know the strength of a $\tfrac{1}{16}$-in. fillet weld 1 in. long. Though this size is below the minimum permissible size given in Table 14-2 the calculated strength of such a weld is useful for determining weld sizes for calculated forces. For a 1-in.-long SMAW weld we have

$$\phi F_w = (0.75)(0.707 \times \tfrac{1}{16} \times 1.0)(0.60 F_{EXX}) = 0.02 F_{EXX}$$

For an E70 electrode ϕF_w is $(0.02)(70) = 1.4$ k/in. for a $\tfrac{1}{16}$-in. weld. If we are designing a fillet weld to carry a factored force of 6.5 k/in. the required weld size will be $6.5/1.4 = 4.64$ sixteenths of an inch or say $\tfrac{5}{16}$ in.

14-13 SHEAR AND TORSION

Fillet welds are frequently loaded with eccentrically applied loads with the result that the welds are subjected to either shear and torsion or to shear and bending. Figure 14-18 is presented in an attempt to show the student the difference between the two situations. Shear and torsion, shown in part (a) of the figure, is the subject of this section while shear and bending shown in part (b) of the figure are discussed in Sections 14-14 and 15-8.

As is the case for eccentrically loaded bolt groups (Section 13-1) the LRFD Specification provides the permissible design strength of welds but does not specify a method of analysis for eccentrically loaded welds. The method to be used is left up to the designer.

Elastic Method   Initially the very conservative elastic method is presented. In this method friction or slip resistance between the connected parts is neglected and the connected parts are assumed to be perfectly rigid.

For this discussion the welded bracket of part (a) of Fig. 14-18 is considered. The pieces being connected are assumed to be completely rigid as they were in bolted connections. The effect of this assumption is that all deformation occurs in the weld. The weld is subjected to a combination of shear and torsion as was the eccentrically loaded bolt group considered in Section 13-1. The force caused by torsion can be computed from the following familiar expression.

$$f = \frac{Td}{J}$$

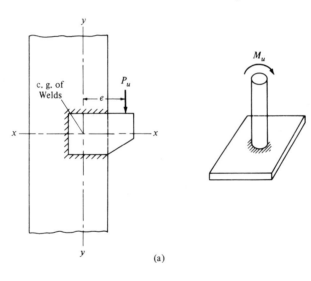

(a)

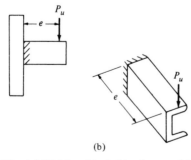

(b)

Figure 14-18   (a) Welds subjected to shear and torsion. (b) Welds subjected to shear and bending.

In this expression T is the torsion, d is the distance from the center of gravity of the weld to the point being considered, and J is the polar moment of inertia of the weld. It is usually more convenient to break the force down into its vertical and horizontal components. In the expressions to follow, h and v are the horizontal and vertical components of the distance d. (These formulas are almost identical to those used for determining stresses in bolt groups subject to torsion.)

$$f_h = \frac{Tv}{J} \qquad f_v = \frac{Th}{J}$$

These components are combined with the usual direct shearing stress which is assumed to equal the reaction divided by the total length of the welds. For design of a weld subject to shear and torsion it is convenient to assume a 1-in. weld and compute the stresses on a weld of that size. Should the assumed weld be overstressed a larger weld is required; if understressed a smaller one is desirable.

Although the calculations will in all probability show the weld to be overstressed or understressed the math does not have to be repeated because a ratio can be set up to give the weld size for which the load would produce a stress exactly equal to the design stress. The student should note that the use of a 1-in. weld simplifies the units because 1 in. of length of weld is 1 in.^{2} of weld and the computed stresses can be said to be either kips per square inch or kips per inch of length. Should the calculations be based on some size other than a 1-in. weld the student will have to be very careful in keeping the units straight particularly in obtaining the final weld size. To further simplify the calculations the welds are assumed to be located at the edges along which the fillet welds are placed rather than at the centers of their effective throats. As the throat dimensions are rather small this assumption changes the results very little. Example 14-9 illustrates the calculations involved in determining the weld size required for a connection subjected to a combination of shear and torsion.

EXAMPLE 14-9

For the bracket shown in Fig. 14-19(a), determine the fillet weld size required if E70 electrodes, the LRFD Specification, and the SMAW process are used.

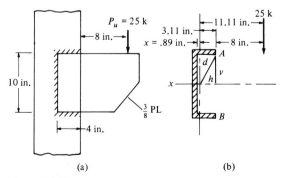

(a)   (b)

Figure 14-19

Solution. Assuming a 1-in. weld as shown in part (b) of Fig. 14-19,

$$A = 18 \text{ in.}^2$$

$$\bar{x} = \frac{(4)(2)(2)}{18} = 0.89 \text{ in.}$$

$$I_x = (\tfrac{1}{12})(1)(10)^3 + (2)(4)(5)^2 = 283.3 \text{ in.}^4$$

$$I_y = (2)(\tfrac{1}{3})(0.89^3 + 3.11^3) + (10)(0.89)^2 = 28.5 \text{ in.}^4$$

$$J = 283.3 + 28.5 = 311.8 \text{ in.}^4$$

The most stressed portions of the weld are the greatest distance from the weld center of gravity [A and B in Fig. 14-19(b)].

$$f_h = \frac{Tv}{J} = \frac{(25 \times 11.11)(5)}{311.8} = 4.45 \text{ k/in.}^2$$

$$f_v = \frac{Th}{J} = \frac{(25 \times 11.11)(3.11)}{311.8} = 2.77 \text{ k/in.}^2$$

$$f_s = f_{shear} = \frac{25}{18} = 1.39 \text{ k/in.}^2$$

$$f_r = f_{resultant} = \sqrt{(2.77 + 1.39)^2 + (4.45)^2} = 6.09 \text{ k/in.}^2$$

Design capacity of a 1-in. fillet weld (E70 electrode) $\phi F_w = (0.75)(0.60 \times 70)(0.707 \times 1.0)(1.0) = 22.27$ k/in.
or $(16)(0.02F_{EXX}) = (16)(0.02 \times 70) = 22.4$ k/in.

$$\text{Weld size required} = \frac{6.09}{22.27} = 0.273 \text{ in.} \qquad (\text{say } \tfrac{5}{16} \text{ in.})$$

Use $\tfrac{5}{16}$-in. fillet weld

Ultimate Strength Method   An ultimate strength analysis of eccentrically loaded welded connections is much more realistic than is the very conservative elastic procedure just described. For the discussion to follow the eccentrically loaded fillet weld of Fig. 14-20 is considered. As for eccentrically bolted connec-

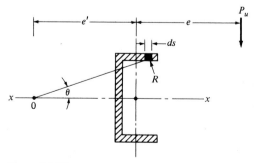

Figure 14-20

tions the load tends to cause a relative rotation and translation between the parts connected by the weld.

Even if the eccentric load is of such a magnitude as to cause the most stressed part of the weld to yield the entire connection will not yield. The load may be increased, the less stressed fibers will begin to resist more of the load, and failure will not occur until all the weld fibers yield. The weld will tend to rotate about its instantaneous center of rotation. The location of this point (which is indicated by the letter O in the figure) is dependent upon the location of the eccentric load, the geometry of the weld, and the deformations of the different elements of the weld.

If the eccentric load P_u is vertical and if the weld is symmetrical about a horizontal axis through its center of gravity the instantaneous center will fall somewhere on the horizontal x axis. Each differential element of the weld will provide a resisting force R. As shown in Fig. 14-20 each of these resisting forces is assumed to act perpendicular to a ray drawn from the instantaneous center to the center of gravity of the weld element in question. The sum of the moments of the resisting forces of all the elements of the weld about point 0 must be equal and opposite to the moment of the eccentric load about the same point.

$$P_u(e' + e) = \Sigma(Rds)(d)$$

$$P_u = \frac{\Sigma(Rds)(d)}{e' + e}$$

Studies have been made to determine the maximum shear forces which eccentrically loaded weld elements can withstand.[3,4] The results which depend on the load-deformation relationship of the weld elements may be represented either with curves or in formula fashion. The ductility of the entire weld is governed by the maximum deformation of the weld element which first reaches its limit (it's probably the element the greatest distance from the weld's instantaneous center). In the expression to follow R is the ultimate shear force for a single weld element (ϕF_W) subjected to eccentric shear while R_{ult} is the ultimate pure shear force for a single weld element.

$$R = R_{ult}(1 - e^{[-k(\Delta/\Delta_o)]})^{k_2}$$

The other items in the expression are given in the "Eccentric Loads on Weld Groups" section of Part 5 of the LRFD Manual. According to the LRFD Specification (J2) the design strength of a fillet weld is limited to $0.6 F_{EXX}$. Thus the values of R obtained from the preceding equation and used in the Manual tables were limited to this maximum value.

We can assume a location of the instantaneous center, determine R values for the different elements of the weld, and compute the value of P_u from the equation given. If this value of P_u does not equal the sum of the vertical components of the R values we will need to assume another location for point 0, and so on.

[3]L. J. Butler, S. Pal, and G. L. Kulak, "Eccentrically Loaded Weld Connections," *Journal of the Structural Division,* vol. 98, no. ST5, May 1972, pp. 989–1005.

[4]G. L. Kulak and Timler, "Tests on Eccentrically Loaded Fillet Welds," Dept. of Civil Engineering, University of Alberta, Edmonton, December 1984.

A numerical example of this tedious trial-and-error process is not presented in this text. The values given in the tables of Part 5 of the LRFD Manual entitled "Eccentric Loads on Weld Groups" were developed by this procedure. Using the tables the ultimate strength P_u of a particular connection can be determined from the expression to follow in which C is a tabular coefficient which includes a ϕ of 0.75, C_1 is a coefficient depending on the electrode number (it's 1.0 for E70XX), D is the weld size in sixteenths of an inch, and l is the length of the vertical weld.

$$P_u = CC_1Dl$$

The Manual includes tables for both vertical and inclined loads (at angles from the vertical of 0°, 45°, and 75°). The user is warned not to interpolate for angles in between these values because the results may be quite unconservative. Therefore, he or she is advised to use the value given for the next lower angle. Should the connection arrangement being considered not be covered by the tables it is suggested that the conservative elastic procedure previously described be used. A simple illustration of these ultimate strength tables is presented in Example 14-10.

EXAMPLE 14-10

Repeat Example 14-9 using the LRFD tables which are based on an ultimate strength analysis. The connection is redrawn in Fig. 14-21.

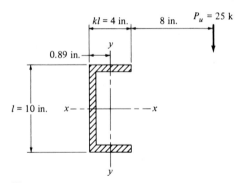

Figure 14-21

Solution

$$e = 11.11 \text{ in.}$$

$$l = 10 \text{ in.}$$

$$a = \frac{11.11}{10} = 1.111$$

$$k = \tfrac{4}{10} = 0.4$$

$C = 0.799$ from Table XXII, Part 5, LRFD Manual

$C_1 = 1.0$ from same table

$$D = \text{weld size required} = \frac{P_u}{CC_1l}$$

$$= \frac{25}{(0.799)(1.0)(10)} = 3.13 \text{ sixteenths}$$

$$= 0.196 \text{ in. (as compared with } 0.274 \text{ in}$$

by the elastic method

$$\underline{\underline{\text{Use } \tfrac{1}{4} \text{ in.}}}$$

14-14 SHEAR AND BENDING

The welds shown in Fig. 14-18(b) and in Fig. 14-22 are subjected to a combination of shear and bending.

For short welds of this type the usual practice is to consider a uniform variation of shearing stress. If, however, the bending stress is assumed to be given by the flexure formula, the shear does not vary uniformly for vertical welds but as a parabola with a maximum value $1\frac{1}{2}$ times the average value. These stress and shear variations are shown in Fig. 14-23.

The student should carefully note that the maximum shearing stresses and the maximum bending stresses occur at different locations. It is, therefore, probably not necessary to combine the two stresses at any one point. If the weld is capable of withstanding the worst shear and the worst moment individually it is probably satisfactory. In Example 14-11, however, a welded connection subjected to shear and bending is designed by the usual practice of assuming a uniform shear distribution in the weld and combining the value vectorially with the maximum bending stress.

EXAMPLE 14-11

Using E70 electrodes, the SMAW process, and the LRFD Specification determine the weld size required for the connection of Fig. 14-22 if $P_u = 45$ k, $e = 2\frac{1}{2}$ in., and $L = 8$ in. Assume that the member thicknesses do not control weld size.

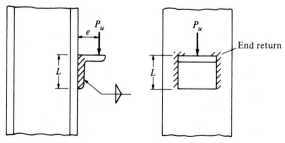

Figure 14-22

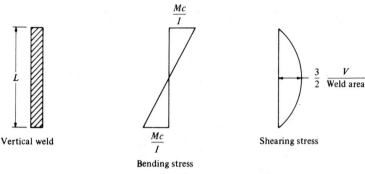

Figure 14-23

Solution

$$f_s = \frac{45}{(2)(8)} = 2.81 \text{ k/in.}$$

$$f = \frac{(45 \times 2.5)(4)}{(\frac{1}{12})(1)(8)^3(2)} = 5.27 \text{ k/in.}$$

$$f_r = \sqrt{(2.81)^2 + (5.27)^2} = 5.97 \text{ k/in.}$$

$$\begin{array}{l} \text{Weld size} \\ \text{required} \end{array} = \frac{5.97}{(0.707)(1)(0.75)(0.60 \times 70)} = 0.268 \text{ in.} \qquad (\text{say } \tfrac{5}{16} \text{ in.})$$

The subject of shear and bending is a very practical one, as it is the situation commonly faced in moment resisting connections. This topic is continued at length in Section 15-10.

PROBLEMS

14-1. A $\frac{5}{16}$-in. fillet weld SMAW process is used to connect the members shown in the accompanying illustration. Determine the design load that can be applied to this connection according to the LRFD Specification if the steel is A36 and E70 electrodes are used. (*Ans. $P_u = 97.2$ k*)

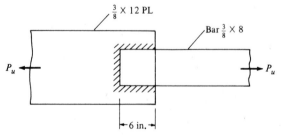

Problem 14-1

14-2. Rework Prob. 14-1 if the SAW process is used.

14-3. Rework Prob. 14-1 if A572 grade 65 steel and E80 electrodes are used. (*Ans.* $P_u = 159.1$ k)

14-4. Design maximum size fillet welds to develop the full strength of the A36 bar shown in the accompanying illustration. Use E70 electrodes and the SMAW process.

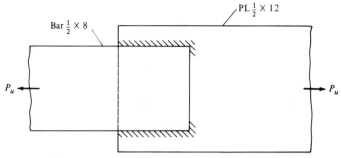

Problem 14-4

14-5. Rework Prob. 14-4 if the SAW process is used. (*Ans.* Use 8-in. welds each side as required by LRFD Section J2.2b including end returns)

14-6. Rework Prob. 14-4 using side welds and a vertical end weld at the end of the $\frac{1}{2} \times 8$ bar. Also use A572 grade 65 steel and E80 electrodes.

14-7. Rework Prob. 14-4 using side welds and welds at the end of the $\frac{1}{2} \times 8$ PL at the connection. (*Ans.* Total $L = 13.32$ in., say 14 in.)

14-8. The $\frac{5}{8} \times 8$-in. PL shown in the accompanying illustration consists of A36 steel and is to be connected to a gusset plate with $\frac{5}{16}$-in. SMAW fillet welds. Determine the length L required to develop the full strength of the bar if E70 electrodes are used.

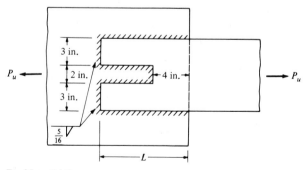

Problem 14-8

14-9. Design maximum size side SMAW fillet welds to develop the full tensile strength of a $6 \times 4 \times \frac{1}{2}$L using E70 electrodes and A36 steel. The member is connected on the 6-in. leg and is subject to alternating loads. (*Ans.* Use $5\frac{1}{2}$-in. and 11-in. side welds)

14-10. Rework Prob. 14-9 using side welds and a weld at the end of the angle.

14-11. Rework Prob. 14-10 using A588 steel and E80 electrodes. (*Ans.* $3\frac{1}{2}$-in. and 10-in. side welds and 6-in. end weld)

14-12. One leg of an $8 \times 8 \times \frac{3}{4}$ angle is to be connected with side welds and a weld at the end of the angle to a plate behind to develop the full capacity of the angle (A36). Balance the SMAW fillet welds around the center of gravity of the angle, use maximum weld size and assume E70 electrodes.

14-13. It is desired to design $\frac{5}{16}$-in. SMAW fillet welds necessary to connect a C10 × 30 made from A36 steel to a $\frac{3}{8}$-in. gusset plate. End, side, and slot welds may be used to develop the full tensile capacity of the channel. No welding is permitted on the back of the channel. Use E70 electrodes. It is assumed that due to space limitations the channel can lap over the gusset plate by a maximum of 10 in. (*Ans.* $\frac{5}{16}$-in. fillet welds and one $1\frac{1}{16} \times 2\frac{1}{2}$ in. slot weld)

14-14. Rework Prob. 14-13 using A572 grade 60 steel and E70 electrodes and $\frac{3}{8}$-in. fillet welds.

14-15. Using the elastic method determine the maximum force per inch to be resisted by the fillet weld shown in the accompanying illustration. (*Ans.* 8.83 k/in.)

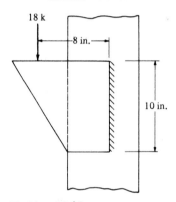

Problem 14-15

14-16. Using the elastic method determine the maximum force to be resisted per inch by the fillet weld shown in the accompanying illustration.

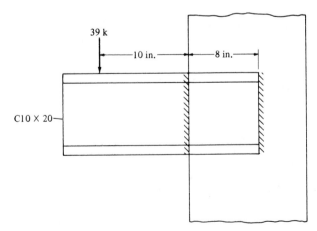

Problem 14-16

14-17. Using the elastic method rework Prob. 14-16 if welds are used on the sides of the channel in addition to those shown in the figure. (*Ans.* 4.36 k/in.)

14-18. Using the elastic method determine the maximum force per inch to be resisted by the fillet welds shown in the accompanying illustration. Also determine the required weld thickness using A36 steel E70 electrode and SMAW welds.

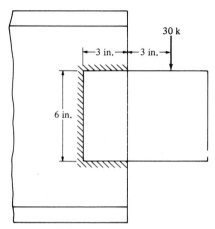

Problem 14-18

14-19. Determine the maximum eccentric load P_u that can be applied to the connection shown in the accompanying illustration if $\frac{1}{4}$-in. SMAW fillet welds are used. Assume plate thickness = $\frac{1}{2}$ in. E70. A36. (a) Use elastic method. (b) Use LRFD tables and ultimate strength method. [*Ans.* (a) P_u = 16.48 k (b) P_u = 23.52 k]

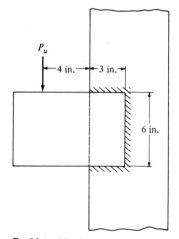

Problem 14-19

14-20. Rework Prob. 14-19 if $\frac{3}{8}$-in. fillet welds are used and the vertical weld is 10 in. high.

14-21. Determine the SMAW fillet weld size required for the connection of Prob. 14-15 if the load is increased to 30 k and the height of the weld is 12 in. A36. E70. (a) Use elastic method (b) use LRFD tables and ultimate strength method. [*Ans.* (a) 0.463 in., say $\frac{1}{2}$ in. (b) 0.309 in., say $\frac{5}{16}$ in.]

14-22. Using E70 electrodes, the SMAW process, and A36 steel determine the fillet weld size required for the bracket shown in the accompanying illustration. (a) Use elastic method. (b) Use LRFD Tables and ultimate strength method.

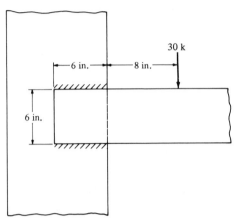

Problem 14-22

14-23. Rework Prob. 14-22 if the load is increased from 30 to 50k and the weld lengths increased from 6 to 8 in. [(a) 0.705 in., say $\frac{3}{4}$ in. (b) 0.476 in., say $\frac{1}{2}$ in.]

14-24. Determine the SMAW fillet weld size required for the connection shown in the accompanying illustration. E70. A36. (a) Use elastic method. (B) Use LRFD tables and ultimate strength method.

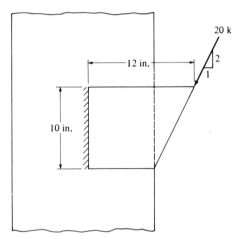

Problem 14-24

14-25. Determine the SMAW fillet weld size required for the connection shown in the accompanying illustration. E70. A36. What angle thickness should be used? (a) Use elastic method. (b) Use LRFD tables and ultimate strength method. [*Ans.* (a) 0.181 in., say $\frac{3}{16}$ in., $\frac{1}{4}$ in. angle t. (b) 0.136 in., say $\frac{3}{16}$ in., $\frac{1}{4}$ in., angle t]

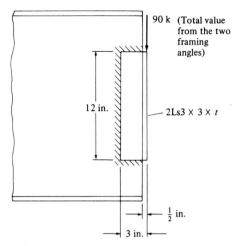

90 k (Total value from the two framing angles)

12 in.

2Ls3 × 3 × t

$\frac{1}{2}$ in.

3 in.

Problem 14-25

14-26. Assuming the SMAW process is to be used, determine the fillet weld size required for the connection shown in the accompanying illustration. Use A36 steel E70 electrodes and the elastic method.

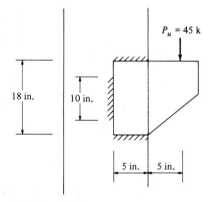

P_u = 45 k

18 in.

10 in.

5 in.   5 in.

Problem 14-26

14-27. Determine the fillet weld size required for the connection shown in the accompanying illustration. A36. E70. The SMAW process is to be used. (a) Use elastic method. (b) Use LRFD tables and ultimate strength method. [*Ans.* (a) 0.705 in., say $\frac{3}{4}$ in. (b) 0.45 in., say $\frac{1}{2}$ in.]

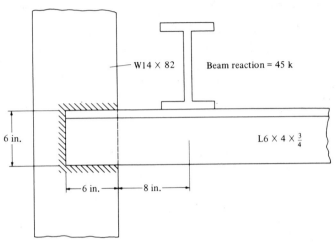

Problem 14-27

14-28. Determine the value of the load P_u that can be applied to the connection shown in the accompanying illustration if $\frac{3}{8}$-in. fillet welds are used. E70.A36. SAW. Use elastic method.

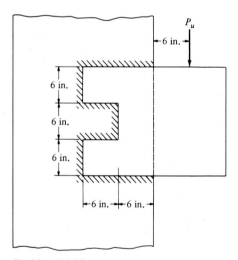

Problem 14-28

14-29. Determine the length of $\frac{1}{4}$-in. SMAW fillet welds 12 in. on center required to connect the cover plates for the section shown in the accompanying illustration at a point where the external shear is 180 k. A36. E70. (*Ans.* 6.08 in., say 6-in. welds, 12 in. o.c.)

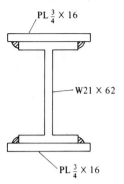

PL $\frac{3}{4}$ × 16

W21 × 62

PL $\frac{3}{4}$ × 16

Problem 14-29

14-30. The welded plate girder shown in the accompanying illustration has an external shear of 720 k at a particular section. Determine the fillet weld size required to fasten the plates to the web if the SMAW process is used. E70.A36.

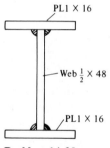

PL1 × 16

Web $\frac{1}{2}$ × 48

PL1 × 16

Problem 14-30

Chapter 15

Building Connections

15-1 SELECTION OF TYPE OF FASTENER

This chapter is concerned with the actual beam-to-beam and beam-to-column connections commonly used in steel buildings. Under present-day steel specifications four types of fasteners are permitted for these connections. These are welds, unfinished bolts, high-strength bolts, and rivets.

Selection of the type of fastener or fasteners to be used for a particular structure usually involves many factors, including requirements of local building codes, relative economy, preference of designer, availability of good welders or riveters, loading conditions (as static or fatigue loadings), preference of fabricator, and equipment available. It is impossible to list a definite set of rules from which the best type of fastener can be selected for any given structure. One can only give a few general statements which may be helpful in making a decision. These are listed as follows.

1. Unfinished bolts are often economical for light structures subject to small static loads and for secondary members (such as purlins, girts, bracing, etc.) in larger structures.
2. Field bolting is very rapid and involves less skilled labor than welding or riveting. The purchase price of high-strength bolts, however, is rather high.
3. If a structure is later to be disassembled, welding is probably ruled out, leaving the job open to bolts.
4. For fatigue loadings fully tensioned high-strength bolts and welds are very good.

5. Welding requires the smallest amounts of steel, probably provides the most attractive-looking joints, and also has the widest range of application to different types of connections.
6. When continuous and rigid fully moment resisting joints are desired welding will probably be selected.
7. Welding is almost universally accepted as being satisfactory for shopwork. For fieldwork it is very popular in some areas of the United States while in a few others it is stymied by the fear that field inspection is rather questionable.
8. Rivets can be rapidly installed in the shop with heavy riveters but nevertheless are very seldom if ever used.
9. For field work, rivets are almost extinct.

A most interesting article entitled "Choosing the Best Structural Fastener" by Henry J. Stetina appeared in the November 1963 issue of *Civil Engineering*. It would be well worth the student's time to read this article.

A valuable reference on the relative economy of various types of beam to beam and beam to column connections is given in a handbook published by the United States Steel Corporation in October 1963 entitled *Building Design Data*.

15-2 TYPES OF BEAM CONNECTIONS

All connections have some restraint, that is, some resistance to changes of the original angles between intersecting members. Depending on the amount of restraint the LRFD Specification (A2.2) classifies connections as being fully restrained (Type FR) and partially restrained (Type PR). These two types of connections are described in more detail as follows:

1. Type FR connections are commonly referred to as rigid or continuous frame connections. They are assumed to be sufficiently rigid or restrained to keep the original angles between members virtually unchanged under load.
2. Type PR connections are those which have insufficient rigidity to keep the original angles virtually unchanged under load. Included in this classification are simple and semirigid connections as described in detail in this section.

A *simple connection* is a Type PR connection for which restraint is ignored. It is assumed to be completely flexible and free to rotate thus having no moment resistance. A *semirigid connection* is a Type PR connection whose resistance to angle change falls somewhere between the simple and rigid types.

From a practical standpoint, since there are no connections that are completely rigid or completely flexible, it is a common practice to classify them on a percentage of moment developed to complete rigidity or complete moment resis-

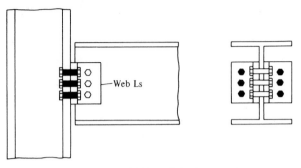

(a) Framed simple connection

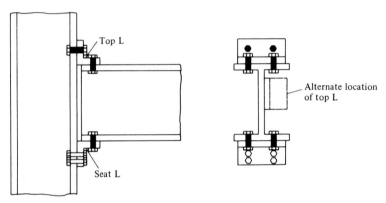

Top L

Alternate location
of top L

Seat L

(b) Seated simple connection

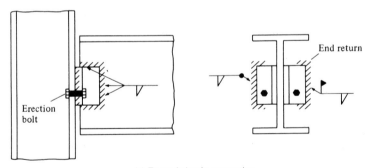

End return

Erection
bolt

(c) Framed simple connection

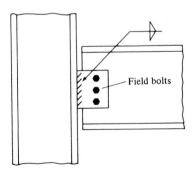

Field bolts

(d) Single plate simple connection

Figure 15-1 Some simple connections. (a) Framed simple connection. (b) Seated simple connection. (c) Framed simple connection. (d) Single-plate simple connection.

tance. (A measure of the rotational characteristics of a particular connection cannot be practically obtained by a theoretical method and it is necessary to run tests and plot curves of the relationships between moments and rotations for each type of connection.) A rough rule is simple connections 0–20 percent rigidity, semirigid 20–90 percent, and rigid above 90 percent.

Each of these three general types of connections is briefly discussed in this section with little mention of the specific types of connectors used. The remainder of the chapter is concerned with detailed designs of these connections using specific types of fasteners. *In this discussion the author probably overemphasizes the semirigid and rigid type connections, because a very large percentage of the building designs with which the average designer works will be assumed to have simple connections.* A few descriptive comments are given in the following paragraphs concerning each of these three types of connections.

Simple connections (Type PR) are quite flexible and are assumed to allow the beam ends to be substantially free to rotate downward under load as true simple beams should. Although simple connections do have some moment resistance (or resistance to end rotation) it is assumed to be negligible and they are assumed to be able to resist shear only. Several types of simple connections are shown in Fig. 15-1. More detailed descriptions of each of these connections and their assumed behavior under load are given in later sections of this chapter. In this figure most of the connections are shown as being made entirely with the same type of fastener while in actual practice two types of fasteners are often used for the same connection. For example, a very common practice is to shop-weld the web angles to the beam web and field-bolt them to the column or girder.

Semirigid connections (Type PR) are those which have appreciable resistance to end rotation thus developing appreciable end moments. In design practice it is quite common for the designer to assume all connections are either simple or rigid with no consideration given to those situations in between, thereby simplifying the analysis. Should he or she make such an assumption for a true

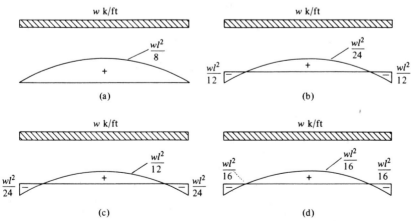

Figure 15-2 (a) Simple connections (0 percent). (b) Rigid connections (100 percent). (c) Semirigid connections (50 percent). (d) Semirigid connections (75 percent).

semirigid connection he or she may miss an opportunity for appreciable moment reductions. To understand this possibility the reader is referred to the moment diagrams shown in Fig. 15-2 for a group of uniformly loaded beams supported with connections having different percentages of rigidity. This figure shows that the maximum moments in a beam vary greatly with different types of end connections. For example, the maximum moment in the semirigid connection of part (d) of the figure is only 50 percent of the maximum moment in the simply supported beam of part (a) and only 75 percent of the maximum moment in the rigidly supported beam of part (b).

Actual semirigid connections are used fairly often but usually no advantage is taken of their moment reducing possibilities in the calculations. Perhaps one factor that keeps the design profession from taking advantage of them more often is the LRFD Specification (Section A2) which states that consideration of a connection as being semirigid is permitted only upon presentation of evidence that it

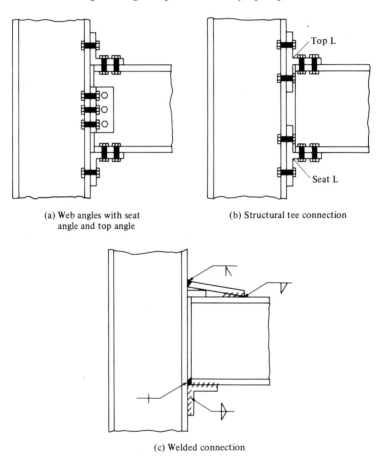

(a) Web angles with seat   (b) Structural tee connection
    angle and top angle

(c) Welded connection

Figure 15-3 Some semirigid connections. These may be rigid if column web stiffeners are not required. (a) Web angles with seat angle and top angle. (b) Structural tee connection. (c) Welded connection.

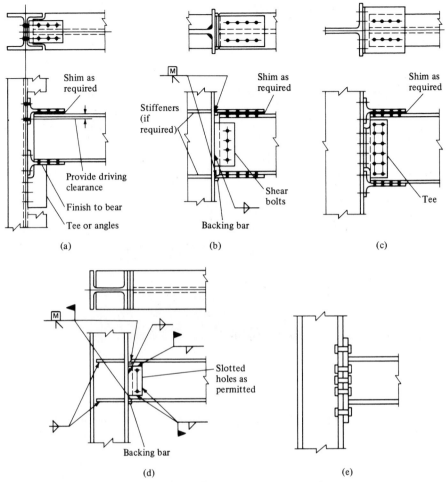

Figure 15-4   Moment-resisting connections.

is capable of providing a certain percentage of the end restraint furnished by a completely rigid connection.

A second deterring factor is the need for a method of analysis which falls in between analysis for simple beams and analysis for statically indeterminate structures with completely rigid joints. The student can see that analysis of a building by moment distribution could be drastically affected if the end connections were assumed to have varying percentages of moment restraint. This subject is discussed in detail by Bruce Johnston and Edward H. Mount in a paper entitled "Analysis of Building Frames with Semi-Rigid Connections."[1] The student is also referred to Chapter 8 of *Advanced Design in Structural Steel* by John E. Lothers (Prentice-Hall, 1960) for an excellent discussion of this subject. Several types of semirigid connections are shown in Fig. 15-3.

[1]*Trans. ASCE* 107 (1942), p. 993.

A semirigid beam-to-column connection, Ainsley Building, Miami, Fla. (Courtesy of the Lincoln Electric Company.)

Rigid connections (Type FR) are those which theoretically allow no rotation at the beam ends and thus transfer 100 percent of the moment of a fixed end. Connections of this type may be used for tall buildings in which wind resistance is developed by providing continuity between the members of the building frame. Several Type FR connections which provide almost 100 percent restraint are shown in Fig. 15-4. It will be noticed in the figure that column web stiffeners may be required for some of these connections to provide sufficient resistance to rotation. The design of these stiffeners is discussed in Section 15-11.

The moment connection shown in part (d) is rather popular with steel fabricators and the end-plate connection of part (e) has also been frequently used in recent years.[2]

15-3 STANDARD BOLTED BEAM CONNECTIONS

Several types of standard bolted connections are shown in Fig. 15-5. These connections are usually designed to resist shear only, as testing has proved this practice to be quite satisfactory. Part (a) of the figure shows a connection between beams with the so-called *framed connection*. This type of connection consists of a

---

[2]J. D. Griffiths, "End-Plate Moment Connections—Their Use and Misuse," *Engineering Journal,* AISC, vol. 21, no. 1 (first quarter, 1984) pp. 32–34.

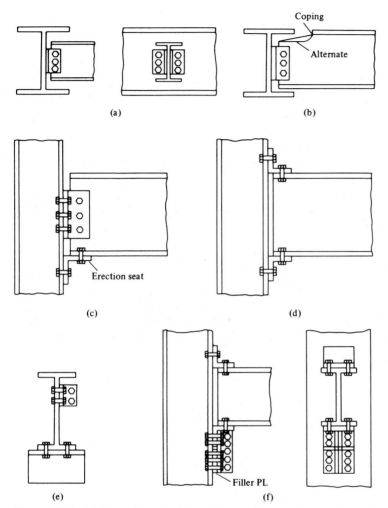

Figure 15-5 (a) Framed connection. (b) Framed connection. (c) Framed connection. (d) Seated connection. (e) Seated connection. (f) Seated connection with stiffener angles.

pair of flexible web angles probably shop-connected to the web of the supported beam and field-connected to the supporting beam or column. When two beams are being connected it is usually necessary to keep their top flanges at the same elevation, with the result that the top flange of one will have to be cut back (called *coping*) as shown in part (b) of the figure. For such connections we must check block shear as discussed in Section 3-7 of this text.

Simple connections of beams to columns can be either framed or seated as shown in Fig. 15-5. In part (c) of the figure a framed connection is shown in which two web angles are connected to the beam web in the shop, after which bolts are placed through the angles and column in the field. It is often convenient to have an angle, called an *erection seat,* to support the beam during erection. Such an angle is shown in the figure.

The seated connection has an angle under the beam similar to the erection seat just mentioned, which is shop-connected to the column. In addition, there is another angle probably on top of the beam which is field-connected to the beam and column. A seated connection of this type is shown in part (d) of the figure. Should space limitation prove a problem above the beam, the top angle may be placed in the optional location shown in part (e) of the figure. The top angle, at either of the locations mentioned, is very helpful in keeping the top flange of the beam from being accidentally twisted out of place during construction.

The amount of load which can be supported by the types of connections shown in parts (c), (d), and (e) of Fig. 15-5 is severely limited by the flexibility or bending strength of the horizontal legs of the seat angles. For heavier loads it is necessary to use stiffened seats such as the one shown in part (f) of the figure.

The majority of these connections are selected by referring to standard tables. The LRFD Manual has excellent tables for selecting bolted or welded beam connections of the types shown in Fig. 15-5. After a rolled-beam section has been selected it is quite convenient to refer to these tables and select one of the standard connections, which will be suitable for the vast majority of cases.

In order to make these standard connections have as little moment resistance as possible the angles used in making up the connections are usually light and flexible. To qualify as simple end supports the ends of the beams should be as free as possible to rotate downward. Figure 15-6 shows the manner in which framed and seated end connections will theoretically deform as the ends of the

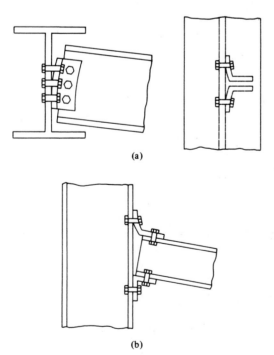

(a)

(b)

Figure 15-6   (a) Bending of framed-beam connection. (b) Bending of seated-beam connection.

beams rotate downward. The designer does not want to do anything that will hamper these deformations.

For the rotations shown in Fig. 15-6 to occur there must be some deformation of the angles. As a matter of fact, if end slopes of the magnitudes which are computed for simple ends are to occur, the angles will actually bend enough to be stressed beyond their yield points. If this situation occurs, they will be permanently bent and the connections will quite closely approach true simple ends. The student should now see why it is desirable to use rather thin angles and large gages for the bolt spacing if flexible simple end connections are the goal of the designer.

These connections do have some resistance to moment. When the ends of the beam begin to rotate downward, the rotation is certainly resisted to some extent by the tension in the top bolts, even if the angles are quite thin and flexible. Neglecting the moment resistance of these connections will cause conservative beam sizes. If moments of any significance are to be resisted, more rigid-type joints need to be provided than are available with the framed and seated connections.

15-4 LRFD MANUAL STANDARD CONNECTION TABLES

In Part 5 of the LRFD Manual a series of tables are presented with which the designer may select several different types of standard connections. There are tables for bolted or welded two-angle framed connections, seated beam connections, stiffened seated beam connections, eccentrically loaded connections, single-angle framed connections, and others.

In the next few sections of this chapter (15-5 through 15-8) a few standard connections are selected using the tables in the Manual. The author hopes that these will be sufficient to introduce the reader to the LRFD tables and enable him or her to make designs using the other tables with little difficulty.

The last three sections (15-9 through 15-11) present design information for some connections not altogether included in the LRFD tables.

15-5 DESIGNS OF STANDARD BOLTED FRAMED CONNECTIONS

For small and low-rise buildings (and that means most buildings) simple framed connections of the types previously shown in parts (a) and (b) of Fig. 15-5 are usually used to connect beams to girders or to columns. The angles used are rather thin (5/8 in. is the arbitrary maximum thickness used in the LRFD Manual) so they will have the necessary flexibility shown in Fig. 15-6. The angles will develop some small moments (supposedly not more than 20 percent of full fixed-end conditions) but they are neglected in design.

The framing angles extend out from the beam web by $\frac{1}{2}$ in. as shown in Fig. 15-7. This protrusion, which is often referred to as the *setback,* is quite useful in fitting members together during steel erection.

Several standard bolted framed connections for simple beams are designed in this section with the tables provided in Part 5 of the LRFD Manual. In these tables the following abbreviations for different bolt conditions are used:

1. A325-SC and A490-SC (slip-critical connections).
2. A325-N and A490-N (bearing-type connections with threads included in the shear planes).
3. A325-X and A490-X (bearing-type connections with threads excluded from shear planes).
4. A307 (common bolts).

To determine the design capacity of bolted framed beam connections it is necessary to check the shearing and bearing capacities of the bolts as well as the shearing and bearing capacities of the framing or connection angles. To make these checks the following LRFD tables are used: I-D (shearing strength of bolts), I-E (bearing strength of connections), I-F (bearing strength of bolts with different edge distances), I-G-1 and I-G-2 (block shearing strengths), II-A and II-B (bolt shearing strengths for bearing-type and slip-critical connections).

It can be seen in Tables II-A and II-B that the standard connection angle lengths (L values in the tables) vary from $5\frac{1}{2}$ to $29\frac{1}{2}$ in. If the bolts are staggered, however, the lengths (L' values in the tables) vary from 7 to 31 in. The minimum center-to-center spacing of the bolts is $2\frac{2}{3}d$ (LRFD Specifications J3.9), but the preferable spacing $3d$ is used in the standard connection tables.

It is thought that the minimum depth of framing angles should be at least equal to one-half of the distance between the web toes of the beam fillets (called the T distances and given in the properties tables of Part 1 of the Manual). This minimum depth is used so as to provide sufficient stability during steel erection.

Example 15-1 presents the design of standard framing angles for a simply supported beam using bearing-type bolts in standard-size holes. In this example the design strengths of the bolts and the angles are taken from the appropriate tables. The author shows how the table values can be checked with the procedures presented in Chapters 12 and 13.

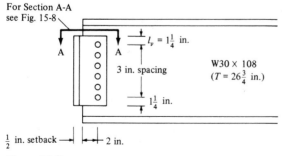

For Section A-A
see Fig. 15-8

$l_v = 1\frac{1}{4}$ in.

3 in. spacing

W30 × 108
($T = 26\frac{3}{4}$ in.)

$1\frac{1}{4}$ in.

$\frac{1}{2}$ in. setback ⟶ ⟵ 2 in.

Figure 15-7

EXAMPLE 15-1

Select a framed simple end connection for the W30×108 ($t_w = 0.545$ in.) shown in Fig. 15-7 if the factored end reaction R_u equals 160 k. The beam consists of A36 steel and $\frac{3}{4}$-in. A325-N bolts are to be used in standard-size 13/16-in. holes.

Solution

Table II-A Shear Strength of Bolts.   From this table for $\frac{3}{4}$-in. A325-N bolts and a reaction of 160 k a 6-row connection with an angle thickness of 5/16 in. and a length $L = 17\frac{1}{2}$ in. is selected. (This length seems satisfactory in comparison with the T of this shape of $26\frac{3}{4}$ in.) These bolts will carry 186 k in shear.

Checking the preceding value: the 6 bolts are in double shear and will resist $(6)(2)(0.44)(0.65 \times 54) = 185.3$ k. ✓

Table II-C Shear Capacity of Connection Angles.   From this table the design shear capacity of the 5/16-in. angles for a six-row connection of $\frac{3}{4}$-in. A325-N bolts is seen to equal 206 k.

This value is based on the shearing strength vertically through the holes for 2 angles as determined from the formula to follow in which ϕ is 0.75 and d is the bolt diameter. Section J4 of the LRFD Specification says that in computing shear capacity we use $d + \frac{1}{16}$ in. rather than $d + \frac{1}{8}$ in. as for tension members.

$$R_u = \phi 2t[(L \text{ or } L') - n(d + 1/16)]0.6\,F_u$$

$$R_u = (0.75)(2 \times 5/16)[(17.5) - (6)(3/4 + 1/16)](0.6)(58) = 205.9 \text{ k} \checkmark$$

Table I-E Bearing Capacity of Bolts.   For one $\frac{3}{4}$-in. A325-N bolt bearing on 1 in. the table gives a value of 78.3 k. As we have 6 bolts bearing on 0.545 in. our total bearing capacity is $(6)(78.3)(0.545) = 256$ k > 160 k.

This value can be checked as follows with the usual bearing expression $\phi\,2.4\,dtF_u$ with $\phi = 0.75$.

$$(6)(0.75)(2.4)(3/4)(0.545)(58) = 256 \text{ k} \checkmark$$

Table I-F Bearing Capacity of Angles.   The top bolt with a $1\frac{1}{4}$-in. edge distance from the top (as per LRFD Table 1.F) can resist 54.4 k for the two angles if it's bearing on 1 in. but it's bearing on $2 \times \frac{5}{16}$ in. of angle thickness.

Top bolt carries $(1)(54.4)(2 \times 5/16) = 34$ k

Other 5 bolts carry $(5)(78.3)(2 \times 5/16) = \underline{244.6 \text{ k}}$

Total $= 278.6$ k > 160 k

This value can be checked as follows with LRFD Formulas J3.1c and J3.1a. For the top bolt bearing on $2 \times \frac{5}{16}$ in. and located $1\frac{1}{4}$ in. from the top edge we have $\phi\,LtF_u = (0.75)(1.25)(2 \times \frac{5}{16})(58)$ plus the strength of the other 5 bolts $(0.75)(2.4)(2 \times \frac{5}{16})(58) = 278.7$ k ✓.

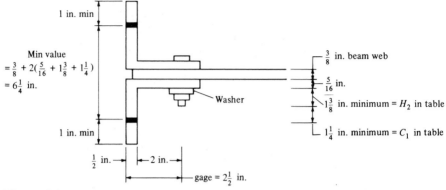

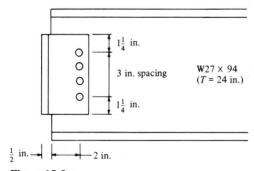

Figure 15-8

To select the lengths of the angle legs it is necessary to study the dimensions given in Fig. 15-8 which is a view along section *A-A* in Fig. 15-7. On the lower right part of the figure are shown the minimum clearances needed for insertion and tightening of the bolts. These are the H_2 and C_1 distances shown and are obtained for $\frac{3}{4}$-in. bolts from the table of "Assembling Clearances" given in Part 5 of the LRFD Manual.

For the legs bolted to the beam web a $2\frac{1}{2}$-in. gage is used. Using a minimum edge distance of 1 in. here we will make these angle legs $3\frac{1}{2}$ in. For the outstanding legs the minimum gage is $\frac{5}{16} + 1\frac{3}{8} + 1\frac{1}{4} = 2\frac{15}{16}$ in. say 3 in. We will make this a 4 in. angle leg.

$$\underline{\text{Use 2Ls } 4 \times 3\tfrac{1}{2} \times \tfrac{5}{8} \times 1 \text{ ft} - 5\tfrac{1}{2} \text{ in.}}$$

The design of another bolted framed connection is presented in Example 15-2. The problem is very close to the one of Example 15-1 except that the strength of the connection is controlled by the shear capacity of the net section of the angles rather than by the shear capacity of the bolts.

EXAMPLE 15-2

Select a framed beam connection for the W27×94 ($t_w = 0.490$ in.) shown in Fig. 15-9. This simply supported beam consists of A36 steel and has a factored end reaction R_u equal to 200 k. The connection is to be made with $\frac{7}{8}$-in. A325-N bolts used in standard $\frac{15}{16}$-in. holes.

Figure 15-9

Solution

    Table II-A Shear Strength of Bolts. From this table we select a 5-row connection with $\frac{3}{8}$-in. angles and length $L = 14\frac{1}{2}$ in. (seems ok with respect to 24-in. T) and a capacity of 211 k. Footnote c in the table indicates, however, that the capacity is controlled by the net shear on the angles.

    Table II-C Shear Capacity of Connection Angles. The design shear capacity of the angles we selected in the last step is equal to 192 k < 200 k. Therefore, we go to a six-row connection which has a capacity of 232 k > 200 k.

    Table I-E Bearing Capacity of Bolts

Bearing capacity of the bolts = (6)(91.3)(0.490) = 268.4 k > 200 k

    Table I-F Bearing Capacity of Angles

Bearing capacity of the bolts = (1)(54.4)(2 × 3/8) =   40.8

$$+ \ (5)(91.3)(2 \times 3/8) = \underline{342.4}$$

$$383.2 \text{ k} > 200 \text{ k}$$

    Use 2Ls 6 × $3\frac{1}{2}$ × $\frac{3}{8}$ × 1 ft $5\frac{1}{2}$ in. (lengths of legs selected as described in solution for Problem 15.1)

15-6 DESIGNS OF STANDARD WELDED FRAMED CONNECTIONS

Table III of Part 5 of the LRFD Manual presents the information needed for designing welded framing angles for beams. This table is normally used where the angles are attached to the beams in the shop and then field-bolted to other members. Should the framing angles be welded to both members the weld values provided in Table IV of the Manual are considered more appropriate.

    The weld used to connect the angles to the beam is called Weld A as shown in Fig. 15-10. If a weld is used to connect the beam to another beam or column that weld is called Weld B as shown in Fig. 15-12.

    For the usual situations 4 × $3\frac{1}{2}$ in. angles are used with the $3\frac{1}{2}$-in. legs connected to the beam webs. The 4-in. outstanding legs will usually accommodate the standard gages for the bolts going into the other members. The angle thick-

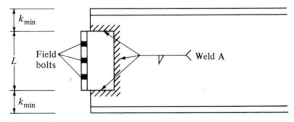

Figure 15-10

ness selected equals the weld size plus $\frac{1}{16}$ in. or the minimum value given in Table II-A for the bolts. The angle lengths are the same as those used for the nonstaggered bolt cases (that is, $5\frac{1}{2}$ through $29\frac{1}{2}$ in.).

The design strengths of the welds to the beam webs (Weld A) given in Table III of the Manual were computed taking into account eccentricities using the instantaneous center method which was briefly discussed in Chapter 14. To select a connection of this type the designer picks a weld size from Table III and then goes to Table II-A to determine the number of bolts required for connection to the other member. This procedure is illustrated in Example 15-3.

EXAMPLE 15-3

Design a framed beam connection to be welded (SMAW) to a W30×90 beam ($t_w = 0.470$ in. and $T = 26\frac{3}{4}$ in.) and then bolted to another member. The factored beam reaction R_u is 210 k, the steel is A36, the weld E70, and the bolts are $\frac{3}{4}$-in. A325-N.

Solution

Table III   Weld A Design Strength.   From this table one of several possibilities is a $\frac{1}{4}$-in. weld $20\frac{1}{2}$ in. long with a design strength of 269 k. The depth or length is compatible with the T value of $26\frac{3}{4}$ in. For a $\frac{1}{4}$-in. weld the minimum angle thickness is $\frac{5}{16}$ in. as per LRFD Specification J2.2. The minimum beam web thickness is 0.55 in., which is greater than the 0.470 in. furnished. The weld capacity is reduced proportionately to the following:

$$\frac{0.470}{0.55}(269) = 230 \text{ k} > 210 \text{ k} \qquad\qquad \text{OK}$$

Table II-A   Shear Strength of Bolts.   The angle length selected is $20\frac{1}{2}$, and this corresponds to a 7-row bolted connection in Table II-A. From this latter table we find that 7 rows of $\frac{3}{4}$-in. A325-N bolts will support 217 k and an angle of $\frac{5}{16}$ in. thickness is required.

Use 2Ls $4 \times 3\frac{1}{2} \times \frac{5}{16} \times 1$ ft $8\frac{1}{2}$ in.

Table IV in the Manual (Part 5) provides the information necessary for designing all-welded standard framed connections, that is, with welds A and B as shown in Fig. 15-11. The design strengths for welds A again (as in Table III) have taken into account the eccentricity of the load but the values given for welds B do not.

Erection bolts are often necessary for erecting these beams. They are probably placed near the bottom of the angles so they will not appreciably reduce the flexibility of the angles. In some situations they may be necessary at the top of these angles to stabilize the joint during erection. Such bolts may be removed at a later date if it is felt that they provide too much restraint against rotation.

Example 15-4, which follows, illustrates the design of a framed beam connection connected with welds A and B.

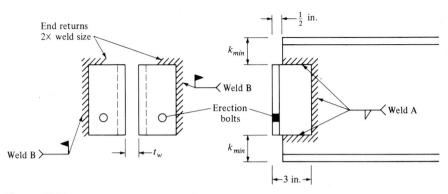

Figure 15-11

EXAMPLE 15-4
Select a standard framed connection with welds A and B for an A W30×90 ($t_w = 0.470$ in., $T = 26\frac{3}{4}$ in.) with a factored reaction equal to 210 k. The welds are E70 and SMAW.

Solution

Table IV   Welds A and B Design Strengths.   For weld A one possibility is a $\frac{1}{4}$-in. weld 20 in. long (compatible with T) which will support 257 k. A $\frac{5}{16}$-in. weld B for this case will support 228 k. The minimum web thickness is 0.55 in. As a result the weld capacity is reduced proportionately as follows:

$$\frac{0.470}{0.55}(257) = 219.6 \text{ k} > 210 \text{ k} \qquad\qquad \text{OK}$$

Use 2 Ls 4 × 3 × $\frac{3}{8}$ × 0 ft 8 in.

15-7 SINGLE-PLATE FRAMING CONNECTIONS

A rather economical type of flexible connection which is being used more and more frequently and which is not presented in the LRFD Manual is the single-plate framing connection which was illustrated in Fig. 15-1 (d). The bolt holes are prepunched in the plate and the web of the beam. Then the plate is shop-welded to the supporting beam or column, and lastly the beam is bolted to the plate in the field. Steel erectors like this kind of connection because of its simplicity. They are particularly pleased with it when there is a beam connected to each side of a girder as shown in Fig. 15-12 (a). All they have to do is bolt the beam webs to the single plate on each side of the girder. Should web clip angles be used for such a connection the bolts will have to pass through the angles on each side of the girder as well as the girder web as shown in part (b) of the figure. This is a litte more difficult field operation.

With the single-plate connection the reaction or shear load is assumed to be distributed equally among the bolts passing through the beam web. It is also

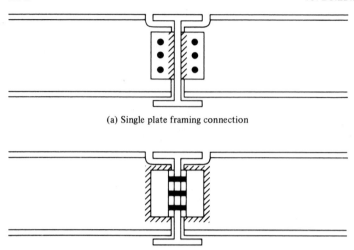

(a) Single plate framing connection

(b) Simple connection with web angles

Figure 15-12   (a) Single-plate framing connection. (b) Simple connection with web angles.

assumed that relatively free rotation occurs between the end of the member and the supporting beam or column. Because of these assumptions this type of connection is often referred to as a "shear tab" connection. Various studies and tests have shown that these connections can develop some end moments depending on the number and size of the bolts and their arrangement, the thicknesses of the plate and beam web, the span-to-depth ratio of the beam, the type of loading, and the flexibility of the supporting element.

An example design of a single-plate framing connection is not presented here because they are not included in the tables of the LRFD Manual. A proposed (*very empirical*) design procedure using service loads is available.[3]

15-8 DESIGNS OF WELDED SEATED BEAM CONNECTIONS

Another type of fairly flexible beam connection is obtained by using a beam seat such as the one shown in Fig. 15-13. Beam seats are obviously of advantage to the worker performing the steel erection. The connections for these angles may be bolts or welds, but only the welded type is considered here because of lack of space. For such a situation the seat angle would usually be shop-welded to the column and field-welded to the beam. When welds are used, the seat angles, which are also called *shelf angles,* may be punched for erection bolts as shown in the figure. These holes can be slotted if desired to permit easy alignment of the members.

[3]R. M. Richard et al. "The Analysis and Design of Single Plate Framing Connections," *Engineering Journal,* AISC, vol. 17, no. 2 (second quarter, 1980), pp. 38–52.

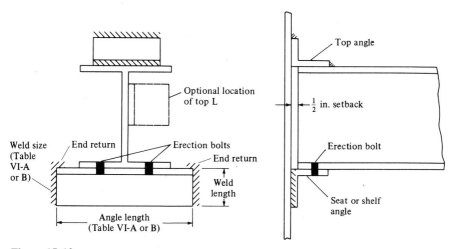

Figure 15-13

A seated connection may be used only when a top angle is used, as shown in Fig. 15-13. This angle, which provides lateral support for the beam, may be placed on top of the beam, or at the optional location shown on the side of the beam in part (a) of the figure. As the top angle is not usually assumed to resist any of the load its size is probably selected by judgment. Fairly flexible angles are used which will bend away from the column or girder to which it is connected when the beam tends to rotate downward under load. This desired situation was illustrated in Fig. 15-6 (b). A common top angle size selected is $4 \times 4 \times \frac{1}{4}$.

As will be seen in the LRFD tables unstiffened seated beam connections of practical sizes can support only fairly light factored loads up to as high as 102 k if 1-in.-thick seat angles of A36 steel are used. For loads of this size two vertical end welds on the seat are sufficient. The top angle is welded only on its toes so that as the beam tends to rotate this thin flexible angle will be free to pull away from the column.

The design loads shown in Tables VI-A and VI-B were calculated on the basis of a $\frac{3}{4}$-in. setback rather than the nominal $\frac{1}{2}$-in. value used for web framing angles. This value is used to provide for possible mill underrun in the beam length. The values given in Table VI-A are based on beams and seat angles both made from A36 steel. The values in Table VI-B are based on A36 seat angles and $F_y = 50$ ksi beams. The numbers in both of these tables represent the maximum loads which can be placed out on the horizontal outstanding legs of the seat angles.

Table VI-C shows the design strengths of the seat angle welds. These values were determined from the elastic (or vector) method which was previously described in Section 14-13 of the last chapter. Example 15-6, which follows, illustrates the selection of a welded unstiffened beam seat and top angles for a beam.

EXAMPLE 15-5

Design a welded unstiffened seated beam connection to support a factored beam reaction of 60 k. The beam is to be a W24×55 ($t_w = \frac{3}{8}$ in.). It is assumed that there is sufficient length along the beam web to install an 8-in.-long seat angle. Use A36 steel and E70 SMAW fillet welds.

Solution

    Table VI-A. For a $\frac{3}{8}$-in. beam web thickness it is necessary to go to an 8-in. outstanding seat angle leg to the value 75.0 k. For this value a 1-in. thick seat angle is required as shown in the table.

    Table VI-C. It is possible to support the 60-k load with a $5 \times 3\frac{1}{2}$ in. angle and a $\frac{5}{8}$-in. weld, or a 6×4 angle with a $\frac{1}{2}$-in. weld, or a 7×4 angle with a $\frac{3}{8}$-in. weld. However, with an 8×4 angle only a one-pass weld size of $\frac{5}{16}$ in. is needed.

$$\text{Use seat L } 8 \times 4 \times 1 \times 0 \text{ ft 8 in.}$$
$$\text{with a top L } 4 \times 4 \times \tfrac{1}{4}$$

15-9 STIFFENED SEATED BEAM CONNECTIONS

When beams are supported by seated connections and when the factored reactions become fairly large (say above 100 k depending on the steel grade) it is necessary to stiffen the seats. These larger reactions cause moments in the outstanding or horizontal legs of the seat angles which cannot be supported by standard thickness angles unless they are stiffened in some manner. Typical stiffened seated connections are shown in Fig. 15-5 (f) and in Fig. 15-14.

    Stiffened seats may be bolted or welded. Bolted seats may be stiffened with a pair of angles as shown in part (a) of Fig. 15-14. Structural tee stiffeners either bolted or welded may be used. A welded one is shown in part (b) of the same figure. Also commonly used are welded two-plate stiffeners such as the one shown in part (c).

    Example designs of both bolted and welded stiffened seat connections are presented in Part 5 of the LRFD Manual. These designs make use of Tables VII and VIII in the Manual.

15-10 DESIGN OF MOMENT-RESISTING CONNECTIONS

In this section moment-resisting connections are considered. It is not the purpose of the author to describe all the different arrangements of bolted and welded moment-resisting connections. He instead attempts to present the basic theory of transferring shear and moment from the beam to another member and provides one simple numerical example. If this information is clearly understood the reader should be able to satisfactorily design other moment-resisting connections whatever the arrangements of the bolts and/or welds.

    The beam connections discussed prior to this section were designed with the object of eliminating most of the moment resistance. For fully continuous struc-

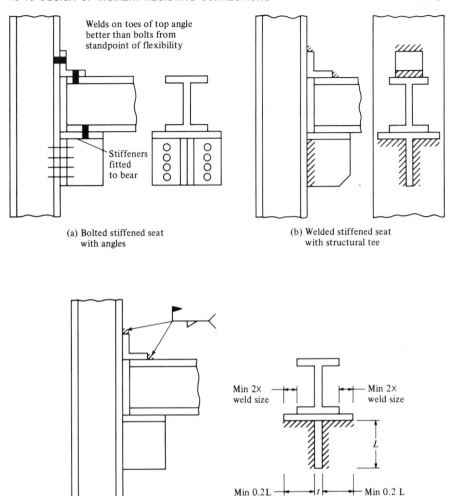

(a) Bolted stiffened seat
with angles

(b) Welded stiffened seat
with structural tee

(c) Welded two-plate stiffened seat

Figure 15-14   (a) Bolted stiffened seat with angles. (b) Welded stiffened seat with structural tee. (c) Welded two-plate stiffened seat.

tures the connections are designed to resist the full calculated moments. Figure 15-15 (a) shows a common type of moment-resisting connection. In the connection shown the tensile force at the top of the beam is transferred by fillet welds to the top plate and by groove welds from the plate to the column. For easier welding the top plate may be tapered as shown in part (b) of the figure. The student may have noticed such tapered plates used for facilitating welding is other situations.

Sometimes the beam flanges are groove-welded flush with the column on one end and connect the beam on the other end with the type of connection just described. This practice is illustrated in Example 15-6.

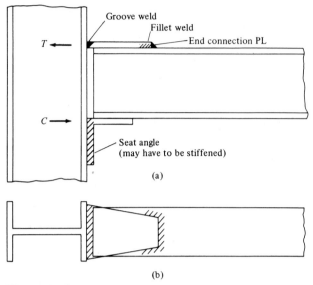

Figure 15-15

EXAMPLE 15-6

Design moment-resistant connections of the type shown in Fig. 15-16 for the ends of a W18×46. The beam, which consists of A36 steel, has factored end reactions of 65 k and end moments of 250 ft-k. SMAW welds and E70 electrodes are to be used. It is assumed that a 6 × 4 × $\frac{3}{4}$-in. seat L has previously been selected.

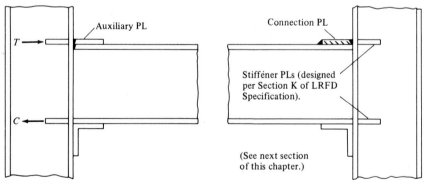

Figure 15-16

Solution

Shear connection: Try $\frac{1}{4}$-in. fillet welds on seat L:

$$\text{Depth required} = \frac{65}{(2)(0.707)(\frac{1}{4})(0.75)(0.6 \times 70)}$$
$$= 5.84 \text{ in. (use 6 in. each side)}$$

Moment connections on flush end: Assuming a bevel butt weld for full flange width:

T = force to be carried

   = moment divided by center-to-center distance of flanges

$$= \frac{12 \times 250}{18.06 - 0.605} = 171.9 \text{ k}$$

Strength of butt weld in tension for full width of flange = $(6.060)(0.605)(0.9)(36) = 118.8$ k. Tension to be resisted by auxiliary PL = $171.9 - 118.8 = 53.1$ k. Assuming a $\frac{3}{8}$-in.-thick PL, its width will equal

$$\frac{53.1}{\frac{3}{8} \times 0.9 \times 36} = 4.37 \text{ in.} \qquad (\text{say } 4\tfrac{1}{2} \text{ in.})$$

Assuming a $\frac{3}{16}$-in. fillet weld on auxiliary PL as shown in Fig. 15-7,

Length of fillet weld $= \dfrac{53.1}{(\frac{3}{16})(0.707)(0.75 \times 0.60 \times 70)} - 4.5 = 12.72$ in.

$$(\text{say } 6\tfrac{1}{2} \text{ in. each side})$$

Allow 1 in. for bevel groove weld.

$$\underline{\text{Use auxiliary PL } \tfrac{3}{8} \times 4\tfrac{1}{2} \times 8,}$$

Moment connection at nonflush end: Assuming T and C a distance apart equal to the beam depth:

$$T = \frac{12 \times 250}{18.06} = 166.1 \text{ k}$$

Assuming top connection has a width something less than the width of the beam flange (say 5 in.), its thickness can be found as follows:

$$t = \frac{166.1}{(5)(0.9 \times 36)} = 1.025 \text{ in.} \qquad (\text{say } 1\tfrac{1}{8} \text{ in.})$$

Assume $\frac{3}{8}$-in. fillet welds of PL:

Required length for weld $= \dfrac{166.1}{(3/8)(0.707)(0.75 \times 0.6 \times 70)} = 19.89$ in.

(see Fig. 15-18)

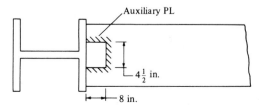

Figure 15-17

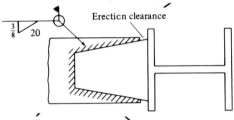

Figure 15-18

15-11 COLUMN WEB STIFFENERS

If a column to which a beam is being connected bends appreciably at the connection the moment resistance of the connection will be reduced no matter how good the connection may be. Furthermore, if the top connection plate in pulling away from the column tends to bend the column flange as shown in part (a) of Fig. 15-19 the middle part of the weld may be greatly overstressed (like the prying action we discussed in Chapter 13 for bolts).

When there is a danger of the column flange bending as described here we must make sure that the desired moment resistance of the connection is provided. This may be done by either using a heavier column with stiffer flanges or by

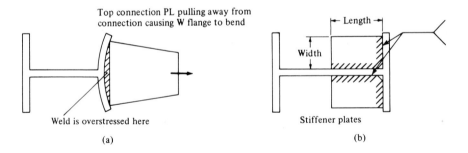

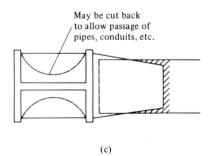

Figure 15-19

introducing column web stiffener plates as shown in part (b) of Fig. 15-19. It is almost always desirable to use a heavier column because column web stiffener plates are quite expensive and a nuisance to use.

Column web stiffener plates are somewhat objectionable to architects as they find it convenient to run pipes and conduits inside their columns but this objection can easily be overcome. First if the connection is to only one column flange the stiffener does not have to run for more than half the column depth as shown in part (b) of Fig. 15-19. If connections are made to both column flanges the column stiffener plates may be cut back to allow the passage of pipes, conduits, etc., as shown in part (c) of the figure.

For this discussion the factored force applied from the beam flange to the column is referred to as P_{bf}. Should its value be larger than any one of the following resisting forces it will be necessary to provide column web stiffeners, says LRFD Specification K1. In the expressions to follow h_c is the depth of the column web between the web toes of the fillets and t_f is the thickness of the beam flange or connection plate from which the concentrated force is being applied.

$$P_{fb} = \text{flange design strength}$$

$$= 5.625 \, t_f^2 F_{yf} \qquad \text{(LRFD Formula K1.1)}$$

$$P_{wi} t_b + P_{wo} = \text{local web yielding strength}$$
$$\text{considered in Chapter 10}$$

$$= F_{yw} t_w (5k + t_b) \qquad \text{(LRFD Formula K1.2)}$$

$$P_{wb} = \text{compression buckling strength of an}$$
$$\text{unstiffened member web}$$

$$= 3690 \left(\frac{t_w}{h_c}\right) t_w^2 \sqrt{F_{yw}} \qquad \text{(LRFD Formula K1.8)}$$

The values of P_{wb}, P_{fb}, and $P_{wi} t_b + P_{wo}$ for the W shapes normally used as columns have been computed and recorded in the column tables of Part 2 of the LRFD Manual for steels with $F_y = 36$ ksi and 50 ksi.

The LRFD Manual presents a set of suggested rules for the design of column web stiffeners. These are as follows:

1. The width of the stiffener plus one-half of the column web thickness should not be less than one-half the width of the beam flange or of the moment connection plate which applies the concentrated force.
2. The stiffener thickness should not be less than $t_b/2$.
3. If there is a moment connection applied to only one flange of the column the length of the stiffener plate does not have to exceed one-half the column depth.
4. The stiffener plate should be welded to the column web with a sufficient strength to carry the force caused by the unbalanced moment on the opposite sides of the column.

For the column given in Example 15-7, which follows, it is necessary to use column web stiffeners or to select a larger column. These two alternatives are considered in the solution.

EXAMPLE 15-7

It is assumed that the column of Example 15-7 is a W12×87 consisting of A36 steel and subjected to 171.9-k C and T forces transferred by the FR type connection through the flanges of the W18×46 beam. It will be found that this column is not satisfactory to resist these forces. (a) Select a larger W12 column section which will be satisfactory. (b) Using a W12×87 column, design column web stiffeners including the stiffener connections using E70 SMAW welds.

Solution

Beam is a W18×46 ($b_f = 6.06$ in., $t_b = t_f = 0.605$ in.)
Column is a W12×87 ($d = 12.53$ in., $t_w = 0.515$ in., $t_f = 0.810$ in.)
Checking to see if forces transferred to column are too large. Using LRFD column tables for the W12×87,

$$P_{fb} = 133 \text{ k} < 171.9 \text{ k} \qquad\qquad\qquad \text{NG}$$

$$P_{wi}t_b + P_{wo} = (19)(0.605) + 139 = 150.5 \text{ k} < 171.9 \text{ k} \qquad \text{NG}$$

$$P_{wb} = 311 \text{ k} > 171.9 \text{ k} \qquad\qquad\qquad \text{OK}$$

∴Either a larger column or column web stiffeners are required.
(a)   Selecting a larger column
By examination a W12×96 will not do because P_{fb} in the table is < 171.9 k.
Try W12×106

$$P_{fb} = 198 \text{ k} > 171.9 \text{ k} \qquad\qquad\qquad \text{OK}$$

$$P_{wi}t_b + P_{wo} = (22)(0.605) + 185 = 198.3 \text{ k} > 171.9 \text{ k} \qquad \text{OK}$$

$$P_{wb} = 518 \text{ k} > 171.9 \text{ k} \qquad\qquad\qquad \text{OK}$$

Use W12×106

(b)   Design of web stiffeners using a W12×87 column and the suggested LRFD rules presented before this example.

$$\text{Required stiffener area} = \frac{171.9 - 133}{36} = 1.08 \text{ in.}^2$$

$$\text{Min. width of stiffener} = \tfrac{1}{3} b_f - \frac{t_w}{2} = \frac{6.06}{3} - \frac{0.515}{2} = 1.76 \text{ in.}$$

$$\text{Min. } t \text{ of stiffeners} = \frac{t_b}{2} = \frac{0.605}{2} = 0.3025 \text{ in.}$$

Required t of stiffeners $= \dfrac{1.08}{1.76} = 0.614$ in.   say $\frac{5}{8}$ in.

Required width $= \dfrac{1.08}{0.625} = 1.728$ in.   say 4 in. for practical conditions

Min. length $= \dfrac{d}{2} - t_f = \dfrac{12.53}{2} - 0.810 = 5.46$ in.   say 6 in.

Use 2 PLs $\frac{5}{8} \times 4 \times 0$ ft 6 in.

Design of welds for stiffener plates
Min. weld size as required by LRFD Specification Table J2.4 $= \frac{1}{4}$ in.

based on the column web t_w of 0.515 in.

Required length of weld $= \dfrac{171.9 - 133}{(0.75 \times 0.60 \times 70)(0.707)(\frac{1}{4})}$

$= 6.99$ in.   say 7 in.

PROBLEMS

15-1. Determine the maximum end reaction that can be transferred through the web angle connection shown in the accompanying illustration. The steel is A36 and the bolts are $\frac{3}{4}$-in. A325-N and are used in standard-size holes. (*Ans. $R_u = 122$ k*)

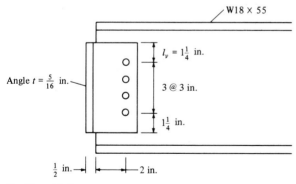

Angle $t = \frac{5}{16}$ in.

W18 × 55

$l_v = 1\frac{1}{4}$ in.

3 @ 3 in.

$1\frac{1}{4}$ in.

$\frac{1}{2}$ in.   2 in.

Problem 15-1

15-2. Repeat Prob. 15-1 if the bolts are $\frac{7}{8}$-in. A325-N.

15-3. Repeat Prob. 15-1 if the bolts are 1-in. A325-X. (*Ans. $R_u = 118$ k*)

15-4 Using the LRFD Manual select a pair of bolted standard web angles for a W33×141 with a dead load service reaction of 150 k and a live load service reaction of 100 k. The bolts are to be $\frac{7}{8}$-in. A325-N in standard-size holes and the steel is A36.

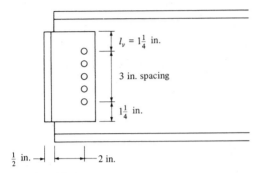

Problem 15-4

15-5. Repeat Prob. 15-4 if 1-in. A325-N bolts are to be used. (*Ans.* 7-row connection 2Ls $7 \times 4 \times \frac{5}{8} \times 1$ ft $8\frac{1}{2}$ in.)

15-6. Repeat Prob. 15-1 if $\frac{3}{4}$-in. A325 S.C. bolts are to be used.

15-7. Design a framed beam connection for a W27×84 to support a factored beam reaction of 250 k. The bolts are to be 1-in. A325-X in standard-size holes, and A36 steel is to be used. The edge distances and bolt spacings are the same as those shown in the sketch for Prob. 15.4. (*Ans.* 6-row connection 2Ls $7 \times 4 \times \frac{5}{8} \times 1$ ft $5\frac{1}{2}$ in.)

15-8. Repeat Prob. 15-1 if the bolts are 1-in. A325-N and are used in $1\frac{1}{16} \times 1\frac{5}{16}$ in. short slots with long axes perpendicular to the transmitted force. Angle *t* is $\frac{5}{8}$ in.

15-9. Design a framed beam connection for a W18×50 to support a factored beam reaction of 65 k. The beam's top flange is to be coped for a 2-in. depth, and $\frac{7}{8}$-in. A325-X bolts in standard-size holes are to be used. A36 steel. (*Ans.* Use 3-row connection with $6 \times 4 \times \frac{5}{8} \times 0$ ft $8\frac{1}{2}$ in. angles)

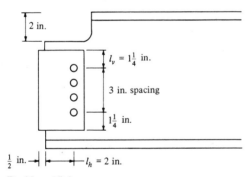

Problem 15-9

15-10. Repeat Prob. 15-7 if the factored beam reaction is to be 285 k and if A242 steel ($F_y = 50$ ksi and $F_u = 70$ ksi) is to be used.

15-11. Select a framed beam connection for a W33×130 beam (A36 steel) with a factored beam reaction of 200 k. The web angles are to be welded with E70 electrodes (weld A in the Manual) and are to be field-connected to the girder with $\frac{7}{8}$-in. A325-N bolts. (*One ans.* 2Ls $4 \times 3\frac{1}{2} \times \frac{5}{16} \times 1$ ft $8\frac{1}{2}$ in., $\frac{1}{4}$-in. weld *A* and 7-row bolt connection to girder)

15-12. Repeat Prob. 15-11 using SMAW shop and field welds (welds A and B in the Manual).

15-13. Select a framed beam connection from the LRFD Manual for a W30×124 consisting of A36 steel (F_y = 36 ksi) using SMAW E70 shop and field welds. The factored beam reaction is 190 k. (*Ans.* 2Ls 4×3×$\frac{3}{8}$×1 ft 6 in. with $\frac{1}{4}$-in. weld *A* and $\frac{5}{16}$-in. weld B)

15-14. Repeat Prob. 15-13 if R_u = 290 k.

15-15. Select a seated beam connection bolted with $\frac{3}{4}$-in. A325-N bolts in standard-size holes for the following data: Beam is W16×67 consisting of A36 steel, R_u is 65 k, and the column gage is $5\frac{1}{2}$ in. (*Ans.* Use L8×4×1×0 ft 8 in., O.S. leg is 4 in. with 6 $\frac{3}{4}$-in. A325-N bolts)

15-16. Repeat Prob. 15-15 if $\frac{7}{8}$-in. A325-X bolts are used to attach the beam flange to the seat using E70 SMAW fillet welds for the other connections.

15-17. Repeat Prob. 15-15 if E70 SMAW fillet welds are used for all connections and if there is sufficient room for an 8-in.-long seat angle. (*Ans.* L8×4×1 with $\frac{5}{16}$-in. weld)

15-18. Design SMAW welded Type FR connections for the ends of a W24×76 to resist an R_u of 120 k and an M_u of 250 ft-k. Use A36 steel and E70 electrodes. Assume the column flange is 16 in. wide. The moment is to be resisted by full-penetration groove welds in the flanges and the shear is to be resisted by welded clip angles along the web. Assume beam was selected for bending with 0.9 F_y.

15-19. The beam shown in the accompanying illustration is assumed to be attached at its ends with Type FR connections. Select the beam assuming full lateral support, A36 steel, and E70 SMAW electordes. Use a connection of the type used in Prob. 15-18. (*Ans.* Use W21×62 with full-groove welds each flange plus auxiliary plate with $\frac{3}{16}$-in. fillet welds)

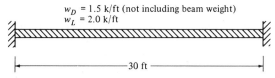

w_D = 1.5 k/ft (not including beam weight)
w_L = 2.0 k/ft

|←————————30 ft ————————→|

Problem 15-19

15-20. The two W16×40 beams shown in the accompanying illustration are to be made continuous across the supporting W24×84 girder. If the factored end reactions are each 80 k and their factored end moments are 70 ft-k, design the connections using E70 electrodes, SMAW welds, and A36 steel.

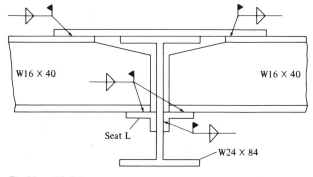

W16 X 40   W16 X 40

Seat L

W24 X 84

Problem 15-20

Chapter 16

Design of Steel Buildings

16-1 INTRODUCTION

The material in this chapter generally pertains to the design of steel buildings from one to several stories in height, while the design of multistory buildings is discussed in Chapter 20. The present discussion applies to apartment houses, office buildings, warehouses, schools, and institutional buildings which are not very tall with respect to their least lateral dimensions. The separating factor between the buildings discussed here and those mentioned in the multistory chapter is the matter of wind forces and not the actual height of the building in question. The usual rule of thumb is that if the height of the building is not greater than twice its least lateral dimension, provision for wind forces is unnecessary. For buildings of these dimensions the walls and partitions probably provide sufficient resistance to wind forces except in unusual circumstances. Following such a rule of thumb, however, is a dangerous engineering practice. Each structure should be considered on the basis of its dimensions, location, surrounding structures, and so on. Then a decision can be made as to whether a definite system of lateral bracing needs to be used.

16-2 TYPES OF STEEL FRAMES USED FOR BUILDINGS

Steel buildings are usually classified as being in one of four groups according to their type of construction. These types are: bearing-wall construction, skeleton construction, long-span construction, and combination steel and concrete framing. More than one of these construction types can be used in the same building. Each of these types is briefly discussed in the following paragraphs.

Bearing-Wall Construction

Bearing-wall construction is the most common type of single-story light commercial construction. The ends of beams or joists or light trusses are supported by the walls that transfer the loads to the foundation. The old practice was to rapidly thicken load-bearing walls as buildings became taller. For instance, the wall on the top floor of a building might be one or two bricks thick while the lower walls might be increased by one brick thickness for each story as we come down the building. As a result this type of construction was usually thought to have an upper economical limit of about 2 or 3 stories. A great deal of research has been conducted for load-bearing construction in recent decades and it has been discovered that thin load-bearing walls may be quite economical for many buildings up to 10 or 20 or even more stories.

The average engineer is not very well versed in the subject of wall-bearing construction with the result that he or she may often specify complete steel or reinforced concrete frames where wall-bearing construction might have been just as satisfactory and at the same time more economical. Wall-bearing construction is not very resistant to seismic loadings and does have an erection disadvantage for buildings of more than one story. For such cases, it is necessary to place the

Coliseum in Spokane, Wash. (Courtesy of Bethlehem Steel Corporation.)

steel members floor by floor as the masons complete their work below thus requiring alternation of the masons and ironworkers.

Bearing plates were probably necessary under the ends of the beams or light trusses which are supported by the masonry walls because of the relatively low bearing strength of the masonry. Although theoretically the beam flanges may on many occasions provide sufficient bearing without bearing plates, the plates are almost always used—particularly where the members are so large and heavy that they must be set by a steel erector. The plates are probably shipped loose and set in the walls by the masons. Setting them in their correct positions and at the correct elevation is a very critical part of the construction. Should they not be properly set there will be some delay in correcting their positions. If a steel erector is used he will probably have to make an extra trip to the job.

When the ends of a beam are enclosed in a masonry wall, some type of wall anchor is desirable to prevent the beam from moving longitudinally with respect to the wall. The usual anchors consist of bent steel bars passing through beam webs. These are called *government anchors* and are shown in Fig. 16-1. Occasionally clip angles attached to the web are used instead of government anchors. These are shown in part (b) of the same figure. Should longitudinal loads of considerable magnitude be anticipated, regular vertical anchor bolts may be used at the beam ends.

For small commercial and industrial buildings bearing-wall construction is quite economical when the clear spans are not greater than roughly 35 or 40 ft. If the clear spans are much greater it becomes necessary to thicken the walls and use pilasters to ensure stability. For these cases it may often be more economical to use intermediate columns if permissible.

Skeleton Construction

In skeleton construction the loads are transmitted to the foundations by a framework of steel beams and columns. The floor slabs, partitions, exterior walls, etc., are all supported by the frame. This type of framing, which can be erected to tremendous heights, is often referred to as beam-and-column construction.

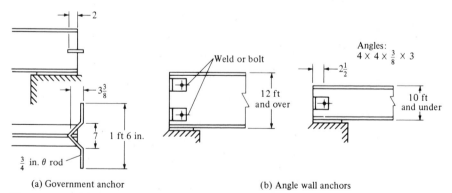

(a) Government anchor   (b) Angle wall anchors

Figure 16-1   (a) Government anchor. (b) Angle wall anchors.

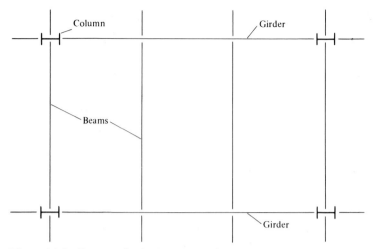

Figure 16-2   Beam and column construction.

In beam-and-column construction the frame usually consists of columns spaced 20, 25, or 30 ft apart with beams and girders framed into them from both directions at each floor level. One very common method of arranging the members is shown in Fig. 16-2. The girders are placed in the longer direction between the columns, while the beams are framed between the girders in the short direction. With various types of floor construction other arrangements of beams and girders may be used.

For skeleton framing the walls are supported by the steel frame and are generally referred to *as nonbearing* or *curtain walls*. The beams supporting the exterior walls are called *spandrel beams*. These beams, which are illustrated in Fig. 16-3, can usually be placed so that they will serve as the lintels for the windows.

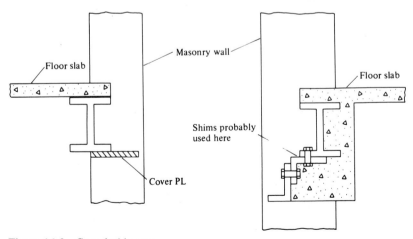

Figure 16-3   Spandrel beams.

Long-Span Steel Structures

When it becomes necessary to use very large spans between columns as for field houses, auditoriums, theaters, hangars, or hotel ballrooms, the usual skeleton construction may not be sufficient. Should the ordinary rolled W sections be insufficient it may be necessary to use coverplated beams, plate girders, box girders, large trusses, arches, rigid frames, and the like. When depth is limited cover-plated beams, plate girders, or box girders may be called upon to do the job. Should depth not be so critical trusses may be satisfactory. For very large spans arches and rigid frames are often used. These various types of structures are referred to as long-span structures. Figure 16-4 shows a few of these types of structures.

Combination Steel and Concrete Framing

A tremendous percentage of the buildings erected today make use of a combination of reinforced concrete and structural steel. If reinforced-concrete columns were used in very tall buildings they would be rather large on the lower floors and take up considerable space. Steel column shapes surrounded by and bonded to reinforced concrete may be used and are referred to as combination or composite

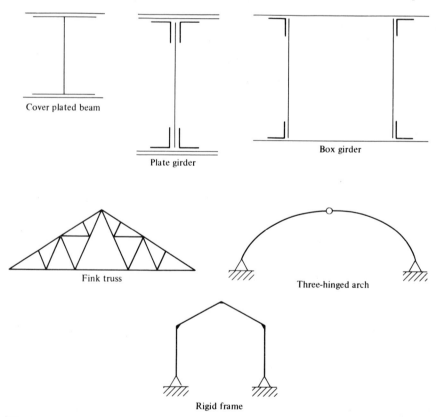

Cover plated beam

Plate girder

Box girder

Fink truss

Three-hinged arch

Rigid frame

Figure 16-4

columns. Chapter 18 is devoted to the design of composite columns where steel sections and reinforced concrete are bonded together in such a manner that they act compositely in resisting loads.

16-3 COMMON TYPES OF FLOOR CONSTRUCTION

Concrete floor slabs of one type or another are used almost universally for steel-frame buildings. They are strong, have excellent fire ratings, and good acoustic ratings. On the other hand, appreciable time and expense is required to provide the formwork necessary for most slabs. Concrete floors are heavy, they need some type of reinforcing bars or mesh included, and there may be a problem involved in making them watertight. Among the many types of concrete floors used today for steel frame buildings are the following.

1. Concrete slabs supported with open-web steel joists (Section 16-4).
2. One-way and two-way reinforced-concrete slabs supported on steel beams (Section 16-5).
3. Concrete slab and steel beam composite floors (Section 16-6).
4. Concrete-pan floors (Section 16-7).
5. Steel decking floors (Section 16-8).
6. Flat slab floors (Section 16-9).
7. Precast concrete slab floors (Section 16-10).

Among the several factors to be considered in selecting the type of floor system to be used for a particular building are: loads to be supported, fire rating desired, sound and heat transmission, dead weight of floor, ceiling situation below (to be flat or have beams exposed), facility of floor for locating conduits, pipes, wiring, etc.; appearance, maintenance required, time required to construct, and depth available for floor.

The student can obtain a great deal of information about these and other construction practices by referring to various engineering magazines and catalogs particularly *Sweet's Catalog File* published by McGraw-Hill Information Systems Company of New York. The author cannot make too strong a recommendation for the student to examine these books to see the tremendous amount of data available there which may be of use in his or her engineering practice. The sections to follow present brief descriptions of the floor systems mentioned in this section along with some discussions of their advantages and chief uses.

16-4 CONCRETE SLABS ON OPEN-WEB STEEL JOISTS

Perhaps the most common type of floor slab in use for small steel-frame buildings is the slab supported by open-web steel joists. The joists are small parallel chord trusses whose members are often made from bars (from whence the common name *bar joist*) or small angles or other rolled shapes. Steel forms or decks are usually attached to the joists by welding or self-drilling, self-tapping screws, then concrete slabs are poured on top. This is one of the lightest types of concrete floors and also one of the most economical ones. A sketch of an open-web joist floor is shown in Fig. 16-5.

Open-web joists are particularly well suited to building floors with relatively light loads and for structures where there is not too much vibration. They have been used a good deal for fairly tall buildings but generally speaking they are better suited for the shorter buildings. They are very satisfactory for supporting floor and roof slabs for schools, apartment houses, hotels, office buildings, restaurant buildings, and other similar low-level buildings.

Open-web joists must be braced laterally to keep them from twisting or buckling and also to keep the floors from being too springy. Lateral support is provided by *bridging,* which consists of continuous horizontal rods fastened to the top and bottom chords of the joists or of diagonal cross bracing. Bridging is desirably used at spaces not exceeding 7 ft on centers.

Open-web joists are easy to handle and are quickly erected. If desired, a ceiling can be attached to the bottom of the joists or suspended therefrom. The open spaces in the webs are admirably suited for placing conduits, ducts, wiring, piping, etc. The joists should be either welded to supporting steel beams or well anchored in the masonry walls. When concrete slabs are placed on top of the joists they are usually from 2 to $2\frac{1}{2}$ in. thick. Nearly all of the many concrete slabs on the market today can be used successfully on top of open-web joists.

Open-web joists are selected from manufacturers' catalogs. The student can learn a great deal about open-web joists by studying various manufacturers' literature. An exceptionally useful reference of this type is the previously mentioned *Sweet's Catalog File.*

The joists are designated as "J-Series" when the material used has a minimum yield point of 36 ksi (248 MPa), and as "H-Series" when the chord material consists of steel with a minimum yield point of 50 ksi (345 MPa). There are also long-span joists "L-J Series" and the deep long-span joists (DLJ Series with $F_y = 36$ ksi) intended for roof construction. The DLH Series are deep long-span joists with minimum yield points of 50 ksi. In the tables the number preceding the joist number is the nominal depth of the joist in inches, the letter or letters (as J or DLJ) indicates the joist series, and the number to the right designates the chord section.

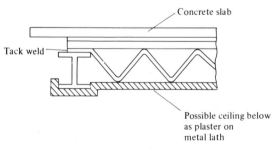

Figure 16-5 Open-web joists.

Short-span joists in the Univac Center of Sperry-Rand Corporation, Blue Bell, Pa. (Courtesy of Bethlehem Steel Corporation.)

The tables provided in the manufacturers' catalogs give the safe uniform loads in pounds per foot which the joists can support. In addition, the maximum end reaction in pounds which a joist can support from the standpoint of shear and the maximum uniform live load in pounds per foot which it can support without having a deflection larger than $\frac{1}{360}$ of the span are provided.

16-5 ONE-WAY AND TWO-WAY REINFORCED-CONCRETE SLABS

One-Way Slabs

A very large number of concrete floor slabs in old office and industrial buildings consisted of one-way slabs about 4 in. thick supported by steel beams 6 to 8 ft on centers. These floors were often referred to as concrete arch floors because at one time brick or tile floors were constructed in approximately the same shape, that is, in the shape of arches with flat tops.

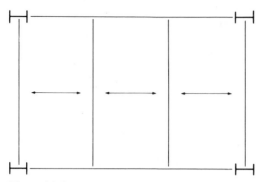

Figure 16-6

A one-way slab is shown in Fig. 16-6. The slab spans in the short direction shown by the arrows in the figure. One-way slabs are usually used when the long direction is two or more times the short direction. In such cases the short span is so much stiffer than the long span that almost all of the load is carried by the short span. The short direction is the main direction of bending and will be the direction of the main reinforcing bars in the concrete, but temperature and shrinkage steel is needed in the other direction.

A typical cross section of a one-way slab floor with supporting steel beams is shown in Fig. 16-7. When steel beams or joists are used to support reinforced-concrete floors it may be necessary to encase them in concrete or other materials to provide the required fire rating. Such a situation is shown in the figure.

It may be necessary to leave steel lath protruding from the bottom flanges or soffits of the beam for the purposes of attaching plastered ceilings. Should such ceilings be required to cover the beam stems, this floor system will lose a great deal of its economy.

One-way slabs have an advantage when it comes to formwork in that the forms can be supported entirely by the steel beams with no vertical shoring needed. They have a disadvantage in that they are much heavier than most of the newer lightweight floor systems. The result is that they are not used as often as formerly for lightly loaded floors; but when a floor to support heavy loads is desired or a rigid floor or a very durable floor, the one-way slab may be a very desirable selection.

Two-Way Concrete Slabs

The two-way concrete slab is used when the slabs are square or nearly so and supporting beams are planned under all four edges. The main reinforcing runs in both directions. Other characteristics are similar to those of the one-way slab.

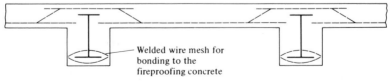

Welded wire mesh for
bonding to the
fireproofing concrete

Figure 16-7

The 104-ft joists for the Bethlehem Catholic High School, Bethlehem, Pa. (Courtesy of Bethlehem Steel Corporation.)

16-6 COMPOSITE FLOORS

Composite floors are those in which the steel beams (rolled sections, coverplated beams, or built-up members) are bonded together with the concrete slabs in such a manner that the two act as a unit in resisting the total loads which the beam sections would otherwise have to resist alone. There can be a saving in sizes of steel beams when composite floors are used because the slabs act as part of the beams.

A particular advantage of composite floors is that they utilize concrete's high compressive strength by keeping all or nearly all of the concrete in compression, and at the same time stress a larger percentage of the steel in tension than is normally the case with steel-frame structures. The result is less steel tonnage in the structure. A further advantage of composite floors is that they can permit an appreciable reduction in total floor thickness which is particularly important in taller buildings.

Two types of composite floor systems are shown in Fig. 16-8. The steel beam can be completely encased in the concrete and the horizontal shear transferred by friction and bond (plus some shear reinforcement if necessary). This type of composite floor, which is usually uneconomical, is shown in part (a) of the figure. A second possibility is shown in part (b), where the steel beam is bonded to the concrete slab with some type of shear connectors. Various types of shear connectors have been used during the past few decades including spiral bars,

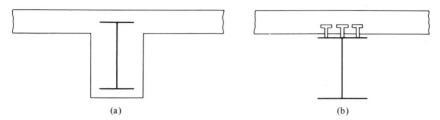

(a)   (b)

Figure 16-8 Composite floors. (a) Steel beam encased in concrete. (b) Steel beam bonded to concrete slab with shear connectors.

channels, angles, studs, etc., but economic considerations usually lead to the use of round studs welded to the top flanges of the beams in place of the other types mentioned. Typical studs are $\frac{1}{2}$ to $\frac{3}{4}$ in. in diameter and 2 to 4 in. in length.

Cover plates may occasionally be welded to the bottom flanges of rolled steel sections. The student can see that with the slab acting as a part of the beam there is quite a large area available on the compressive side of the beam. By adding plates to the tensile flange a little better balance is obtained. Chapter 17 is devoted entirely to the design of composite floors.

16-7 CONCRETE-PAN FLOORS

There are several types of pan floors which are constructed by placing concrete on removable pan molds. (Some special light corrugated pans are also available which can be left in place.) Rows of the pans are arranged on wooden floor forms and the concrete is placed over the top of them producing a floor cross section similar to the one shown in Fig. 16-9. Joists are formed between the pans, giving a tee-beam type floor.

One-piece metal dome pans. (Courtesy of Gateway Erectors, Inc.)

Temple Plaza parking facility, Salt Lake City, Utah. (Courtesy of Ceco Steel Products Corporation.)

These floors, which are suitable for fairly heavy loads, are appreciably lighter than the one-way and two-way concrete slab floors. Although fairly light, they require a good deal of formwork including appreciable shoring underneath the stems. Labor is thus higher than for many floors but saving due to weight reduction and reuse of standard-size pans may make them competitive. If suspending ceilings are required pan floors will have a decided economic disadvantage.

Two-way construction is available—that is, with ribs or stems running in both directions. Pans with closed ends are used and the result is a waffle type floor. This latter type floor is usually used when the floor panels are square or nearly so. Two-way construction can be obtained for reasonably economical prices and makes a very attractive ceiling below and one which has fairly good acoustical properties.

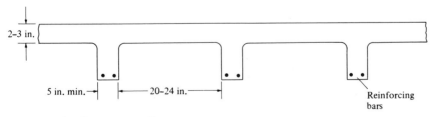

Figure 16-9   Concrete-pan floor.

16-8 STEEL-DECKING FLOORS

Typical cross sections of steel-decking floors are shown in Fig. 16-10. Only two types are shown in the figure but several other variations are available. Today formed steel decking with a concrete topping is by far the most common type of floor system used for office and apartment buildings. They are also popular for hotels and other buildings where the loads are not very large.

A particular advantage of steel-decking floors is that as soon as the decking is placed a working platform is available for the workers. The light steel sheets are quite strong and can span up to 20 ft or more. Due to the considerable strength of the decking the concrete does not have to be particularly strong. This fact permits the use of lightweight concrete as thin as 2 or $2\frac{1}{2}$ in.

The cells in the decking can be conveniently used for placing conduits, pipes, and wiring. The steel is probably galvanized and if exposed underneath can be left as it comes from the manufacturer or painted as desired. Should fire resistance be necessary, a suspended ceiling with metal lath and plaster may be used. The same is necessary if a flat ceiling is required below for the types of decking shown in Fig. 16-10 but decking is available with a flat soffit.

16-9 FLAT SLABS

Formerly flat slab floors were limited to reinforced-concrete buildings but today it is possible to use them in steel-frame buildings. A flat slab is a slab that is reinforced in two or more directions and transfers its loads to the supporting columns without the use of beams and girders protruding below. The supporting concrete beams and girders are made so wide that they are the same depth as the slab.

Flat slabs are of great value when the panels are approximately square,

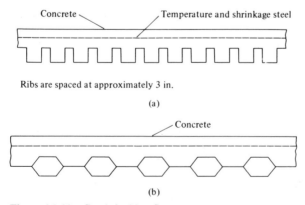

Ribs are spaced at approximately 3 in.

(a)

(b)

Figure 16-10   Steel-decking floors.

when more headroom is desired than is provided with the normal beam and girder floors, when heavy loads are anticipated, and when it is desired to place the windows as near to the tops of the walls as possible. Another advantage is the flat ceiling produced for the floor below. Although the large amounts of reinforcing steel required cause additional costs, the simple formwork cuts expenses decidedly. The significance of simple formwork will be understood when it is realized that over one-half of the cost of the average poured concrete floor slab is in the formwork.

For some reinforced-concrete frame buildings with flat slab floors, it is necessary to flare out the tops of the columns forming column capitals and perhaps thicken the slab around the column with the so-called drop panels. These items, which are shown in Fig. 16-11, may be necessary to prevent shear failures in the slab around the column.

It is possible today in steel-frame buildings to use short steel cantilever beams connected to the steel columns and embedded in the slabs. These beams serve the purposes of the flared columns and drop panels in ordinary flat-slab construction. This latter arrangement is often called a *steel grillage* or *column head*. The flat slab is not a very satisfactory type of floor system for the usual tall building where lateral forces (wind or earthquake) are appreciable, because protruding beams and girders are desirable to serve as part of the lateral bracing system.

16-10 PRECAST CONCRETE FLOORS

Precast concrete sections are more commonly associated with roofs than they are with floor slabs but their use for floors is increasing. They are quickly erected and reduce the need for formwork. Lightweight aggregates are generally used in the concrete making the sections light and easy to handle. Some of the aggregates used make the slabs nailable and easily cut and fitted on the job. For floor slabs with their fairly heavy loads, the aggregates should be of a quality that will not greatly reduce the strength of the resulting concrete.

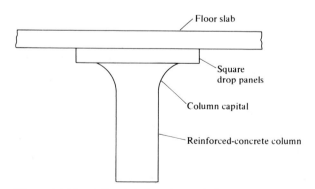

Figure 16-11   A flat slab floor for a reinforced-concrete building.

Interns' living quarters, Good Samaritan Hospital, Dayton, Ohio. (Courtesy of the Flexi-core Company, Inc.)

The reader is again referred to *Sweet's Catalog File* at this point. In these catalogs a great amount of information is available on the various types of precast floor slabs on the market today. A few of the common types of precast floor slabs available are listed after this paragraph and a cross section of each type mentioned is presented in Fig. 16-12. Due to slight variations in the upper surfaces of precast sections it is necessary to use a mortar topping of 1 to 2 in. before asphalt tile or other floor coverings can be installed.

1. *Precast concrete planks* are roughly 2 to 3 in. thick, 12 to 24 in. wide, and are placed on joists up to 5 or 6 ft on center. They probably have tongue and groove edges.
2. *Hollow-cored slabs* are sections of roughly 6- to 8-in. depth and 12- to 18-in. widths, which have hollow perhaps circular cores formed in them in a longitudinal direction, thus reducing their weights by approximately 50 percent (as compared with solid slabs of the same dimensions). These sections, which can be used for spans of from roughly 10 to 25 ft, may be prestressed.
3. *Prefabricated concrete block systems* are slabs made by tying together precast concrete blocks with steel rods (may also be prestressed). These

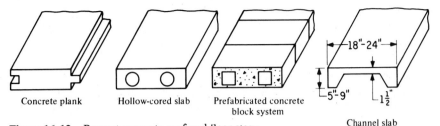

Concrete plank   Hollow-cored slab   Prefabricated concrete   Channel slab
                                                   block system

Figure 16-12   Precast concrete roof and floor slaps.

very seldom used floors vary from 4 to 8 in. deep, and can be used for
spans of roughly 8 to 32 ft.

4. *Channel slabs* are used for spans of approximately 8 to 24 ft. Rough
   dimensions of these slabs are given in Fig. 16-12.

16-11 TYPES OF ROOF CONSTRUCTION

The types of roof construction commonly used for steel-frame buildings include
concrete slabs on open-web joists, steel-decking roofs, and various types of precast concrete slabs. For industrial buildings the predominant roof system in use
today is one which involves the use of a cold rolled steel deck. In addition, the
other types mentioned in the preceding sections on floor types are occasionally
used but they usually are unable to compete economically. Among the factors to
be considered in selecting the specific types of roof construction are strength,
weight, span, insulation, acoustics, appearance below, and type of roof covering
to be used.

The major differences between floor-slab selection and roof-slab selection
probably occur in the considerations of strength and insulation. The loads applied
to roofs are generally much smaller than those applied to floor slabs, thus permitting the use of many types of lightweight aggregate concretes which may be
appreciably weaker. Roof slabs should have good insulation properties or they
will have to have insulation materials placed on them and covered by the roofing.
Among the many types of lightweight aggregates used are wood fibers, zonolite,
foams, sawdust, gypsum, expanded shale, etc. Although some of these materials
decidedly reduce concrete strengths, they provide very light roof decks with
excellent insulating properties.

Precast slabs made with these aggregates are light, quickly erected, have
good insulating properties, and usually can be sawed and nailed. For poured concrete slabs on open-web joists several lightweight aggregates work very well (zonolite, foams, gypsum, etc.), and the resulting concrete can easily be pumped up to
the roofs, thereby facilitating construction. By replacing the aggregates with certain foams, concrete can be produced which is so light it will float in water.
Needless to say the strength of the resulting concrete is quite low.

Steel decking with similar thin slabs of lightweight and insulating concrete placed on top make very good and economical roof decks. A competitive variation consists of steel decking with rigid insulation board placed on top followed by the regular roofing material. The other concrete-slab types are hard pressed to compete economically with these types for lightly loaded roofs. Should other types of poured concrete decks be used, the labor will be much higher. Furthermore, if an insulating type of lightweight aggregate is not used it will be necessary to place some type of insulating board on top of the slab before the roofing is applied.

16-12 EXTERIOR WALLS AND INTERIOR PARTITIONS

Exterior Walls

The purposes of exterior walls are to provide resistance to atmospheric conditions including insulation against heat and cold, satisfactory sound-absorption and light-refraction characteristics, sufficient strength, and fire ratings. They should have satisfactory appearance and be reasonably economical.

For many years exterior walls were constructed of some type of masonry, glass, or corrugated sheeting. In recent decades, however, the number of satisfactory materials for exterior walls has increased tremendously. Available and commonly used today are precast concrete panels, insulated metal sheeting, and many other prefabricated units. Of increasing popularity are the light prefabricated sandwich panels which consist of three layers. The exterior surface is made from aluminum, stainless steel, ceramic or plastic materials, and others. The center of the panel consists of some type of insulating material such as fiberglass or fiberboard, while the interior surface is probably made from metal, plaster, masonry, or some other attractive material.

Interior Partitions

The main purpose of interior partitions is to divide the inside space of a building into rooms. Their selection is based on appearance, fire rating, weight, or acoustical properties. Partitions are referred to as being bearing or nonbearing. These two types are described in the following two paragraphs.

1. *Bearing partitions* are those that support gravity loads in addition to their own weights and are thus permanently fixed in position. They can be constructed from wood or steel studs or from masonry units and faced with plywood, plaster, wallboard, or other material.

2. *Nonbearing partitions* are those that do not support any loads in addition to their own weights and thus may be fixed or movable. Selection of the type of material to be used for a partition is based on the answers to the following questions: Is the partition to be fixed or movable? Is it to be transparent or opaque? Is it to extend all the way to the ceiling? Is it to be used to conceal piping and electrical conduits? Are there fire-rating and acoustical requirements? Perhaps the more common types are made of metal, masonry, or concrete. For design of the floor slabs in a building that has movable partitions, some allowance should be given to the fact that the partitions may be moved. Probably the usual practice is to increase the floor-design live load by 15 or 20 psf over the entire floor area.

16-13 FIREPROOFING OF STRUCTURAL STEEL

Although structural steel members are incombustible their strength is tremendously reduced at temperatures normally reached in fires when the other materials of a building burn. Many disastrous fires have occurred in empty buildings where the only fuel for the fires was the buildings themselves. It is true that steel is an excellent heat conductor and nonfireproofed steel members may transmit enough heat from one burning compartment of a building to ignite materials with which they are in contact in adjoining sections of the building.

Steels (particularly those with rather high carbon contents) may actually increase a little in strength as they are heated from room temperature up to approximately 600°F. As temperatures are raised into the 800° to 1000°F range steel strengths are drastically reduced and at 1200°F they have little strength left.

The fire resistance of structural steel members can be greatly increased by applying to them fire protective covers such as concrete, gypsum, mineral fiber sprays, special paints, and others. The thickness and kind of fireproofing used depends on the type of structure, the degree of fire hazard, and economics.

In the past, concrete was commonly used for fire protection. Although concrete is not a particularly good insulation material it is very satisfactory when applied in thickness of $1\frac{1}{2}$ to 2 or more in. due to its mass. Furthermore the water in concrete (16 to 20 percent when fully hydrated) improves its fireproofing qualities appreciably. This is due to the fact that the boiling off of the water from the concrete requires a great deal of heat. It is true, however, that in very intense fires the boiling off of the water may cause severe cracking and spalling of the concrete.

Though concrete is an everyday construction material and though in mass it is quite a satisfactory fireproofing material its installation cost is extremely high and its weight is large. As a result, for most steel construction, sprayed-on fireproofing materials have almost completely replaced concrete.

The spray-on materials usually consist of either mineral fibers or cementious fireproofing materials. The mineral fibers formerly used were comprised of asbestos. Today, however, due to the health hazards associated with this material, its use has been discontinued and other fibers are now used by manufacturers. The cementious fireproofing materials are probably composed of gypsum, perlite, vermiculite, and others. Sometimes, when plastered ceilings are required in a building, it is possible to hang the ceilings and light gage furring channels by wires from the floor systems above and use the plaster as the fireproofing.

The cost of fireproofing structural steel buildings is high and hurts steel in its economic competition with other materials. As a result, a great deal of research is being conducted by the steel industry on imaginative new fireproofing methods. Among these ideas is the coating of steel members with expansive and insulative paints. When heated to certain temperatures these paints will char, foam and expand, forming an insulative shield around the members.

Other techniques involve the isolation of some steel members outside of the building where they will not be subject to damaging fire exposure; the circulation of liquid coolants inside box or tube shaped members in the building. It is probable that major advances will be made with these and other fireproofing ideas in the near future.[1]

---

[1]W. A. Rains, "A New Era in Fire Protective Coatings for Steel," *Civil Engineering,* September, 1976, *ASCE*: New York, pp. 80–83.

Chapter 17

Composite Beams

17-1 COMPOSITE CONSTRUCTION

When a concrete slab is supported by steel beams and there is no provision for shear transfer between the two the result is a noncomposite section. There is no doubt in noncomposite construction but that the load causes the slab to deflect along with the beam, with the result that some of the load is carried by the slab. Unless, however, there is a great deal of bond between the two (as would be the case where the steel beam is completely encased in the concrete, or where a system of mechanical shear connectors is provided) the load carried by the slab is small and may be neglected.

For many years steel beams and reinforced-concrete slabs were used together with no consideration being made for any composite effect. In recent decades, however, it has been shown that a great strengthening effect can be obtained by tying the two together to act as a unit in resisting loads. Steel beams and concrete slabs joined together compositely can often support a one-third, one-half, or even greater increase in load than could the steel beams alone in noncomposite action.

Composite construction for highway bridges was given the green light by the adoption of the 1944 AASHTO Specifications which approved the method. Since about 1950 the use of composite bridge floors has rapidly increased until today they are commonplace all over the United States. In these bridges the longitudinal shears are transferred from the stringers to the reinforced-concrete slab or deck with shear connectors (to be described in Section 17.5) causing the slab or deck to assist in carrying the bending moments. This type of section is shown in part (a) of Fig. 17-1.

The first approval for composite building floors was given by the 1952

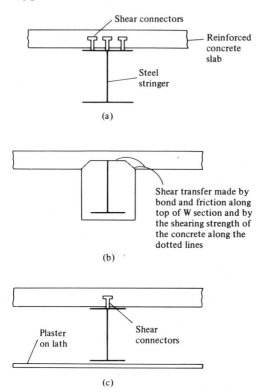

Figure 17-1 (a) Composite bridge floor with shear connectors. (b) Encased section for building floors. (c) Building floors with shear connectors.

AISC Specification and today they are very common. These floors may either be encased in concrete (very rare due to expense) as shown in part (b) of Fig. 17-1 or be nonencased with shear connectors as shown in part (c) of the figure. The larger percentage of composite building floors being built today are of the nonencased type. If the steel sections are encased in concrete the shear transfer is made by bond and friction between the beam and the concrete and by the shearing strength of the concrete along the dotted lines shown in part (b) of Fig. 17-1. If more shearing strength is required, some type of steel reinforcing is provided along the sections indicated with the dotted lines.

Today formed steel deck is very commonly used for composite floor systems. This type of construction is illustrated in Fig. 17-2.

17-2 ADVANTAGES OF COMPOSITE CONSTRUCTION

The floor slab in composite construction acts not only as a slab for resisting the live loads but also as an integral part of the beam. It actually serves as a large cover plate for the upper flange of the steel beam, appreciably increasing the beam's strength.

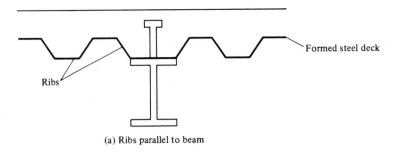

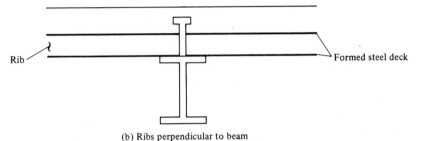

(b) Ribs perpendicular to beam

Figure 17-2 Composite sections using formed steel deck.

A particular advantage of composite floors is that they make use of concrete's high compressive strength by putting a large part of the slab in compression. At the same time a larger percentage of the steel is kept in tension (also advantageous) than is normally the case in steel-frame structures. The result is less steel tonnage required for the same loads and spans (or longer spans for the same sections). Composite sections have greater stiffness than noncomposite sections and they have smaller deflections—perhaps only 20 to 30 percent as large.

Bridge on John F. Kennedy Expressway, Chicago, Ill. (Courtesy of Robert McCullough, Chicago, Ill.)

Furthermore, tests have shown that the ability of a composite structure to take overload is decidedly greater than for a noncomposite structure.

A further advantage of composite construction is the possibility of having smaller overall floor depths—a fact of particular importance for tall buildings. Smaller floor depths permit reduced building heights with the consequent advantages of smaller costs for walls, plumbing, wiring, ducts, elevators, and foundations. Another important advantage available with reduced beam depths is a saving in fireproofing costs for the beams. A disadvantage for composite construction is the cost of furnishing and installing the shear connectors. This extra cost will usually exceed the cost reductions mentioned when spans are short and lightly loaded.

17-3 DISCUSSION OF SHORING

After the steel beams are erected the concrete slab is placed on them. The formwork, wet concrete, and other construction loads must, therefore, be supported by the beams or by temporary shoring. Should no shoring be used, the steel beams must support all of these loads as well as their own weights. Most specifications say that after the concrete has gained 75 percent of its 28-day strength the section has become composite and all loads applied thereafter may be considered to be supported by the composite section. When shoring is used it supports the wet concrete and the other construction loads. It does not really support the weight of the steel beams unless they are given an initial upward deflection (probably impractical). When the shoring is removed (after the concrete gains at least 75 percent of its 28-day strength) the weight of the slab is transferred to the composite section, not just to the steel beams. The student can see that if shoring is used it will be possible to use lighter and thus cheaper steel beams. The question then arises, "Will the saving in steel cost be greater than the extra cost of shoring?" Probably the answer is "No." The usual decision is to use heavier steel beams and do without shoring for several reasons. These include the following.

1. Apart from reasons of economy, the use of shoring is a tricky operation, particularly where settlement of the shoring is possible, as is often the case in bridge construction.

2. Both theory and load tests show that the ultimate strengths of composite sections of the same sizes are the same whether shoring is used or not. If lighter steel beams are selected for a particular span because shoring is used, the result is a smaller ultimate strength.

3. Another disadvantage of shoring is that after the concrete hardens and the shoring is removed the slab will participate in composite action in supporting the dead loads. The slab will be placed in compression by these long-term loads and will have substantial creep and shrinkage parallel to the beams. The result will be a great decrease in the stress in the slab with a corresponding increase in the steel stresses. The probable consequence is that most of the dead load will be supported by the steel beams anyway and composite action will really apply only to the live loads as though shoring had not been used.

17-4 EFFECTIVE FLANGE WIDTHS

There is a problem involved in estimating how much of the slab acts as part of the beam. Should the beams be rather closely spaced the bending stresses in the slab will be fairly uniformly distributed across the compression zone. If, however, the distances between beams are large, bending stresses will vary quite a bit and non-linearly across the flange. The farther a particular part of the slab or flange is away from the steel beam the smaller will be its bending stress. Specifications attempt to handle this problem by replacing the actual slab with a narrower or effective slab which has a constant stress. This equivalent slab is deemed to support the same total compression as is supported by the actual slab. The effective width of the slab b_e is shown in Fig. 17-3.

The portion of the slab or flange which can be considered to participate in the composite beam action is controlled by the specifications. LRFD Specification I3.1 states that the effective width of the concrete slab on each side of the beam center line is to be taken equal to the least of the values to follow. This same set of rules applies whether the slab exists on one or both sides of the beam.

1. One-eighth of the span of the beam measured center to center of supports for both simple and continuous spans.
2. One-half of the distance from the beam center line to the center line of the adjacent beam.
3. The distance from the beam center line to the edge of the slab.

The AASHTO requirements for determining effective flange widths are somewhat different. The maximum total flange width may not exceed one-fourth of the beam span, twelve times the least thickness of the slab, or the distance center to center of the beams. Should the slab exist on only one side of the beam its effective width may not exceed one-twelfth of the beam span, six times the slab thickness, or one-half of the distance from the center line of the beam to the center line of the adjacent beam.

17-5 SHEAR TRANSFER

The concrete slabs may rest directly on top of the steel beams or the beams may be completely encased in concrete for fireproofing purposes. This latter case,

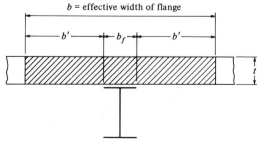

Figure 17-3

however, is very expensive and is rarely used. The longitudinal shear can be transferred between the two by bond and shear and possibly some type of shear reinforcing if needed when the beams are encased. When not encased, mechanical connectors must transfer the load. Fireproofing is not necessary for bridges and the slab is placed on top of the steel beams. Bridges are subject to heavy impactive loads and the bond between the beams and the deck, which is easily broken, is considered negligible. For this reason shear connectors are designed to resist all of the shear between bridge slabs and beams.

Various types of shear connectors have been tried including spiral bars, channels, zees, angles, and studs. Several of these types of connectors are shown in Fig. 17-4. Economic considerations have usually led to the use of round studs welded to the top flanges of the beams. These studs are available in diameters from $\frac{1}{2}$ to 1 in. and in lengths from 2 to 8 in. but the LRFD Specification (I5.1) states that their length may not be less than 4 stud diameters.

They actually consist of rounded steel bars welded on one end to the steel beams. The other end is upset to prevent vertical separation of the slab from the beam. These studs can be quickly attached to the steel beams with stud-welding guns by semiskilled workers.

Shop installation of shear connectors is initially more economical but there is a growing tendency to use field installation. There are two major reasons for this trend: the connectors may easily be damaged during transportation and setting of the beams, and they serve as a hindrance to the workers walking along the top flanges during the early phases of construction.

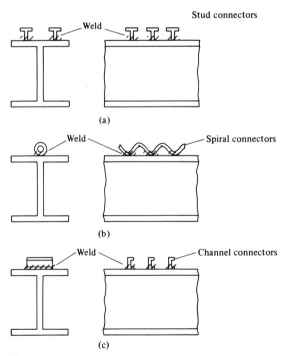

Figure 17-4   Shear connectors.

Channel-section shear connectors, Grand Rapids, Mich. (Courtesy of the Lincoln Electric Company.)

When a composite beam is being tested, failure will probably occur with a crushing of the concrete. At that time it seems reasonable to assume that the concrete and steel will both have reached a plastic condition.

For this discussion reference is made to Fig. 17-5. Should the neutral axis fall in the slab the maximum horizontal shear (or horizontal force on the plane between the concrete and the steel) is said to be $A_s F_y$; and if the neutral axis is in the steel section, the maximum horizontal shear is considered to equal $0.85 f'_c A_c$. (For the student unfamiliar with the strength design theory for reinforced con-

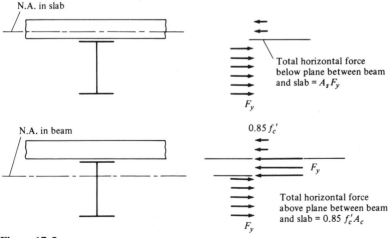

Figure 17-5

crete, the average stress at failure on the compression side of a concrete beam is usually assumed to be $0.85 f'_c$.)

From this information expressions for $\Sigma \, Q_n$ (the shear to be taken by the connectors) can be determined. The LRFD (I5.2) says that for composite action the total horizontal shear between the points of maximum positive moment and zero moment is to be taken as the least of the following where $\Sigma \, Q_n$ is the total nominal strength of the shear connectors provided as described in Section 17-7.

(a) $0.85 f'_c A_c$
(b) $A_s F_y$
(c) $\Sigma \, Q_n$

17-6 PARTIALLY COMPOSITE BEAMS

For this discussion it is assumed that we need to select a steel section which when made composite with the concrete slab will have a design strength of 450 ft-k. It is further assumed that when we select a section from the Manual it has (when made composite with the slab) an M_u equal to 510 ft-k. If we now provide shear connectors for full composite action the section will have a design strength of 510 ft-k. But we need only 450 ft-k.

It seems very logical to assume that we may decide to provide only a sufficient number of connectors to develop a design strength of 450 ft-k. In this way we can reduce the number of connectors and save some money (perhaps a good bit if we repeat this section a good many times in the structure). The resulting section is a *partially composite section*, one which does not have a sufficient number of connectors to develop the full flexural strength of the composite beam. We will encounter this situation in Examples 17-3 and 17-4.

It is usually felt that the total strength of the shear connectors used in a particular beam should not be less than 25 percent of the shearing strength required for full composite action ($A_s F_y$). Otherwise our calculations may not accurately show the stiffness and strength of a composite section.

17-7 STRENGTH OF SHEAR CONNECTORS

For composite sections it is permissible to use normal weight stone concrete (made with aggregates conforming to ASTM C33) or lightweight concrete weighing not less than 90 lb/ft³ (made with rotary kiln-produced aggregates conforming to ASTM C330).

The LRFD Specification provides strength values for headed steel studs not less than 4 diameters in length after installation and for hot-rolled steel channels. *They do not, however, give resistance factors for the strength of shear connectors.* This is because they feel that the factor used for determining the flexural strength of the concrete is sufficient to account for variations in concrete strength including those variations which are associated with shear connectors.

Stud Shear Connectors

The nominal shear strength in kips of one stud shear connector embedded in a solid concrete slab is to be determined with the following expression from LRFD Specification I5.3. In this expression A_{sc} is the cross-sectional area of the shank of the connector in square inches and f'_c is the specified compressive stress of the concrete in ksi. E_c is the modulus of elasticity of the concrete in ksi and equals $w^{1.5} \sqrt{f'_c}$ in which w is the unit weight of the concrete in lb/ft³. Finally F_u is the specified minimum tensile strength of the steel stud in ksi.

$$Q_n = 0.5 \, A_{sc} \sqrt{f'_c E_c} \lessgtr A_{sc} F_u$$

Table 17-1 (which is Table 4.1 from Part 4 of the LRFD Manual) shows a set of Q_n values calculated with the preceding equation for 3/4-in. headed studs made from A36 steel and embedded in concrete slabs with several different f'_c values and concrete weights of 115 and 145 lb/ft³.

Channel Shear Connectors

The nominal shear strength in kips for one channel shear connector is to be determined from the following expression from LRFD Specification I5.4 in which t_f and t_w are respectively the flange and web thicknesses of the channel and L_c is its length. All of these values are to be used in inches.

$$Q_n = 0.3 \, (t_f + 0.5 \, t_w) \, L_c \sqrt{f'_c E_c}$$

Other Connectors

Should other types of shear connectors be used the LRFD Specification (I6) states that their nominal strengths are to be determined with suitable tests.

Stud Shear Connectors in the Ribs of Steel Decks

When shear connectors are placed in the ribs of steel decks the LRFD Specification (I3.5) states that their nominal capacity as determined by the appropriate

TABLE 17-1 NOMINAL STUD SHEAR STRENGTH, Q_n (KIPS) FOR ³/₄-IN. HEADED STUDS

f'_c ksi	w lb/ft³	Q_n kips
3.0	115	17.7
3.0	145	21.0
3.5	115	19.8
3.5	145	23.6
4.0	115	21.9
4.0	145	26.1

Q_n expression may have to be reduced. Two reduction factors are provided one of which is for the case where deck ribs are perpendicular to the steel beams and the other for the case where the ribs are parallel to the steel sections. These factors account for the effects of connector spacings and for varying rib dimensions. If the appropriate factor is computed to be less than 1.0 it must be multiplied by Q_n.

17-8 NUMBER, SPACING, AND COVER REQUIREMENTS FOR SHEAR CONNECTORS

The number of shear connectors to be used between the point of maximum moment and each adjacent point of zero moment equals the horizontal force to be resisted divided by the nominal strength of one connector Q_n.

Spacing of Connectors

Tests of composite beams with shear connectors spaced uniformly and of composite beams with the same number of connectors spaced in variation with the statical shear show little differences as to ultimate strengths and as to deflections at working loads. These results are the case as long as the total number of connectors is sufficient to develop the shear on both sides of the point of maximum moment. As a result the LRFD Specification (I5.6) permits uniform spacings of connectors on each side of the maximum moment point except that the number of connectors placed between a concentrated load and the nearest point of zero moment must be sufficient to develop the maximum moment at the concentrated load.

Maximum and Minimum Spacings

Except for formed steel decks the minimum center-to-center spacing of shear connectors along the longitudinal axes of composite beams permitted by the LRFD Specification (I5.6) is 6 diameters while the minimum value transverse to the longitudinal axis is 4 diameters. Within the ribs of formed steel decks the minimum permissible spacing is 4 diameters in either direction.

When the flanges of steel beams are rather narrow it may be difficult to achieve the minimum transverse spacings described here. For such situations the studs may be staggered. Figure 17-6 shows possible arrangements.

If the deck ribs are parallel to the axis of the steel beam and more connectors are required than can be placed within the rib the LRFD Commentary (I5.6) permits the splitting of the deck so that adequate room is made available.

Shear connectors must be capable of resisting both horizontal and vertical movement because there is a tendency for the slab and beam to separate vertically as well as to slip horizontally. The upset heads of stud shear connectors help to prevent vertical separation. The LRFD Specification (I5.6) states that the maximum spacing of shear connectors may not exceed 8 times the total slab thickness.

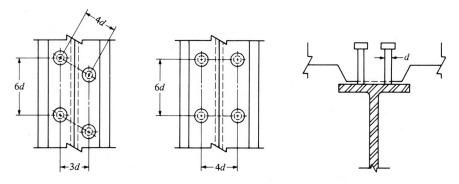

Figure 17-6   Connector arrangements.

Cover Requirements

The LRFD Specification (I5.6) requires that there must be at least 1 in. of lateral concrete cover provided for shear connectors. This rule does not apply to connectors used within the ribs of formed steel decks because tests have shown that strengths are not reduced even when studs are placed as close as possible to the ribs.

When studs are not placed directly over beam webs, they have a tendency to tear out of the beam flanges before their full shear capacity is reached. To keep this situation from occurring, the LRFD Specification (I5.6) requires that the diameter of the studs may not be greater than 2.5 times the flange thickness of the beam to which they are welded.

When formed steel deck is used, the steel beam must be connected to the concrete slab with stud shear connectors with diameters not larger than 3/4 in. These may be welded through the deck or directly to the steel beam. After their installation they must extend for at least $1\frac{1}{2}$ in. above the top of the steel deck and the concrete slab thickness above the steel deck may not be less than 2 in. (LRFD Specification I3.5a).

17-9 MOMENT CAPACITY OF COMPOSITE SECTIONS

The nominal flexural strength of a composite beam in the positive moment region may be controlled by the plastic strength of the section or by the strength of the concrete slab or by the strength of the shear connectors. Furthermore if the web is very slender and if a large portion of the web is in compression web buckling may possibly limit the nominal strength of the member.

Little research has been done on the subject of web buckling for composite sections, and for this reason the LRFD Specification (I3.2) has conservatively applied the same rules to composite section webs as to plain steel webs. The positive flexural strength ($\phi_b M_n$ with $\phi_b = 0.85$) of a composite section is to be determined assuming a plastic stress distribution if $h_c/t_w \leq 640/\sqrt{F_{yf}}$. In this expression h_c is the distance between the web toes of the fillet, that is, d-$2k$, t_w is the web thickness, and F_{yf} is the yield stress of the beam flange. All of the rolled W, S, M, HP, and C shapes in the Manual meet this requirement for F_y values up to 65 ksi.

Should h_c/t_w be greater than $640/\sqrt{F_{yf}}$ the value of $\phi_b M_n$ with $\phi_b = 0.90$ is to be determined by superimposing the elastic stresses. The effects of shoring must be considered for these calculations.

The nominal moment capacity of composite sections as determined by load tests can be estimated very accurately with the plastic theory. With this theory the steel section at failure is assumed to be fully yielded and a part of the concrete slab which is on the compression side of the neutral axis is assumed to be stressed to $0.85 f'_c$. If any part of the slab is on the tensile side of the neutral axis it is assumed to be cracked and incapable of carrying stress.

The plastic neutral axis (PNA) may fall in the slab or in the flange of the steel section or in its web. Each of these cases is discussed in this section.

Neutral Axis in Concrete Slab

The concrete slab compression stresses vary somewhat from the PNA out to the top of the slab. For convenience in calculations, however, they are assumed to be uniform with a value of $0.85 f'_c$ over an area of depth a and width b_e determined as described in Section 17-4. (This distribution is selected to provide a stress block having the same total compression C and the same center of gravity for the total force as we have in the actual slab.)

The value of a can be determined from the following expression where the total tension in the steel section is set equal to the total compression in the slab.

$$A_s F_y = 0.85 f'_c a b_e$$

$$a = \frac{A_s F_y}{0.85 f'_c b_e}$$

If a is equal to or less than the slab thickness, the PNA will fall in the slab and the nominal or plastic moment capacity of the composite section may be written as the total tension T or the total compression times the distance between their centers of gravity. Reference is made here to Fig. 17-7.

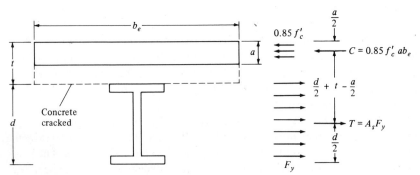

Figure 17-7   Plastic neutral axis in the slab.

Example 17-1, which follows, illustrates the calculation of $M_u = \phi_b M_p = \phi_b M_n$ for a composite section where the PNA falls within the slab.

EXAMPLE 17-1

Compute $M_u = \phi_b M_p = \phi_b M_n$ for the composite section shown in Fig. 17-8 if $f_c' = 4$ ksi and $F_y = 36$ ksi.

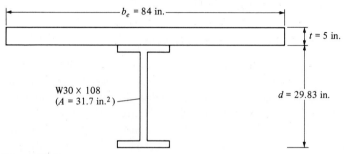

$b_e = 84$ in.

$t = 5$ in.

W30 × 108
($A = 31.7$ in.²)

$d = 29.83$ in.

Figure 17-8

Solution

$$a = \frac{A_s F_y}{0.85 f_c' b_e} = \frac{(31.7)(36)}{(0.85)(4)(84)} = 4.00 \text{ in.} < t$$

$$M_n = M_p = A_s F_y \left(\frac{d}{2} + t - \frac{a}{2} \right)$$

$$= (31.7)(36) \left(\frac{29.83}{2} + 5 - \frac{4.0}{2} \right)$$

$$= 20{,}445 \text{ in.-k} = 1704 \text{ ft-k}$$

$$M_u = \phi_b M_n = (0.85)(1704) = \underline{\underline{1448.4 \text{ ft-k}}}$$

Note: If the reader refers to Part 4 of the LRFD Manual he or she can determine the M_u value for this composite beam with reference being made to Fig. 17-9. For this discussion it is assumed that the PNA is located at the top of the steel flange (TFL) or down in the steel shape. In the LRFD tables

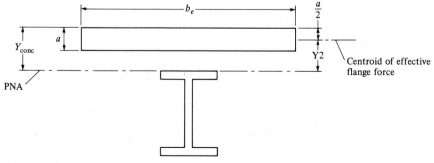

b_e

$\frac{a}{2}$

Y_{conc}

a

Y2

Centroid of effective flange force

PNA

Figure 17-9

Y1 represents the distance from the PNA to the top of the beam flange while Y2 represents the distance from the centroid of the effective concrete flange force to the top flange of the beam.

With the PNA for the preceding example being located at the top of the beam flange from page 4-19 of the Manual with $Y2 = 5 - 4/2 = 3$ in. and for a $W30 \times 108$ the value of $M_u = \phi_b M_p = 1450$ ft-k.

Neutral Axis in Top Flange of Steel Beam

If a is calculated as previously described and is greater than the slab thickness t the PNA will fall down in the steel section. If this happens it will be necessary to find out whether the PNA is in the flange or below the flange. Suppose we assume that it's at the base of the flange. We can calculate the total compressive force C above $= 0.85 f'_c b_e t + A_f F_y$ where A_f is the area of the flange and the total tensile force below $T = F_y(A_s - A_f)$. If C is greater than T the PNA will be in the flange. If $C < T$ the PNA is below the flange.

Assuming we find the PNA is in the flange we can find its location letting $\bar{y}$ be the distance to the PNA measured from the top of the top flange by equating C and T as follows:

$$0.85 f'_c b_e t + F_y b_f \bar{y} = F_y A_s - F_y b_f \bar{y}$$

From which $\bar{y}$ is

$$\bar{y} = \frac{F_y A_s - 0.85 f'_c b_e t}{2 F_y b_f}$$

Then the nominal or plastic moment capacity of the section can be determined from the expression to follow with reference being made to Fig. 17-10. Taking moments about the PNA we get:

$$M_p = M_n = 0.85 f'_c b_e t \left(\frac{t}{2} + \bar{y}\right) + 2 F_y b_f \bar{y} \left(\frac{\bar{y}}{2}\right) + F_y A_s \left(\frac{d}{2} - \bar{y}\right)$$

Example 17-2, which follows, illustrates the calculation of $M_u = \phi_b M_p = \phi_b M_n$ for a composite section in which the PNA falls in the flange.

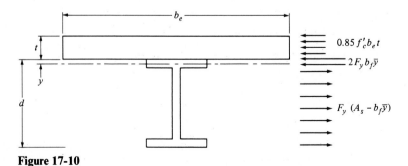

Figure 17-10

EXAMPLE 17-2

Compute $M_u = \phi_b M_p = \phi_b M_n$ for the composite section shown in Fig. 17-11 if A36 steel is used and if f_c' is 4 ksi.

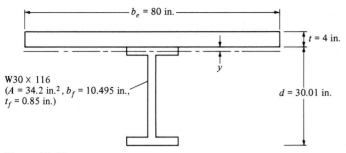

Figure 17-11

Solution

Is PNA located at top of steel flange?

$$a = \frac{(36)(34.2)}{(0.85)(4)(80)} = 4.53 \text{ in.} > 4.00 \text{ in.}$$

∴ PNA is located down in steel section.

Is PNA in flange or in web? Here we assume it is at base of steel flange.

$$C = (0.85)(4)(80)(4) + (36)(10.495)(0.85) = 1409.1 \text{ k}$$

$$T = (36)(34.2 - 10.495 \times 0.85) = 910 \text{ k}$$

Since $C > T$ the PNA falls in the steel flange and can be located as follows:

$$\bar{y} = \frac{(36)(34.2) - (0.85)(4)(80)(4)}{(2)(36)(10.495)} = 0.190 \text{ in.}$$

Then

$$M_n = M_p = (0.85)(4)(80)(4)\left(\frac{4}{2} + 0.190\right)$$

$$+ (2)(36)(10.495)(0.190)\left(\frac{0.190}{2}\right)$$

$$+ (36)(34.2)\left(\frac{30.01}{2} - 0.190\right)$$

$$= 20{,}637 \text{ in.-k} = 1719.7 \text{ ft-k}$$

$$M_u = \phi_b M_p = \phi_b M_n = (0.85)(1719.7) = \underline{\underline{1461.8 \text{ ft-k}}}$$

By interpolation in the LRFD Composite Design tables with Y1 = 0.19 in. and Y2 = 2.0 in. we get $1480 - (0.19/0.21)(1480 - 1460) = 1461.9$ ft-k

If we have a partially composite section with $\Sigma\,Q_n$ less than $A_s F_y$, the PNA will be down in the shape, and if in the flange the value of M_u can be determined with the equation used in Example 17-2. In the composite design tables presented in the Manual values of $\Sigma\,Q_n$ and M_u are shown for seven different PNA locations—the top of the flange, quarter points in the flange, bottom of flange, and two points down in the web. Straight-line interpolation may be used for numbers in between the tabulated values.

Neutral Axis in Web of Steel Section

If for a particular composite section we find that a is larger than the slab thickness and if we then assume the PNA is located at the bottom of the steel flange and we calculate C and T and find T is larger than C the PNA will fall in the web. We can go through calculations similar to the ones we used for the case where the PNA was located in the flange. Space is not taken to show such calculations because the Composite Design tables in the Manual cover most common cases. In addition the LRFD Commentary (I3.2) presents formulas for determining M_u for all these situations.

17-10 DESIGN OF COMPOSITE SECTIONS

Composite construction is of particular advantage economically when loads are heavy, spans are long, and beams are spaced at fairly large intervals. For steel-building frames, composite construction is economical for spans varying roughly from 25 to 50 ft, with particular advantage in the longer spans. For bridges, simple spans have been economically constructed up to approximately 120 ft and continuous spans 50 or 60 ft longer. Composite bridges are generally economic for simple spans greater than about 40 ft and for continuous spans greater than about 60 ft.

Occasionally cover plates are welded to the bottom flanges of steel beams with improved economy. The student can see that with the slab acting as part of the beam there is a very large compressive area available and by adding cover plates to the tensile flange a little better balance is obtained.

In tall buildings where headroom is a problem it is desirable to use the minimum overall floor thicknesses possible. For buildings, minimum depth-span ratios of approximately $\frac{1}{24}$ are recommended if the loads are fairly static and $\frac{1}{20}$ if the loads are of such a nature as to cause appreciable vibration. The thicknesses of the floor slabs are known (from the concrete design) and the depths of the steel beams can be fairly well estimated from these ratios.

Before we attempt some composite designs several additional remarks relating to lateral bracing, shoring, estimated steel beam weights, and lower bound moments of inertia are discussed in the paragraphs to follow.

Lateral Bracing

After the concrete slab hardens it will provide sufficient lateral bracing for the compression flange of the steel beam. However, during the construction phase before the concrete hardens lateral bracing may be insufficient and its design strength may have to be reduced depending on the estimated unbraced length. When steel formed decking or concrete forms are attached to the beam's compression flange they will usually provide sufficient lateral bracing. The designer must be very careful in consideration of lateral bracing for fully encased beams.

Beams with Shoring

If beams are shored during construction we will assume that all loads are resisted by the composite section after the shoring is removed.

Beams without Shoring

If temporary shoring is not used during construction the steel beam alone must be able to support all the loads before the concrete is sufficiently hardened to provide composite action.

    Without shoring the wet concrete loads tend to cause large beam deflections which may cause us to build thicker slabs where the beam deflections are large. This situation can be counteracted by cambering the beams.

    The LRFD Specification does not provide any extra margin against yield stresses occurring in beams during construction of unshored composite floors. Assuming satisfactory lateral bracing is provided the Specification (F1.2) states that the maximum factored moment may not exceed $0.90F_yZ$. The 0.90 in effect keeps the maximum factored moment to a value about equal to the yield moment F_yS.

    To calculate the moment to be resisted during construction it seems logical to count the wet concrete as a live load and to also include some extra live load as perhaps 20 psf to account for construction activities.

Estimated Steel Beam Weight

As illustrated in Example 17-3 it may sometimes be useful to make an estimate of the weight of the steel beam. The LRFD Manual provides the following empirical formula for this purpose in its Part 4.

$$\text{Estimated beam weight} = \left[\frac{12 M_u}{(d/2 + Y_{con} - a/2)\phi F_y} \right] 3.4$$

where M_u = required flexural strength of composite section
        d = nominal steel beam depth
     Y_{con} = distance from top of steel beam to top of concrete slab
        a = effective concrete slab thickness (which can be conservatively estimated as somewhere in the range of about 2 in.)
       ϕ = 0.85

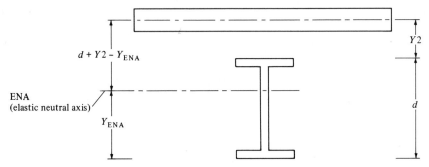

Figure 17-12

Lower Bound Moment of Inertia

To calculate the service load deflections for composite sections a table of lower bound moment of inertia values is presented in Part 4 of the Manual. These values are computed from the area of the steel beam and an equivalent concrete area of $\Sigma\, Q_n/F_y$. The remainder of the concrete flange is not used in these calculations. This means that if we have partially composite sections the value of the lower bound moment of inertia will reflect this situation because $\Sigma\, Q_n$ will be smaller. The lower bound moment of inertia is computed with the following expression, reference being made to Fig. 17-12 where Y_{ENA} is the distance from the bottom of the beam to the elastic neutral axis (ENA).

$$I = I_x + A_s\left(Y_{ENA} - \frac{d}{2}\right)^2 + \left(\frac{\Sigma\, Q_n}{F_y}\right)(d + Y_2 - Y_{ENA})^2$$

Extra Reinforcing

For building design calculations the spans are often considered to be simply supported, but the steel beams do not generally have perfectly simple ends. The result of this situation is that some negative moment may occur at the beam ends with possible cracking of the slab above. To prevent or minimize cracking, some extra steel can be placed in the top of the slab extending 2 or 3 ft out into the slab. The amount of steel added is in addition to the temperature and shrinkage requirements specified by the American Concrete Institute.[1]

Example Problems

Examples 17-3 and 17-4 illustrate the designs of two composite sections, the first being for unshored construction and the second for shored construction.

EXAMPLE 17-3

Beams 10 ft on center with 36-ft simple spans are to be selected to support a 4-in-deep lightweight concrete slab on a 3-in.-deep formed steel deck with

---

[1]*Building Code Requirements for Reinforced Concrete* (ACI std. 318-77). Detroit: American Concrete Institute, 1983, Section 7.12.

no shoring. The ribs for the steel deck which are perpendicular to the beam center lines have average widths of 6 in. If the service dead load (including the beam weight) is to be 0.78 k/ft of length of the beams and the service live load is 1.2 k/ft, (a) select the beams, (b) determine the number of $\frac{3}{4}$-in.-diameter headed studs required, (c) compute the service live load deflection, and (d) check the beam shear. Other data: A36 steel, $f_c' = 3$ ksi, and concrete weight 115 lb/ft³.

Solution

Loads and moments

$$w_u = (1.2)(0.78) + (1.6)(1.2) = 2.856 \text{ k/ft}$$

$$M_u = \frac{(2.856)(36)^2}{8} = 462.7 \text{ ft-k}$$

Effective flange width b_e

$$b_e = (2)(\tfrac{1}{8} \times 36 \times 12) = 108 \text{ in.} \leftarrow$$

$$b_e = (2)(5 \times 12) = 120 \text{ in.}$$

(a) Select W section

$$Y_{\text{con}} = \text{distance from top of slab to top}$$

of steel flange $= 4 + 3 = 7$ in.

Assume $a = 2$ in. (It's usually quite small, particularly for relatively light sections.)

Y1 is distance from PNA to top flange $= 0$ in.

Y2 is the distance from the center of gravity of the concrete flange force to the top flange of the beam $= 7 - a/2 = 7 - 2/2 = 6$ in.

Looking through the composite tables of the Manual with $M_u = 462.7$ ft-k, Y1 = 0, and Y2 = 6 in. we can see that several W18s (55 lb, 50 lb, and 46 lb) seem reasonable.

We can use LRFD estimated weight formula for an 18-in.-deep section.

$$wt = \left[\frac{(12)(462.7)}{(18/2 + 7 - 2/2)(0.85)(36)} \right] 3.4 = 41.12 \text{ lb}$$

Try W18×40 ($A = 11.8$ in.²) though it seems a little small compared with estimated weight.

Assume $\Sigma Q_n = A_s F_y = (11.8)(36) = 424.8$ k

$$\text{Thus } a \text{ required} = \frac{\Sigma Q_n}{0.85 f_c' b_e} = \frac{424.8}{(0.85)(3)(108)} = 1.54 \text{ in.}$$

$$Y1 = 0$$

$$Y2 = 7.0 - \frac{1.54}{2} = 6.23 \text{ in.}$$

M_u from Manual by interpolation $= 450 + \left(\dfrac{0.23}{0.50}\right)(465 - 450)$

$$= 456.9 \text{ ft-k} < 462.7 \text{ ft-k} \qquad\qquad \text{NG}$$

Try W18×46 ($A = 13.5$ in.^{2}, $d = 18.06$ in., $t_w = 0.360$ in.)

Now this time we will go to the case where Y1 is the largest possible to provide an M_u of about 462.7 ft-k with Y2 = 6 in. because $\Sigma\, Q_n$ will be smaller and fewer shear connectors will be required. This will occur with Y1 = 0.30 in. where $M_u = 460$ ft-k if Y2 = 6 in.

$$\Sigma\, Q_n = 354 \text{ k}$$

$$a = \frac{354}{(0.85)(3)(108)} = 1.285 \text{ in.}$$

$$\text{Y2} = 7 - \frac{1.285}{2} = 6.36 \text{ in.}$$

$$M_u = 460 + \left(\frac{0.36}{0.50}\right)\left(472 - 460\right)$$

$$= 468.6 \text{ ft-k} > 462.7 \text{ ft-k} \qquad\qquad \text{OK}$$

$$\underline{\text{Use W18×46}}$$

(b)   Design of studs

Since we have steel formed deck we must compute a stud reduction factor (SRF) as required by Specification I3.5(b).

$$h_r = 3 \text{ in.}$$

Assume $H_s = 5.0$ in. (can't be $> h_r + 3 = 6$ in.)
Assume 1 stud in a rib at beam intersection $= N_r$

$$w_r = 6 \text{ in.}$$

$$\text{SRF} = \frac{0.85}{\sqrt{1}}\left(\frac{6}{3}\right)\left[\frac{5.0}{3} - 1.0\right] = 1.13 > 1.0$$

$$\therefore \text{ No reduction is necessary.}$$

Q_n from Table 17.1 for $\frac{3}{4}$-in. headed studs, $f'_c = 3$ ksi, and concrete weight of 115 lb/ft^3 = 17.7 k

$$\Sigma\, Q_n \text{ for W18×46 with Y1 of 0.30 in.} = 354 \text{ k}$$

$$\text{Total no. of connectors required} = \frac{2\Sigma\, Q_m}{Q_m} = \frac{(2)(354)}{17.7} = 40$$

Use 40-$\frac{3}{4}$-in. studs (20 on each side of point of maximum moment which is ℄ here)

(c) Compute LL deflection

Assume maximum permissible LL deflection

$$\frac{1}{360} \text{ span} = \left(\frac{1}{360}\right)(12 \times 36) = 1.2 \text{ in.}$$

$C_1 = 161$ from Fig. 10.4 of textbook (or Fig. 3.3 of Manual)

$$M_{LL} = \frac{(1.62)(36)^2}{8} = 194.4 \text{ ft-k}$$

I_x = lower bound moment of inertia from tables in Part 4 of Manual

$$= 2000 + \left(\frac{0.36}{0.50}\right)(2080 - 2000) = 2057.6 \text{ in.}^4$$

$$\Delta_{LL} = \frac{ML^2}{C_1 I_x} = \frac{(194.4)(36)^2}{(161)(2057.6)} = 0.760 \text{ in.} < 1.2 \text{ in.} \qquad \text{OK}$$

(d) Check beam shear

$$V_u = \frac{(36)(2.856)}{2} = 51.4 \text{ k}$$

$$\phi V_n = \phi 0.6 F_y A_w \text{ for W}18 \times 46$$

$$= (0.9)(0.6)(36)(18.06)(0.360)$$

$$= 126.4 \text{ k} > 51.4 \text{ k} \qquad \text{OK}$$

Use W18×46 with 40 $\frac{3}{4}$-in. headed studs

EXAMPLE 17-4

Using the same data as for Example 17-3 except that $F_y = 50$ ksi and unshored construction is to be used, perform the following tasks:

(a) Select steel beam.

(b) If the stud reduction factor for metal decks is = 1.0, determine the number of $\frac{3}{4}$-in. headed studs required.

(c) Check the beam strength before the concrete hardens.

(d) Compute service load deflection before concrete hardens. Assume a construction live load of 20 psf.

(e) Determine the service live load deflection after composite action is available.

(f) Check shear.

(g) Select a steel section to carry all the loads if no shear connectors are used and compute its total service load deflection.

Solution

From Example 17-3

$$M_u = 462.7 \text{ ft-k}$$

$$b_e = 108 \text{ in.}$$

(a)   Select W section

$$Y_{con} = 4 + 3 = 7 \text{ in.}$$

$$\text{Assume } a = 2 \text{ in.}$$

$$Y1 = 0$$

$$Y2 = 7 - \tfrac{2}{2} = 6 \text{ in.}$$

Try W18×35 ($A = 10.3 \text{ in.}^2$, $d = 17.70 \text{ in.}$, $t_w = 0.30 \text{ in.}$)

$$\text{Assume } \Sigma \, Q_n = (10.3)(50) = 515 \text{ k}$$

$$a = \frac{515}{(0.85)(3)(108)} = 1.87 \text{ in.}$$

$$Y2 = 7 - \frac{1.87}{2} = 6.065 \text{ in.}$$

M_u from Manual by interpolation

$$= 542 + \left(\frac{0.065}{0.50}\right)(560 - 542) = 544.3 \text{ ft-k} > 462.7 \text{ k} \qquad \text{OK}$$

(b)   Design of studs

Q_n from stud table $= 17.7 \text{ k}$
We can go down in W18×35 table to $Y1 = 0.21 \text{ in.}$ and still obtain
the M_u required (it's between 487 and 500 ft-k)

$$\Sigma \, Q_n = 388 \text{ k}$$

$$\text{Number of connectors required} = \frac{2 \times 388}{17.7} = 43.8$$

Use 22 connectors each side of $\mathbf{\mathring{L}}$

(c)   Check strength of W section before concrete hardens

Assume the wet concrete is a LL during construction and also add a
20 psf construction live load

Concrete wt $= (\tfrac{4}{12})(145)(10) = 483 \text{ lb/ft}$

Other dead loads $= 0.780 - 0.483 = 0.297 \text{ k/ft}$

$w_u = (1.2)(0.297) + (1.6)(0.020 \times 10 + 0.483) = 1.449 \text{ k/ft}$

$$M_u = \frac{(1.449)(36)^2}{8} = 234.7 \text{ ft-k}$$

$\phi_b M_p$ from beam tables for W18×35 $= 249 \text{ ft-k} > 234.7 \text{ ft-k}$ \qquad OK

(d) Service load deflection before concrete hardens

$$C_1 = 161$$

I_x for W18×35 = 510 in.^{4} (not lower bound I)

Use $w_D = 0.78 + (10)(0.02) = 0.98$ k/ft

$$M_{DL} = \frac{(0.98)(36)^2}{8} = 158.8 \text{ ft-k}$$

$$\Delta_{DL} = \frac{(158.8)(36)^2}{(161)(510)} = 2.51 \text{ in.}$$

(We might camber beam for this deflection.)

(e) Service LL deflection after composite action is available

$$M_L = \frac{(1.2)(36)^2}{8} = 194.4 \text{ ft-k}$$

Lower bound I with Y1 = 0.21 in. and Y2 = 6.065 in.

$$I = 1490 + \left(\frac{0.065}{0.50}\right)(1550 - 1490) = 1498 \text{ in.}^4$$

$$\Delta_{LL} = \frac{(194.4)(36)^2}{(161)(1498)} = 1.04 \text{ in.} < \frac{L}{360} = 1.2 \text{ in.} \qquad \text{OK}$$

(f) Check shear

$$V_u = \frac{(36)(2.856)}{2} = 51.4 \text{ k}$$

$$\phi V_n = (0.9)(0.6)(50)(17.70)(0.30)$$

$$= 143.4 \text{ k} > 51.4 \text{ k} \qquad \text{OK}$$

(g) Select section if we have no composite action

$$M_u = 462.7 \text{ ft-k} = \phi_b M_p$$

Requires a W24×55 ($I_x = 1350$ in.^{4}) from Load Factor Design Selection Table in Manual

$$\Delta = \frac{(462.7)(36)^2}{(161)(1350)} = 2.76 \text{ in.} > \frac{L}{360} = 1.2 \text{ in.}$$

Min. I to limit deflection to 1.2 in. $= \left(\frac{2.76}{1.2}\right)(1350) = 3105 \text{ in.}^4$

Use W27×94 (as compared with W18×35 if a composite section is used)

17-11 CONTINUOUS COMPOSITE SECTIONS

The LRFD Specification (I3.2) permits the use of continuous composite sections. The flexural strength of a composite section in a negative moment region may be considered to equal $\phi_b M_n$ for the steel section alone or it may be based upon the plastic strength of a composite section considered to be made up of the steel beam and the longitudinal reinforcement in the slab. For this latter method to be used the following conditions must be met:

1. The steel section must be compact and adequately braced.
2. The slab must be connected to the steel beams in the negative moment region with shear connectors.
3. The longitudinal reinforcing in the slab parallel to the steel beam and within the effective width of the slab must have adequate development lengths. (Development length is a term used in reinforced-concrete design and refers to the length which reinforcing bars have to be extended or embedded in the concrete to properly anchor them or develop their stresses by means of bond between the bars and the concrete.)

For a particular beam the total horizontal shear force between the point of zero moment and the point of maximum negative moments is to be taken as the smaller of $A_r F_{yr}$ and ΣQ_n where A_r is the cross-sectional area of the properly developed reinforcing and F_{yr} is the yield stress of the bars. The plastic stress distribution for negative moment in a composite section is illustrated in Fig. 17-13.

Continuous-welded plate girders in Henry Jefferson County, Iowa. (Courtesy of the Lincoln Electric Company.)

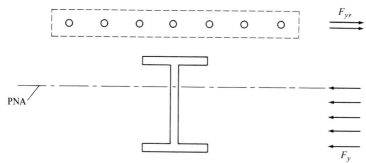

Figure 17-13 Stress distribution in negative moment range.

17-12 DESIGN OF CONCRETE ENCASED SECTIONS

For fireproofing purposes the steel beams in building floors may occasionally be completely encased in concrete. Under certain conditions the horizontal shears between the slabs and beams can be considered to be transferred by natural bond and friction between the two. The LRFD says that for this transfer to be permissible the encasing concrete must be placed integrally with the slab concrete and must cover the steel by at least 2 in. on the sides and bottom (or soffit). It is further required that the top of the steel section be at least $1\frac{1}{2}$ in. below the top of the slab and 2 in. above the bottom of the slab. Finally, the encasing concrete must have adequate mesh or other reinforcing for its full depth and across the soffit of the beam to prevent spalling of the concrete. The exact amount, which is not specified by the LRFD, can be very nominal in size (from "chicken wire size" on up).

1. With one method the design strength of the encased section may be based on the plastic moment capacity $\phi_b M_p$ of the steel section alone.
2. By another method the design strength is based on the first yield of the tension flange assuming composite action between the concrete which is in compression and the steel section.

If the second method is used and we have unshored construction the stresses in the steel section caused by the wet concrete and other construction loads are calculated. Then the stresses in the composite section caused by loads applied after the concrete hardens are computed. These stresses are superimposed on the first set of stresses. The total stresses so computed may not exceed $\phi_b F_y$ with $\phi_b = 0.9$. If we have shored construction all of the loads may be assumed to be supported by the composite section and the stresses computed accordingly. For stress calculations the properties of a composite section are computed with the transformed area method. In this method the cross-sectional area of one of the two materials is replaced or transformed into an equivalent area of the other. For composite design, it is customary to replace the concrete with an equivalent area of steel, whereas the reverse procedure is used in the working stress design method for reinforced-concrete design.

In the transformed area procedure the concrete and steel are assumed to be bonded tightly together so that their strains will be the same at equal distances from the neutral axis. The unit stress in either material can then be said to equal its strain times its modulus of elasticity (ϵE_c for the concrete or ϵE_s for the steel). The unit stress in the steel is then $\epsilon E_s/\epsilon E_c = E_s/E_c$ times as great as the corresponding unit stress in the concrete. The E_s/E_c ratio is referred to as the modular ratio n; therefore, n in.^{2} of concrete are required to resist the same total stress as 1 in.^{2} of steel; and the cross-sectional area of the slab (A_c) is replaced with a transformed or equivalent area of steel equal to A_c/n.

The American Concrete Institute (ACI) Building Code suggests that the following expression may be used for calculating the modulus of elasticity of concrete weighing from 90 to 155 lb/ft^3.

$$E_c = w^{1.5}33\sqrt{f_c'}$$

In this expression w is the weight of the concrete in pounds per cubic foot and f_c' is the 28-day compressive strength in pounds per square inch.

There are no slenderness limitations required by the LRFD Specification for either of the two methods because the encasement is effective in preventing both local and lateral buckling.

In Example 17-5, which follows, stresses are computed with the elastic theory assuming composite action as described for the second method. Notice that the author has divided the effective width of the slab by n to transform the concrete slab into an equivalent area of steel.

EXAMPLE 17-5

Review the encased beam section shown in Fig. 17-14 if no shoring is used and the following data are assumed:

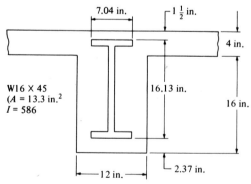

Figure 17-14

Simple span = 36 ft

Service dead load = 0.50 k/ft before concrete hardens plus an

additional 0.25 k/ft after concrete hardens

Construction live loads = 0.2 k/ft

Service live load = 1.0 k/ft after concrete hardens

Effective flange width b_e = 60 in. and n = 9

F_y = 50 ksi

Solution

Calculated Properties of Composite Section: Neglecting concrete area below flange,

$$A = 13.3 + \frac{(4)(60)}{9} = 39.96 \text{ in.}^2$$

$$y_b = \frac{(13.3)(10.44) + (26.66)(18)}{39.96} = 15.50 \text{ in.}$$

$$I = 586 + (13.3)(5.06)^2 + (\tfrac{1}{12})(\tfrac{60}{9})(4)^3 + (26.66)(2.5)^2$$

$$= 1129 \text{ in.}^4$$

Stresses before concrete hardens
    Assume wet concrete is a live load

$$w_u = (1.6)(0.5 + 0.2) = 1.12 \text{ k/ft}$$

$$M_u = \frac{(1.12)(36)^2}{8} = 181.4 \text{ ft-k}$$

$$f_t = \frac{(12)(181.4)(8.065)}{586} = 29.96 \text{ ksi}$$

$$< \phi_b F_y = (0.9)(50) = 45 \text{ ksi} \qquad \text{OK}$$

Stresses after concrete hardens

$$w_u = (1.2)(0.25) + (1.6)(1.0) = 1.9 \text{ k/ft}$$

$$M_u = \frac{(1.9)(36)^2}{8} = 307.8 \text{ ft-k}$$

$$f_t = \frac{(12)(307.8)(15.50 - 2.37)}{1129} = 42.96 \text{ ksi}$$

$$\text{Total } f_t = 29.96 + 42.96 = 72.92 \text{ ksi} > 45 \text{ ksi} \qquad \text{NG}$$

The LRFD Specification does not provide instructions relating to longitudinal shear in encased sections. The author, however, feels the topic should be considered. The critical sections for resistance to this shear in an encased section are shown in Fig. 17-15. The total resistance to the longitudinal shear can be estimated to equal the bond along the top of the steel shape (line 2–3 in the figure) plus the shearing resistance of the concrete along lines 1–2 and 3–4.

A typical value used for the allowable bond between the steel and concrete

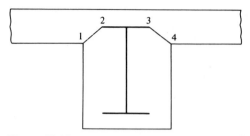

Figure 17-15

is 0.03 f_c' while a common allowable shear in the concrete as along sections 1–2 and 3–4 is 0.12 f_c'. Both of these values are for service load conditions. (A much smaller shear allowable is used in Example 17-6.) Should the longitudinal shear be greater than the sum of the shearing and bond resistance along the sections mentioned, some type of reinforcing will be necessary. The usual types of shear connectors on top of the beam flange will probably not be of much value because relatively large deformations will probably have to occur before much load can be applied to the connectors. By the time that much deformation occurs, the natural bond between the steel and concrete will probably have broken. Some type of shear reinforcing placed along lines 1–2 and 3–4 will probably be the most effective method of increasing the shearing resistance. The longitudinal shear will have to be extremely large to require this shear reinforcing.

Example 17-6 illustrates the calculations involved in reviewing the design of an encased section for shear. Notice in this problem that since no shoring is used the only longitudinal shear to be resisted is that produced by the loads applied after the concrete hardens. The ultimate-strength theory is today considered not to apply to these encased sections due to the lack of shear connectors, and longitudinal shears are computed by the familiar VQ/I expression.

EXAMPLE 17-6

Review the encased beam section shown in Fig. 17-16 for longitudinal shear. No shoring is used and the following data are assumed, for service load conditions.

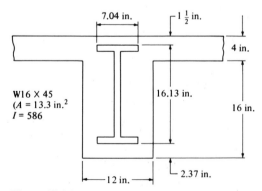

Figure 17-16

$$\text{Allowable bond} = 90 \text{ psi}$$

$$\text{Allowable shear} = 180 \text{ psi}$$

$$\text{Assume service } LL \text{ maximum external shear} = 22 \text{ k}$$

$$\text{Effective flange width} = 60 \text{ in.}$$

$$n = 9$$

Solution. Calculated Properties of Composite Section: Neglecting concrete area below flange,
From solution of Example 17.5

$$A = 39.96 \text{ in}^2.$$

$$y_b = 15.50 \text{ in.}$$

$$I = 1129 \text{ in}^4.$$

Q for area above section 1–2–3–4 in Fig. 17-17,

$$Q = \tfrac{1}{9}[(4)(60)(2.5) - (7.04 \times 2.50)(1.75)$$

$$- (2)(\tfrac{1}{2} \times 2.48 \times 2.5)(1.33)]$$

$$= 62.3 \text{ in.}^3$$

Checking Longitudinal Shear

$$v = \frac{VQ}{I} = \frac{(22,000)(62.3)}{1129} = 1214 \text{ lb/in.}$$

$$\text{Allowable bond} = (7.04)(90) = 634 \text{ lb/in.}$$

$$\text{Allowable shear} = (2)(3.52)(180) = \underline{1270 \text{ lb/in.}}$$

$$\text{Total shear resistance} = 1904 \text{ lb/in.} > 1214 \text{ lb/in.} \quad \text{OK}$$

For buildings continuous composite construction with encased sections is permissible. For continuous construction the positive moments are handled exactly as has been illustrated by the preceding examples. For negative moments, however, the transformed section is taken as shown in Fig. 17-18. The cross-hatched area represents the concrete in compression and all concrete on the tensile side of the neutral axis (that is, above the axis) is neglected. Notice also the

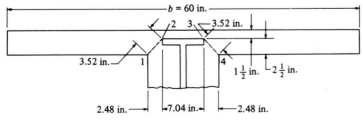

Figure 17-17

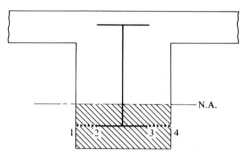

Figure 17-18

fact that the resistance to longitudinal shear is provided along line 1–2–3–4 in the figure.

PROBLEMS

17-1. Determine M_u for the section shown assuming sufficient shear connectors are provided to ensure full composite section. Solve problem using the procedure presented in Section 17-9 and check answer with tables in Manual. A36 steel. $f'_c = 3$ ksi. (*Ans.* 522.4 ft-k)

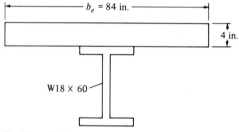

Problem 17-1

17-2. Repeat Prob. 17-1 if $F_y = 50$ ksi.

17-3. Repeat Prob. 17-1 using tables in Manual if it is considered to be partially composite and if ΣQ_n is 350 k. (*Ans.* 482.8 ft-k)

17-4. Determine M_u for the section shown if A36 steel and sufficient shear connectors are used to guarantee full composite action. Use formulas and check with Manual. $f'_c = 4$ ksi.

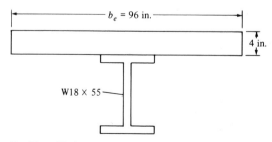

Problem 17-4

17-5. Repeat Prob. 17-4 if $F_y = 50$ ksi (*Ans.* 677.9 ft-k)

17-6. Compute M_u for the composite section shown if A36 steel is used and sufficient shear connectors are used to provide full composite action. A 3 1/4-in. concrete slab is supported by 3-in.-deep composite metal deck ribs perpendicular to beam. $f'_c = 4$ ksi. Check answer with Manual.

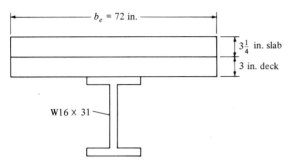

$b_e = 72$ in.

$3\frac{1}{4}$ in. slab

3 in. deck

W16 × 31

Problem 17-6

17-7. Repeat Prob. 17-6 using Manual tables if ΣQ_n of the connectors is 241 k. (*Ans.* 282.2 ft-k)

17-8. Using the Composite Design Tables of the LRFD Manual, A36 steel, a 145 lb/ft³ concrete slab with $f'_c = 3$ ksi and shored construction select the steel section, design 3/4-in. headed studs, calculate live load service deflection, and check the shear if the service live load is 100 psf.

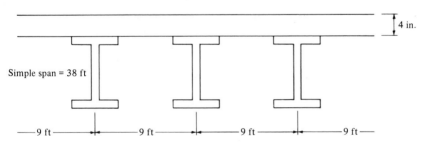

4 in.

Simple span = 38 ft

—9 ft— —9 ft— —9 ft— —9 ft—

Problem 17-8

17-9. Repeat Prob. 17-8 if $F_y = 50$ ksi. (*Ans.* W16×36 with 30 studs)

17-10. Repeat Prob. 17-9 if it's unshored. Calculate deflection during construction for wet concrete plus 20 psf live load for construction activities.

17-11. Select an A36 steel section to support a service dead load of 200 psf and a service live load of 100 psf. The beams are to have 36-ft simple spans and are to be spaced 8 ft 6 in. on center. Construction is shored, concrete weighs 115 lb/ft³, f'_c is 3.5 ksi, a metal deck with ribs perpendicular to the steel beams is used together with a 4-in. concrete slab. The ribs are 3 in. deep and have average widths of 6 in. Design 3/4-in. headed studs and calculate live load deflection. (*Ans.* W21×57 with 46 studs)

17-12. Using the LRFD Manual and A36 steel design a nonencased shored composite section for the simple span beams shown in the accompanying figure if a 4-in. con-

crete slab (145 lb/ft^3) with $f'_c = 3$ ksi is used. The total service dead load is 0.6 k/ft of length of the beam and the service live load is 1.25 k/ft.
(a) Select the beams.
(b) Determine the number of 3/4-in.-diameter headed studs required.
(c) Compute the service live load deflection.
(d) Check the beam shear.

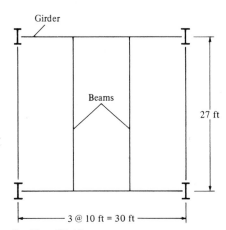

Problem 17-12

17-13. Repeat Prob. 17-12 if $F_y = 50$ ksi and $f'_c = 4$ ksi. (*Ans.* W14×26 with 22 studs.)

17-14. A36 beams 9 ft on center and spanning 40 ft are to be selected to support a 4-in.-deep lightweight concrete slab ($f'_c = 3$ ksi, weight 115 lb/ft^3) on a 3-in.-deep formed steel deck with shoring. The ribs for the steel deck, which are perpendicular to the steel beams, have average widths of 6 in. If the total service dead load is to be 0.80 k/ft of length of the beams and the service live load is 1.25 k/ft, (a) select the beams, (b) determine the number of 3/4-in.-diameter headed studs required, (c) compute the service live load deflection, and check the beam shear.

17-15. Repeat Prob. 17-14 if $F_y = 50$ ksi. (*Ans.* W18×40 with 58 studs)

17-16. Repeat Prob. 17-14 if spans are 34 ft.

17-17. Repeat Prob. 17-16 if $F_y = 50$ ksi. (*Ans.* W16×31 with 52 studs)

17-18. Using the same data as for Prob. 17-14 except that unshored construction is to be used, perform the following tasks:
(a) Select the steel beam.
(b) If the stud reduction factor for metal decks is 1.0 determine the number of $\frac{3}{4}$-in. headed studs required.
(c) Check the beam strength before the concrete hardens.
(d) Compute service load deflection before the concrete hardens assuming a construction live load of 25 psf.
(e) Determine the service load deflection after composite action is available.
(f) Check shear.

17-19. Repeat Prob. 17-18 if $F_y = 50$ ksi. (*Ans.* W18×40 with 58 studs. But deflection is high before concrete hardens.)

17-20. Using the transformed area method compute the stresses in the encased section shown in the accompanying illustration if no shoring is used. The section is assumed to be used for a simple span of 30 ft and to have a service dead uniform load of 30 psf applied after composite action develops and a service live uniform load of 120 psf. Assume $n = 9$, A36 steel, $f'_c = 3$ ksi, and concrete weighing 150 lb/ft^3.

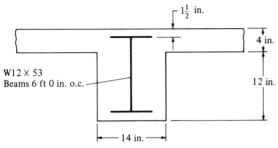

Problem 17-20

Chapter 18

Composite Columns

18-1 INTRODUCTION

Composite columns are constructed with rolled or built-up steel shapes, encased in concrete or with concrete placed inside steel pipes or tubes. The resulting members are able to support significantly higher loads than can reinforced-concrete columns of the same sizes.

Several composite columns are shown in Fig. 18-1. In part (a) of the figure a W shape embedded in concrete is shown. The cross sections, which are usually square or rectangular, have one or more longitudinal bars placed in each corner. In addition lateral ties are wrapped around the longitudinal bars at frequent vertical intervals. Ties are effective in increasing column strengths. They prevent the longitudinal bars from being displaced during construction and they resist the tendency of these same bars to buckle outward under load, which would cause breaking or spalling off of the outer concrete cover. It will be noted that these ties are all open and U-shaped. Otherwise they could not be installed because the steel column shapes will have always been erected at an earlier time. In parts (b) and (c) of the figure steel pipe and steel tubing sections filled with concrete are shown.

18-2 ADVANTAGES OF COMPOSITE COLUMNS

For quite a few decades structural steel shapes have been used in combination with plain or reinforced concrete. Originally the encasing concrete was used to provide only fire and corrosion protection for the steel with no consideration given to its strengthening effects. During the last 20 to 30 years, however, the

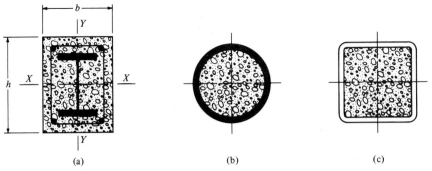

Figure 18-1 Composite columns

development and increasing popularity of composite frame construction has encouraged designers to include the strength of the concrete in their calculations.[1,2]

Composite columns may be practically used for low-rise and high-rise buildings. For the low-rise types such as warehouses, parking garages, and so on the steel columns are often encased in concrete for the sake of appearance or for protection of the steel from fire, corrosion, and from vehicles in garages. If for structures like these we are going to encase the steel in concrete anyway we may as well take advantage of the concrete and use smaller steel shapes.

For high-rise buildings the sizes of composite columns are often considerably smaller than is required for reinforced-concrete columns to support the same loads. The results with composite designs are appreciable savings of valuable floor space. Closely spaced composite steel-concrete columns connected with spandrel beams may be used around the outsides of high-rise buildings to resist lateral loads by the tubular concept (to be described in Chapter 20). Very large composite columns are sometimes placed on the corners of high-rise buildings to increase lateral resisting moments. Also steel sections embedded within reinforced-concrete shear walls may be used in the central core of high-rise buildings.

With composite construction the bare steel sections support the initial loads including the weights of the structure, the gravity and lateral loads occurring during construction, and the concrete later cast around the W shapes or inside the tube shapes. The concrete and steel are combined in such a fashion that the advantages of both materials are used in the composite sections. For instance, the reinforced concrete enables the building frame to more easily limit swaying or lateral deflections. At the same time the light weight and strength of the steel shapes permit the use of smaller and lighter foundations.[3]

Composite high-rise structures are erected in a rather efficient manner.

[1]D. Belford, "Composite Steel Concrete Building Frame," *Civil Engineering* (New York: ASCE, July 1972), pp. 61–65.

[2]Fazlur R. Kahn, "Recent Structural Systems in Steel for High Rise Buildings," BCSA Conference on Steel in Architecture, November 24–26, 1969.

[3]L.G. Griffis, "Design of Encased W-shape Composite Columns," *Proceedings 1988 National Steel Construction Conference* (AISC, Chicago, June 8–11, 1988), pp. 20-1–20-28.

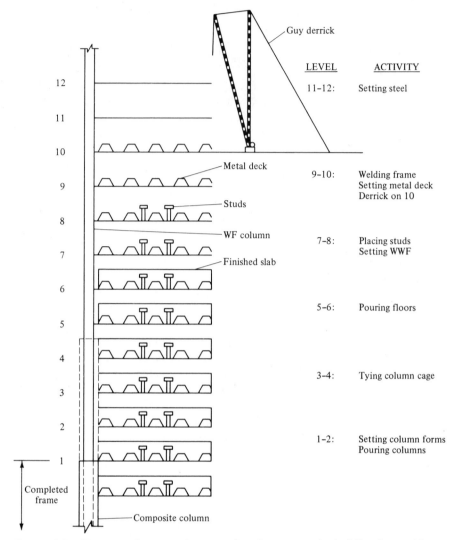

Figure 18-2 Sequence of construction operations for a composite building frame. (Courtesy of AISC.)

There is quite a vertical spread of construction going on at any one time with numerous trades working simultaneously. This situation, which is pictured in Fig. 18-2, is briefly described here.[4]

1. One group of workers may be erecting the steel beams and columns for one or two stories on top of the frame.
2. Two or three stories below another group will be setting the metal decking for the floors.

---

[4]Griffis, op. cit.

3. A few stories below another group may be placing the concrete for the floor slabs.
4. This continues as we go down the building with one group, tying the column reinforcing bars in cages, while below them others are placing the column forms, placing the column concrete, and so on.

18-3 DISADVANTAGES OF COMPOSITE COLUMNS

As described in the preceding section composite sections have several important advantages. They, however, also have a few disadvantages. One particular problem with their use in high-rise buildings is the difficulty of controlling their rates and amounts of shortening in relation to shear walls and perhaps adjacent plain steel columns. The accurate estimation of these items is made quite difficult by the different types and stages of construction activities going on simultaneously over a large number of building stories.

If composite columns are used around the outside of a high-rise building and plain steel sections are used in the building core (or if we have shear walls) the creep in the composite sections can be a problem. The results may be concrete floors that are not very level. Some erectors make very careful elevation measurements at column splices and then try to make appropriate adjustments with steel shims to try to even out the differences between measured elevations and computed elevations.

Another problem with composite columns is the lack of knowledge available concerning the mechanical bond between the concrete and the steel shapes. This is particularly important for the transfer of moments through beam-column joints. It is feared that if large cyclical strain reversals were to occur at such a joint (as in a seismic area) there could be a severe breakdown of the joint.[5]

18-4 LATERAL BRACING

Resistance to lateral loads for the usual structural steel or reinforced-concrete high-rise building is provided as the floors are being constructed. For instance, diagonal bracing or moment-resisting joints may be provided for each floor as a structural steel building frame is being constructed. In a similar manner the needed lateral strength of a reinforced-concrete frame may be provided by the moment resistance obtained with monolithic construction of its members and/or by shear walls.

For composite construction the desired lateral strength of a building is not obtained until the concrete has been placed around or inside the erected steel members and has sufficiently hardened. This situation is probably being achieved 10 to 18 stories behind the steel erection (see Fig. 18-2).

[5]Griffis, op. cit.

As we have mentioned the steel fabricator is used to erecting a steel frame and providing the necessary wind bracing as the floors are erected. The steel frames used for high-rise composite buildings, however, do not usually have such bracing and the frames will not have the desired lateral strength. This strength will be obtained only after the concrete is placed and cured for many building stories. Thus the engineer of record for a composite high-rise building must clearly state the lateral force conditions and what is to be done about them during erection.[6]

18-5 SPECIFICATIONS FOR COMPOSITE COLUMNS

Composite columns can theoretically be constructed with cross sections which are square, rectangular, round, triangular, or any other shape. Practically, however, they are usually square or rectangular with one reinforcing bar in each column corner. This arrangement enables us to use reasonably simple connections from the exterior spandrel beams and floor beams to the steel shapes in the columns without unduly interfering with the vertical reinforcing.

The LRFD Specification does not provide detailed requirements for reinforcing bar spacings, splices, and so on. It therefore seems logical that the requirements in this regard of the ACI 318 Code[7] should be followed for situations not clearly covered by the LRFD Specification.

Section I of the LRFD Specification does provide detailed requirements pertaining to cross-sectional areas of steel shapes, concrete strengths, tie areas and spacings for the vertical reinforcing bars, and so on. This information is listed and briefly discussed in the paragraphs to follow:

1. *The total cross-sectional area of the steel section or sections may not be less than 4 percent of the gross column area.* If the steel percentage is less than 4 percent the member is classified as a reinforced-concrete column, and its design must be handled by the *Building Code Requirements for Reinforced Concrete* of the American Concrete Institute.

2. *When a steel core is encased in concrete the encasement must be reinforced with longitudinal load-carrying bars (which must be run continuously at framed levels) and with lateral ties spaced no farther apart than 2/3 times the least dimension of the composite member. The area of the ties must not be less than 0.007 sq. in. per inch of bar spacing. There must be at least 1 1/2 in. clear cover of concrete outside of any steel (ties or longitudinal bars).* The cover is needed for the protection of the steel from fire or corrosion. The amount of longitudinal and transverse reinforcing required in the encasement is thought to be sufficient to prevent severe spalling of the concrete surface from occurring during a fire.

[6]Griffis, op. cit.

[7]American Concrete Institute, *Building Code Requirements for Reinforced Concrete,* ACI 318-83 (Detroit: 1983).

3. *The specified compression strength of the concrete* f'$_c$ *must be at least 3 ksi but not more than 8 ksi if normal weight concrete is used. For lightweight concrete it may not be less than 4 ksi or more than 8 ksi.* The upper limit of 8 ksi is provided because sufficient test data were not available for composite columns with higher-strength concretes at the time the specification was prepared. The lower limit of *f'$_c$* was specified for the purpose of ensuring the use of good quality but readily available concrete and for the purpose of making sure that adequate quality control is used. This might not be the case if a lower grade of concrete were specified.

4. *The yield stresses of the steel sections and reinforcing bars used may not be greater than 55 ksi. If a steel with a yield stress greater than 55 ksi is actually used in a composite column only 55 ksi may be used in the calculations.* One major objective in composite design is to prevent local buckling of the longitudinal reinforcing bars and the contained steel section. To achieve this objective the covering concrete must not be allowed to break or spall. It was assumed by the writers of the LRFD Specification that this concrete is in danger of breaking or spalling if its strain reaches 0.0018. If we take this strain and multiply it by E_s we get (0.0018) (29,000) $\approx$ 55 ksi. Thus 55 ksi is a limit state for stress in the reinforcing.

5. *The minimum permissible wall thickness of steel tubing filled with concrete is equal to* b $\sqrt{F_y/3E}$ *for each face of width b. The minimum thickness for pipe sections of outside diameter* D *is* D $\sqrt{F_y/8E}$. These values are the same as those given in the 1983 ACI Code. It is desired that designers use steel pipe or tubing of sufficient thickness so that they will not buckle before yielding.

6. *When composite columns contain more than one steel shape those shapes have to be connected together with lacing, tie plates, or batten plates so that local buckling of the individual shapes is not possible before the concrete hardens.* After the concrete hardens it is assumed that all the parts of the column act together as a unit in resisting load.

7. It is necessary to avoid the overstressing of either the concrete or the structural steel at connections. As a consequence the LRFD Specification I2.4 requires that *the part of the design strength of axially loaded composite columns resisted by concrete must be developed by direct bearing at connections. Should the supporting concrete be wider on one or more sides than the loaded area and be otherwise restrained against lateral expansion on the remaining side or sides the design strength of the concrete is to be computed as being equal to 1.7ϕ_cf'$_c$ A$_B$ with ϕ_c = 0.6 for bearing on concrete where A$_B$ is the loaded area.*

18-6 AXIAL DESIGN STRENGTHS OF COMPOSITE COLUMNS

The contribution of each component of a composite column to its overall strength is difficult if not impossible to determine. The amount of flexural concrete cracking varies throughout the height of the column. The concrete is not nearly as homogeneous as is the steel, and furthermore the modulus of elasticity of the concrete varies with time and under the action of long-term or sustained loads. The effective lengths of composite columns in the rigid monolithic structures in which they are frequently used cannot be determined very well. The contribution of the concrete to the total stiffness of a composite column varies depending on whether it is placed inside a tube or whether it's on the outside of a W section where its stiffness contribution is less.[8]

The preceding paragraph presented some of the reasons why it is difficult to develop a useful theoretical formula for the design of composite columns. As a result a set of empirical formulas is presented in the LRFD Specification for the design of composite columns.

The design strengths of composite columns ($\phi_c P_n$ with $\phi_c = 0.85$ and $P_n = A_g F_{cr}$) are determined much as are the design strengths of plain steel columns. The formulas to be used for composite columns for F_{cr}, the critical stress, are the same except that the areas, radii of gyration, yield stresses, and moduli of elasticity are modified in an attempt to account for composite behavior. The column expressions given in Section E2 of the Specification and previously described in Chapter 5 of this text are listed as follows:

If $\lambda_c \leq 1.5$

$$F_{cr} = (0.658^{\lambda_c^2}) F_y \qquad \text{(LRFD Formula E2.2)}$$

If $\lambda_c > 1.5$

$$F_{cr} = \left(\frac{0.877}{\lambda_c^2}\right) F_y \qquad \text{(LRFD Formula E2.3)}$$

where

$$\lambda_c = \frac{KL}{r\pi} \sqrt{\frac{F_y}{E}} \qquad \text{(LRFD Formula E2.4)}$$

The modifications made in these formulas are as follows:

1. *Replace A_g with A_s* where A_s is the area of the steel shape, tube, or pipe but not including any reinforcing bars.
2. *Replace r with r_m* where r_m is the radius of gyration of the steel shapes, pipes, or tubes. For steel shapes encased in concrete it may not be less than 0.3 times the overall thickness of the composite member in the plane of buckling.
3. *Replace F_y with the modified yield stress F_{my} and E with the modified modulus of elasticity F_m.* These values follow:

$$F_{my} = F_y + c_1 F_{yr}\left(\frac{A_r}{A_s}\right) + c_2 f_c'\left(\frac{A_c}{A_s}\right) \qquad \text{(LRFD Formula I2.1)}$$

$$E_m = E + c_3 E_c\left(\frac{A_c}{A_s}\right) \qquad \text{(LRFD Formula I2.2)}$$

---

[8]Task Group 20, Structural Stability Council, "A Specification for the Design of Steel-Concrete Composite Columns," *Engineering Journal,* AISC 16, no. 4 (fourth quarter, 1979), pp. 101–115.

In these expressions for F_{my} and E_m the following abbreviations are used:

1. A_c, A_s, and A_r are respectively the areas of the concrete, steel section, and reinforcing bars.
2. E and E_c are the moduli of elasticity of the steel and concrete.
3. F_y and F_{yr} are the specified minimum yield stresses of the steel section and the reinforcing bars.
4. c_1, c_2, and c_3 are numerical coefficients. For concrete-filled pipes and tubing $c_1 = 1.0$, $c_2 = 0.85$, and $c_3 = 0.4$. For concrete encased shapes $c_1 = 0.7$, $c_2 = 0.6$, and $c_3 = 0.2$.

The value of $\phi_c P_n$ for a composite column with a W section embedded in concrete is calculated in Example 18-1 in accordance with these LRFD requirements.

EXAMPLE 18-1

Compute the value of $\phi_c P_n$ for the composite column shown in Fig. 18-3 if A36 steel, 3.5 ksi concrete with a weight of 145 lb/ft^3, and a KL of 12 ft are used.

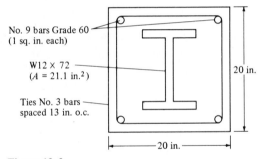

No. 9 bars Grade 60
(1 sq. in. each)

W12 × 72
(A = 21.1 in.^{2})

Ties No. 3 bars
spaced 13 in. o.c.

20 in.

20 in.

Figure 18-3

Solution

Note: As required by LRFD Specification (I2.1) F_{yr} is used as 55 ksi and not 60 ksi.

$$F_{my} = F_y + c_1 F_{yr}\left(\frac{A_r}{A_s}\right) + c_2 f'_c\left(\frac{A_c}{A_s}\right)$$

$$= 36 + (0.7)(55)\left(\frac{4}{21.1}\right) + (0.6)(3.5)\left(\frac{400 - 25.1}{21.1}\right)$$

$$= 80.61 \text{ ksi}$$

$$E_c = w^{1.5}\sqrt{f'_c} = 145^{1.5}\sqrt{3.5} = 3266.5 \text{ ksi}$$

$$E_m = E + c_3 E_c\left(\frac{A_c}{A_s}\right)$$

$$= 29{,}000 + (0.2)(3266.5)\left(\frac{400 - 25.1}{21.1}\right)$$

$$= 29{,}000 + 11{,}608 = 40{,}608 \text{ ksi}$$

Replace F_y with 80.61 ksi, E with 40,608 ksi, and A_g with A_s = 21.1 in.^{2}

r_y of w $12 \times 72 = 3.04$ in., may not be less than $(0.3)(20)$ = 6.0 in.

$$\lambda_c = \frac{KL}{r\pi}\sqrt{\frac{F_y}{E}}$$

$$= \frac{(12)(12)}{6.0\pi}\sqrt{\frac{80.61}{40,608}} = 0.340$$

$$F_{cr} = (0.658^{\lambda_c^2})F_y = (0.658^{0.340^2})80.61$$

$$= 76.80 \text{ ksi}$$

$$\phi_c P_n = \phi_c F_{cr} A_s = (0.85)(76.80)(21.1)$$

$$= \underline{\underline{1377.4}}$$

(LFRD table described in Section 18-4 gives 1380-k)

18-7 LRFD TABLES

The LRFD Manual in Part 4 presents a series of tables for determining the design axial strengths of various square and rectangular encased W shapes as well as for numerous pipes and structural tubes filled with concrete. The tables are set up in exactly the same fashion as they were for axially loaded plain steel columns in Part 2 of the Manual. The strengths are given with respect to the minor axes for a range of $K_y L_y$ values.

The tables were prepared for normal weight concretes and include encased W shapes with F_y values of 36 and 50 ksi and f'_c values of 3.5, 5, and 8 ksi; 36- and 50-ksi pipes filled with 3.5- and 5-ksi concrete; and 46-ksi tubing filled with 3.5- and 5-ksi concrete. The reinforcing bars used for the encased sections are all grade 60 (or 60 ksi).

Examples 18-2 and 18-3, which follow, illustrate the applications of these tables.

EXAMPLE 18-2

(a)   Using A36 steel and 3.5-ksi concrete select a square composite column with a W shape encased in concrete to support a factored load P_u = 940 k if KL = 14 ft.

(b)   Repeat part (a) if a plain W shape is used instead of a composite column.

Solution

(a)   Possible selections from Part 4 of the Manual include:

18 in. $\times$ 18 in. column with W10$\times$45 (971 k)

16 in. $\times$ 16 in. column with W8$\times$67 (1000 k)

<u>Use 18 in. $\times$ 18 in. column with W10$\times$45</u>

(b)   Possible selections from Part 2 of Manual include:

$$W14 \times 120$$

$$W12 \times 136$$

$$\text{Use } W14 \times 120$$

EXAMPLE 18-3

An 18-in.-square composite column with an encased $W14 \times 48$ ($F_y = 36$ ksi, $f'_c = 5$ ksi) has a $K_x L_x = 26$ ft and a $K_y L_y = 16$ ft. Using the composite column tables determine $P_u = \phi_c P_n$.

..Solution.   Entering the tables with $K_y L_y = 16$ ft, $P_u = \phi_c P_n$ is found to equal 1330 k. However, $K_x L_x = 26$ ft and $r_x/r_y = 1.22$.

$$\frac{K_x L_x}{r_x/r_y} = \frac{26}{1.22} = 21.31 \text{ ft} > k_y L_y$$

$$\therefore K_x L_x \text{ controls}$$

Reentering tables with $K_y L_y = 21.31$ ft we find P_u is found by interpolation to equal

$$1230 - \left(\frac{1.31}{2}\right)(1230 - 1180) = \underline{\underline{1197 \text{ k}}}$$

18-8 FLEXURAL DESIGN STRENGTHS OF COMPOSITE COLUMNS

The nominal flexural strength of composite columns is computed assuming a plastic distribution of stresses. We can locate the plastic neutral axis by equating the tensile forces on one side of the member to the compression force on the other side. On the tensile side there will be reinforcing bars and part of the embedded steel section stressed to their yield stresses. On the compression side there will be a compressive force equal to $0.85 f'_c$ times the area of an equivalent stress block. The equivalent stress block will have a width equal to the column width and a depth equal to β_1 times the distance to the PNA. (The value of β_1 is provided by the ACI Code.) The nominal flexural strength M'_n then equals the sum of the moments of the axial forces about the PNA.

Values of $\phi_b M_{nx}$ and $\phi_b M_{ny}$ are shown for each of the composite columns in Part 4 of the Manual. These values will be needed for analyzing beam-columns as described in the next section.

18-9 AXIAL LOAD AND BENDING EQUATION

The following interaction formulas are used to check plain steel members subject to axial load and bending.

If $P_u/\phi P_n \geq 0.2$

$$\frac{P_u}{\phi P_n} + \frac{8}{9}\left(\frac{M_{ux}}{\phi_b M_{ny}} + \frac{M_{uy}}{\phi_b M_{ny}}\right) \leq 1.0 \quad \text{(LRFD Formula H1.1a)}$$

If $P_u/\phi P_n < 0.2$

$$\frac{P_u}{2\phi P_n} + \left(\frac{M_{ux}}{\phi_b M_{nx}} + \frac{M_{uy}}{\phi_b M_{ny}}\right) \leq 1.0 \quad \text{(LRFD Formula H1.1b)}$$

These formulas and their application were discussed in Chapter 11. That presentation included definitions of the various values needed to calculate M_{ux} and $M_{uy}(B_1, M_{nt}, B_2,$ and $M_{1t})$.

These same interaction equations are used to determine the adequacy of composite beam-columns except that some of the terms are modified. These modifications follow:

1. The Euler elastic buckling loads P_{ex} and P_{ey} which are used in the calculations of the bending factors B_1 and B_2 are to be determined with the following expression in which F_{my} is the modified yield stress which was defined in Section 18-3. The values of P_{ex} and P_{ey} times the square of the appropriate effective length in feet divided by 10^4 are given in the tables for each of the composite columns.

$$P_e = \frac{A_s F_{my}}{\lambda_c^2}$$

2. The resistance factor ϕ_b is to be used as it is in composite beams where it equals 0.85 if $h_c/t_w \leq 640/\sqrt{F_{yf}}$ and a plastic stress distribution is used to compute M_n; or it is taken as 0.9 if $h_c/t_w > 640/\sqrt{F_{yf}}$ and M_n determined by superimposing the elastic stresses.
3. The column slenderness parameter λ_c is to be modified as it was for determining the design strengths of axially loaded composite columns in Section 18-3.

18-10 DESIGN OF COMPOSITE COLUMNS SUBJECT TO AXIAL LOAD AND BENDING

This section is devoted to the design of composite columns to resist axial loads and moments. The procedure is a trial-and-error one involving the selection of a trial section, the application of the appropriate interaction formula, probably the selection of another trial section, the application of the formula, and so on until a satisfactory column is obtained.

A perfectly satisfactory design can be made as described above, but the process can involve quite a few trials if the first estimate is not too good. For this reason a rough method of estimating sizes is presented in the next few paragraphs. This method will usually enable the designer to make a fairly good first size estimate and thus reduce the number of trial designs that have to be made.

For this discussion it is assumed that a composite column is to be designed to support a certain P_u and a certain M_{ux} with M_{uy} equal to zero. It is further assumed that LRFD Formula H1-1a applies to the member. If M_{uy} is zero the formula becomes

$$\frac{P_u}{\phi P_n} + \frac{8}{9} \frac{M_{ux}}{\phi_b M_{nx}} \le 1.0$$

The designer may estimate the final values of the two parts of this equation. He or she may very well assume the two parts are equal.

$$\frac{P_u}{\phi P_n} = 0.5 \quad \text{and} \quad \frac{8}{9} \frac{M_{ux}}{\phi_b M_{nx}} = 0.5$$

For this discussion it is assumed that a composite column with an encased section with $KL = 12$ ft consisting of A36 steel and 3.5-ksi concrete is to be selected to resist $P_u = 500$ k and $M_{ux} = 100$ ft-k. Estimated values of ϕP_n and $\phi_b M_{nx}$ can be determined as follows:

$$\frac{P_u}{\phi P_n} = 0.5 \qquad \frac{8}{9} \frac{M_{ux}}{\phi_b M_{nx}} = 0.5$$

$$\frac{500}{\phi P_n} = 0.5 \qquad \frac{8}{9} \frac{100}{\phi_b M_{nx}} = 0.5$$

$$\phi P_n = 1000 \text{ k} \qquad \phi_b M_{nx} = 177.8 \text{ ft-k}$$

The designer may then go to the composite tables and try a section that has ϕP_n and $\phi_b M_{nx}$ values somewhere in the range of these values. For instance, an 18 in. × 18 in. encased section with a W10×49 has a ϕP_n of 1020 k while a 16 in. × 16 in. encased section with a W8×40 has a $\phi_b M_{nx}$ equal to 194 ft-k. As a first trial the designer may like to try a section somewhere in between these two in the tables as perhaps the 18 in. × 18 in. with a W10×39 or a 16 in. × 16 in. column with a W8×67. Example 18-4 presents the complete design of this particular member.

EXAMPLE 18-4

Select a composite column with an encased W section to resist $P_u = 500$ k and $M_{ux} = 100$ ft-k. The column which is braced against sidesway or lateral translation of its ends is to consist of A36 steel and 3.5-ksi concrete. $KL = 12$ ft and $C_m = 0.85$.

Solution. As described in the paragraphs before this example assume rough values of $\phi P_n = 1000$ k and $\phi_b M_{nx} = 177.8$ ft-k and try an 18 in. × 18 in. composite column with an encased W10×39 (for which $\phi P_n = 942$ k, $\phi_b M_{nx} = 245$ ft-k, and $P_e(K_x L_x)^2/10^4 = 107$ ft²-k).

$$P_e = \frac{(107)(10)^4}{(12)^2} = 7430.6 \text{ k}$$

with LRFD Formula H1-3

$$B_1 = \frac{C_m}{1 - P_u/P_e} = \frac{0.85}{1 - 1000/7430.6} = 0.982 \qquad \underline{\text{Use } 1.0}$$

and

$$M_{ux} = (1.0)(100) = 100 \text{ ft-k}$$

$$\frac{P_u}{\phi P_n} = \frac{500}{942} > 0.2$$

∴ Must use LRFD Formula H1-1a

$$\frac{P_u}{\phi P_n} + \frac{8}{9} \frac{M_{ux}}{\phi_b M_{nx}} \leq 1.0$$

$$\frac{500}{942} + \frac{8}{9} \frac{100}{245} = 0.531 + 0.363 = 0.894 < 1.0 \qquad \text{OK}$$

Further calculations show that the 16 in. × 16 in. column with an encased W8×48 will also be satisfactory.

Use 18 in. × 18 in. column with encased W10×39, 4 No. 8 longitudinal bars and No. 3 ties spaced 12 in. o.c.

For Example 18-4 the proportions between the axial load P_u and the bending moment M_{ux} seemed fairly normal. Should the ratio between the two be somewhat different as where we have a rather large moment in comparison with the axial load, we might change the estimated values in the interaction equation and have a little better luck in the first trials. For instance, if the moment M_{ux} is rather large in comparison with P_u we might assume the following approximate values for the interaction formula and backfigure the values of ϕP_n and $\phi_b M_{nx}$ to look up in the tables.

$$\frac{P_u}{\phi P_n} = 0.3 \qquad \frac{8}{9} \frac{M_{ux}}{\phi_b M_{nx}} = 0.7$$

The rough procedure used here to estimate beam-column sizes doesn't always work as well as it does for the author's two enclosed examples (18-4 and 18-5). Frequently several trials will still have to be made.

EXAMPLE 18-5

It is desired to design an encased composite column with A36 steel and 5-ksi concrete to resist a P_u of 600 k and a M_{ux} of 500 ft-k. Assume $KL = 14$ ft and $C_m = 0.85$.

Solution. Assuming that Formula H1-1a will apply. As the moment is quite large in comparison with the axial load the following values are assumed:

$$\frac{P_u}{\phi P_n} = 0.3 \qquad \frac{8}{9}\frac{M_{ux}}{\phi_b M_{nx}} = 0.7$$

$$\frac{600}{\phi P_n} = 0.3 \qquad \frac{8}{9}\frac{500}{\phi_b M_{nx}} = 0.7$$

$$\phi P_n = 2000 \text{ k} \qquad \phi_b M_{nx} = 634.9 \text{ ft-k}$$

Try a 22 in. × 22 in. section with an encased W14×90 ($\phi P_n = 2000$ k, $\phi_b M_{nx} = 691$ ft-k, and $[P_{ex}(K_x L_x)^2]/10^4 = 335$ ft²-k)

$$P_e = \frac{(335)(10)^4}{(14)^2} = 17,092 \text{ k}$$

$$B_1 = \frac{0.85}{1 - 600/17,092} = 0.881 \qquad \underline{\text{Use } 1.0}$$

$$M_{ux} = (1.0)(500) = 500 \text{ ft-k}$$

$$\frac{P_u}{\phi P_n} = \frac{600}{2000} > 0.2$$

∴ Must use LRFD Formula H1-1a

$$\frac{600}{2000} + \frac{8}{9}\frac{500}{691} = 0.3 + 0.643 = 0.943 \qquad \text{OK}$$

Try a 22 in. × 22 in. section with an encased W14×82 ($\phi P_n = 1940$ k, $\phi_b M_{nx} = 663$ ft-k, and $[P_{ex}(K_x L_x)^2]/10^4 = 315$ ft²-k)

$$P_e = \frac{(315)(10)^4}{(14)^2} = 16,071 \text{ k}$$

$$B_1 = \frac{0.85}{1 - 600/16,071} = 0.883 \qquad \underline{\text{Use } 1.0}$$

$$M_{ux} = (1.0)(500) = 500 \text{ ft-k}$$

$$\frac{P_u}{\phi P_n} = \frac{600}{1940} > 0.2$$

∴ Must use LRFD Formula H1-1a

$$\frac{600}{1940} + \frac{8}{9}\frac{500}{663} = 0.979 < 1.0 \qquad \text{OK}$$

Use 22 in. × 22 in. section with an encased W14×82, 4 No. 10 longitudinal bars and No. 3 ties at 14 in. o.c.

18-11 LOAD TRANSFER AT FOOTINGS AND OTHER CONNECTIONS

The LRFD Specification (I2.4) states that the design strength of axially loaded composite columns must be developed by direct bearing at connections. If the supporting concrete area is wider than the column on one or more sides and otherwise restrained against lateral expansion on the remaining sides, the design bearing strength of the concrete equals $1.7 \, \phi_c f'_c A_B$ where ϕ_c is 0.60 and A_B is the loaded area.

A small steel base plate is usually provided at the base of a composite column. Its purpose is to accommodate the anchor bolts needed to anchor the embedded steel shape to the footing for the loads occurring during the erection of the structure before the encasing concrete hardens and composite action is developed. This plate desirably will be sufficiently small so as to be out of the way of the dowels needed for the reinforced-concrete part of the column.[9]

The LRFD Specification does not provide details for the design of these dowels but a procedure similar to the one provided by the ACI 318 Code seems to be a good one. If the column P_u exceeds $1.7 \, \phi_c f'_c A_B$ the excess load should be resisted by dowels. If P_u does not exceed $1.7\phi_c f'_c A_B$ it would appear that no dowels are needed. For such a situation the ACI Code (Sections 15.8.2.1 and 15.8.2.3) states that a minimum area of dowels equal to 0.005 times the cross-sectional area of the column must be used and the diameter of those dowels must not exceed the diameter of the column bars by more than 0.15 in. This diameter requirement ensures sufficient tying together of the column and footing over the whole contact area. The use of a very few large dowels spaced far apart might not do this very well.

PROBLEMS

For all problems grade 60 reinforcing bars and 145 lb/ft^3 concrete are used.

18-1 to 18-3. Determine $\phi_c P_n$ for the composite columns shown using the appropriate LRFD formulas. The steel shapes are A36 and f'_c is 3.5 ksi for the concrete. Column effective lengths, W section sizes, and reinforcing bar numbers are shown with the figures.

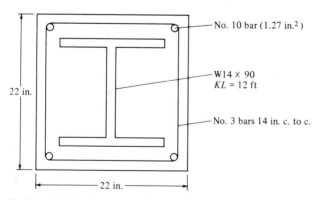

No. 10 bar (1.27 in.^{2})

W14 × 90
$KL = 12$ ft

No. 3 bars 14 in. c. to c.

22 in.

22 in.

Problem 18-1   (*Ans.* 1715 k)

[9]Griffis, op. cit.

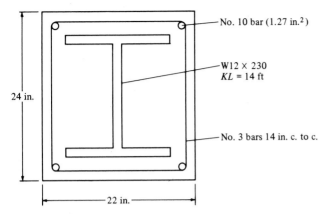

24 in.

22 in.

No. 10 bar (1.27 in.²)

W12 × 230
KL = 14 ft

No. 3 bars 14 in. c. to c.

Problem 18-2

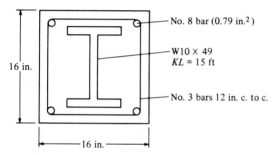

16 in.

16 in.

No. 8 bar (0.79 in.²)

W10 × 49
KL = 15 ft

No. 3 bars 12 in. c. to c.

Problem 18-3   (*Ans.* 989 k)

18-4. Repeat Prob. 18-2 using $F_y = 50$ ksi.

18-5. Repeat Prob. 18-3 using $F_y = 50$ ksi and $f'_c = 5$ ksi. (*Ans.* 1319.5 k)

18-6. Repeat Probs. 18-1, 18-2, and 18-3 using the tables in the LRFD Manual.

18-7 and 18-8. Compute $\phi_c P_n$ for the concrete-filled steel pipe composite columns shown using the appropriate LRFD formulas. The pipes are made from A36 steel and f'_c is 3.5 ksi for the concrete. The section sizes and effective lengths are shown with the figures.

12 in. pipe (65.42 lb/ft, $t_{wall} = 0.500$ in.)
KL = 16 ft

Problem 18-7   (*Ans.* 763 k)

6 in. pipe (53.16 lb/ft, $t_{wall} = 0.864$ in.)
KL = 15 ft

Problem 18-8

18-9. Repeat Probs. 18-7 and 18-8 using the LRFD tables. (*Ans.* 764 k and 345 k respectively)

18-10 and 18-11. *Determine $\phi_c P_n$ for the concrete-filled square structural tubes shown using the appropriate LRFD formulas. The tubes are made from steel with $F_y = 46$ ksi and concrete with $f'_c = 3.5$ ksi. (Ans. 117 k)*

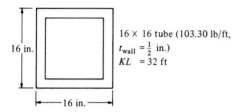

16 × 16 tube (103.30 lb/ft,
$t_{\text{wall}} = \frac{1}{2}$ in.)
$KL = 32$ ft

Problem 18-10

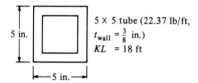

5 × 5 tube (22.37 lb/ft,
$t_{\text{wall}} = \frac{3}{8}$ in.)
$KL = 18$ ft

Problem 18-11   (*Ans.* 117 k)

18-12. Repeat Probs. 18-10 and 18-11 using the tables in the LRFD Manual.

18.13. A 20 in. × 20 in. composite column ($f'_c = 3.5$ ksi) with an encased W12×58 (A36) and with one No. 9 bar (1.00 in.^{2}) in each corner is subjected to $P_u = 1200$ k and $M_{ux} = 125$ ft-k. If the member has a $KL = 14$ ft and is braced against sidesway or lateral translation at its ends and if $C_m = 0.85$, is it satisfactory? (*Ans.* 1.24 > 1.00 NG)

18.14. A 16 in. × 16 in. tube (103.30 lb/ft, $t = \frac{1}{2}$ in., and $F_y = 46$ ksi) is encased with 3.5-ksi concrete. If the member has a KL of 16 ft and is braced against sidesway with $C_m = 0.85$ and if P_u is 800 k and M_{ux} is 100 ft-k, is it satisfactory?

18-15. Select a composite column ($f'_c = 3.5$ ksi) with an encased W section (A36 steel) to resist $P_u = 1500$ k and $M_{ux} = 150$ ft-k. The column which is to have a $KL = 15$ ft is to be braced against lateral translation at its ends and is to have $C_m = 0.85$. (*One ans:* 22 in. × 22 in. column with a W14×109, 4 No. 10 longitudinal bars and No. 3 ties, 14 in. o.c.)

18-16. Select a composite column ($f'_c = 5$ ksi) with an encased W14 section (A36 steel) to resist $P_u = 1800$ k and $M_{ux} = 200$ ft-k. Assume $KL = 12$ ft and $C_m = 0.85$. The column is braced against sidesway at its ends.

Chapter 19

Built-up Beams, Built-up Wide-Flange Sections, and Plate Girders

19-1 COVER-PLATED BEAMS

Should the largest available W section be insufficient to support the loads antic-ipated for a certain span, several possible alternatives may be taken. Perhaps the most economical solution involves the use of a higher-strength steel W section. If this is not feasible, we may make use of one of the following: (1) two or more regular W sections side by side (an expensive solution), (2) a cover-plated beam, (3) a built-up wide-flange section, (4) a plate girder, or (5) a steel truss. This section is concerned with the cover-plated beam alternative, the built-up wide flange is presented in Section 19-2, and the remainder of the chapter is devoted to plate girders.

In addition to being practical for cases where the moments to be resisted are slightly in excess of those which can be supported by the deepest W sections, there are other useful applications for cover-plated beams. On some occasions the total depth may be so limited that the resisting moments of W sections of the specified depth are too small. For instance, the architect may show a certain maximum depth for beams in his drawings for a building. In a bridge, beam depths may be limited by clearance requirements. Cover-plated beams will fre-quently prove to be a very satisfactory solution for situations like these. Further-more, there may be economical uses for cover-plated beams where the depth is not limited and where there are standard W sections available to support the loads. A smaller W section than required by the maximum moment can be selected and have cover plates attached to its flanges. These plates can be cut off where the moments are smaller with resulting saving of steel. Applications of this type are quite common for continuous beams.

Should the depth be fixed and a cover-plated beam seem to be a feasible

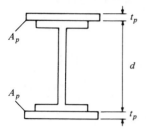

Figure 19-1

solution, the usual procedure will be to select a standard section with a depth which leaves room for top and bottom cover plates. Then the cover plate sizes can be selected.

For this discussion reference is made to Fig. 19-1. For the derivation to follow Z is the plastic modulus for the entire built-up section, Z_W is the plastic modulus for the W section, d is the depth of the W section, t_p is the thickness of one cover plate, and A_p is the area of one cover plate. An expression for the required area of one plate can be developed as follows:

$$Z \text{ required} = \frac{M_u}{\phi_b F_y}$$

The total Z of the built-up section must at least equal the Z required. It will be furnished by the W shape and the cover plates as follows:

$$Z \text{ required} = Z_W + Z_{\text{plates}}$$

$$= Z_W + 2 A_p \left(\frac{d}{2} + \frac{t_p}{2} \right)$$

$$A_p = \frac{Z \text{ required} - Z_W}{d + t_p}$$

Example 19-1 illustrates the design of a cover-plated beam. Quite a few satisfactory solutions involving different W sections and varying size cover plates are available other than the one made in this problem.

EXAMPLE 19-1

Select a beam limited to a maximum depth of 29.50 in. for the loads and span of Fig. 19-2. A36 steel is used and the beam is assumed to have full lateral bracing for its compression flange.

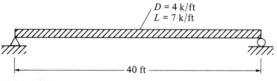

$D = 4$ k/ft
$L = 7$ k/ft

40 ft

Figure 19-2

Solution

$$\text{Assume beam wt} = 350 \text{ lb/ft}$$

$$w_u = (1.2)(4.35) + (1.6)(7) = 16.42 \text{ k/ft}$$

$$M_u = \frac{(16.42)(40)^2}{8} = 3284 \text{ ft-k}$$

$$Z \text{ required} = \frac{M_u}{\phi_b F_y} = \frac{(12)(3284)}{(0.9)(36)} = 1216.3 \text{ in.}^3$$

No single W section in A36 steel of depth 29.50 in. or less will be able to provide this M_u. Therefore, a cover-plated beam will be selected.

Try a W27×217 ($d = 28.43$ in., $Z_X = 708$ in.³, $b_f = 14.115$ in.)

Assume 1-in.-thick plates

$$A_p = \frac{Z \text{ required} - Z_W}{d + t_p} = \frac{1216.3 - 708}{28.43 + 1.0} = 17.27 \text{ in.}^2$$

Use W27×217 with one PL1×18 each flange

19-2 BUILT-UP WIDE-FLANGE SECTIONS

In this section the author initially defines the terms *built-up wide-flange sections* and *plate girders* because the reader may quite understandably find it difficult to distinguish between the two.

In Chapter G of the LRFD Specification a clear distinction is made between beams (whether rolled shapes or built-up wide-flange sections) and plate girders. It is stated that plate girders are to be distinguished from beams on the basis of web slenderness. This slenderness is measured by the ratio h_c/t_w where h_c is the assumed web depth for stability and equals twice the distance from the neutral axis to the inside face of the compression flange less the fillet or corner radius and t_w is the web thickness.

To be defined as a beam a section may be a rolled shape or a welded one but it may not have stiffeners, says the LRFD, and its h_c/t_w ratio may not be greater than $970/\sqrt{F_{yf}}$ where F_{yf} is the specified minimum yield stress of the flange. On the other hand, a plate girder may or may not have stiffeners but its h_c/t_w ratio must be greater than $970/\sqrt{F_{yf}}$. (In this expression the flange yield point F_{yf} is used rather than the web yield point F_{yw} because for hybrid girders inelastic buckling of the web due to bending is dependent upon the strain in the flange.) Reference is made here to Fig. 19-3.

With the LRFD Specification overall economy can often be obtained when built-up wide-flange sections are used instead of plate girders. With these sections the webs selected are sufficiently thick to carry shear without buckling. Even though these sections with their unstiffened webs will be heavier than plate girders for the same spans and loads their overall costs will often be less because

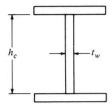

Figure 19-3   A built-up wide-flange or a plate girder? It's a plate girder if $h_c/t_w > 970/\sqrt{F_{yf}}$ whether or not it has stiffeners. If $h_c/t_w \leq 970/\sqrt{F_{yf}}$ and no stiffeners are used, it's a built-up wide-flange section.

of their smaller fabrication costs. In addition design calculations will be appreciably reduced.

According to LRFD Specification F1.1 plastic analysis is permitted for compact beams and girders if they meet certain conditions. They must be singly or doubly symmetric and loaded in the plane of symmetry. When bent about their major axes the laterally unbraced lengths of their compression flanges at plastic hinge locations associated with failure mechanisms must not exceed certain values as given in Chapter F of the Specification.

According to Table B5.1 of the LRFD Specification a section is compact if

$$\frac{b_f}{2\,t_f} \leq 65\,\sqrt{F_y} \quad \text{and} \quad \frac{h_c}{t_w} \leq \frac{640}{\sqrt{F_{yf}}}$$

After a compact web size is selected for a built-up wide-flange member the next step is to select the flange size. For this discussion reference is made to Fig. 19-4. The total design bending strength of the girder shown equals the bending strength of its web plus the bending strength of its flanges.

The plastic modulus of the entire girder equals the statical moment of the compression and tension areas of the web about the neutral axis plus the statical moment of the areas of both flanges about the neutral axis. After such an expression is written it can be solved for the required area of one of the flanges.

$$\text{Total } Z \text{ required for girder} = \frac{M_u}{\phi_b F_y}$$

$$Z \text{ furnished} = (t_w)\left(\frac{h}{2}\right)\left(\frac{h}{4}\right)(2) + (A_f)\left(\frac{h}{2} + \frac{t_f}{2}\right)(2)$$

$$\frac{M_u}{\phi_b F_y} = \frac{t_w h^2}{4} + A_p(h + t_f)$$

$$A_p \text{ required} = \frac{M_u}{\phi_b F_y(h + t_f)} - \frac{t_w h^2}{4(h + t_f)}$$

EXAMPLE 19-2

Design a 60-in.-deep welded built-up wide-flange section with no intermediate stiffeners for a 70-ft simple span to support a service dead load of 1 k/ft and a service live load of 2 k/ft. The section is to be framed between columns (thus no end bearing stiffeners are required as described in Section

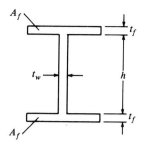

Figure 19-4

19-8) and is to have full lateral bracing for its compression flange. The design is to be made with A36 steel, SMAW fillet welds, and E70 electrodes.

Solution

Maximum shear and moment

$$\text{Assume beam wt} = 200 \text{ lb/ft}$$

$$w_u = (1.2)(1.2) + (1.6)(2) = 4.64 \text{ k/ft}$$

$$M_u = \frac{(4.64)(70)^2}{8} = 2842 \text{ ft-k}$$

$$V_u = (35)(4.64) = 162.4 \text{ k}$$

Design for a compact web and flange

$$Z \text{ required} = \frac{M_u}{\phi F_y} = \frac{(12)(2842)}{(0.9)(36)} = 1052.6 \text{ in.}^3$$

Web size required

For web to be compact h_c/t_w must be

$$\leq \frac{640}{\sqrt{F_{yf}}} = \frac{640}{\sqrt{36}} = 107$$

Assume $h = h_c = 60 - 2 = 58$ in.

$$\text{Min. } t_w = \frac{58}{107} = 0.542 \text{ in.} \qquad \underline{\text{say } \tfrac{9}{16} \text{ in.}}$$

Try $\tfrac{9}{16} \times 58$ web ($A_w = 32.62$ in.^{2})

$$\frac{h}{t_w} = \frac{58}{9/16} = 103.1$$

Since this value is $> 418/\sqrt{F_{yf}} = 70$ transverse stiffeners may be needed. However, if $V_u \leq \phi V_n$ as determined from LRFD Formulas F2-3 or A-G3-3 stiffeners will not be necessary. (This topic is discussed in Section 19-5.)

Instead of substituting into the above-mentioned formulas it is possible to determine the value of $\phi_v V_n / A_w$ from Table 10-36 in Part 6 of the Manual with $a/h > 3.0$ and $h/t_w = 103.1$. (a is the clear distance between transverse stiffeners.)

$$\frac{\phi_v V_n}{A_w} = 11.2 \text{ ksi}$$

$$\frac{V_u}{A_w} = \frac{162.4}{32.62} = 4.98 \text{ ksi} < 11.2 \text{ ksi}$$

$\therefore$ Intermediate stiffeners are not required

Flange area

$$A_f = \frac{M_u}{\phi_b F_y (h + t_p)} - \frac{t_w h^2}{4(h + t_f)}$$

Assume $\frac{3}{4}$-in. plates

$$A_f = \frac{(12)(2842)}{(0.9)(36)(58 + 0.75)} - \frac{(9/16)(58)^2}{(4)(58 + 0.75)} = 9.87 \text{ in.}^2$$

Try $\frac{3}{4} \times 14$ plate each flange (10.5 in.^{2}). Is flange compact?

$$\frac{b_f}{2 t_f} = \frac{14}{(2)(0.75)} = 9.33 < \frac{65}{\sqrt{36}} = 10.8 \qquad \text{YES}$$

Check section

$$Z = (29)(\tfrac{9}{16})(14.5)(2) + (14)(0.75)(29.375)(2)$$

$$= 1090 \text{ in.}^3 > 1052.6 \text{ in.}^3 \qquad \text{OK}$$

Use $\frac{9}{16} \times 58$ web and one $\frac{3}{4} \times 14$ plate each flange

Comments on Solution

1. On some occasions the tables entitled "Built-Up Wide-Flange Sections" in Part 3 of the Manual may be very useful in designing these types of sections.
2. Should the section selected be noncompact the beam will be designed as per the requirements of appendix F1.7 of the LRFD Specification.
3. As the reader goes through Example 19-2 (and subsequent examples in this chapter) he or she will probably find that the flow chart presented on pages 3-104 through 3-107 of the Manual for the design of welded girders and welded plate girders will be very useful in following the necessary design steps. At first glance this chart may seem a little overpowering, but after the reader goes through an example problem or two he or she will begin to realize how useful the chart is and how well it was prepared.

19-3 INTRODUCTION TO PLATE GIRDERS

Plate girders are large I-shaped sections built up from plates and perhaps rolled sections. (Also see Fig. 19-3 of this chapter.) They usually have design strengths somewhere between those of rolled beams and steel trusses. Several possible arrangements are shown in Fig. 19-5. Rather obsolete bolted or riveted girders are shown in parts (a) and (b) of the figure, while several welded types are shown in parts (c) through (f). *Since nearly all plate girders constructed today are welded (although they may make use of bolted field splices) this chapter is devoted almost exclusively to welded girders.*

The welded girder of part (d) of Fig. 19-5 is arranged to reduce overhead welding as compared with the girder of part (c), but in so doing may be creating a little worse corrosion situation if the girder is exposed to the weather. A box girder, illustrated in part (g), is occasionally used where moments are large and depths are quite limited. Box girders also have great resistance to torsion and lateral buckling.

Plates and shapes can be arranged to form plate girders of almost any reasonable proportions. This fact may seem to give them a great advantage for all situations, but for the smaller sizes the advantage is usually canceled by the higher fabrication costs. For example, it is possible to replace a W36 with a plate girder roughly twice as deep which will require considerably less steel and will have much smaller deflections; however, the higher fabrication costs will almost always rule out such a possibility.

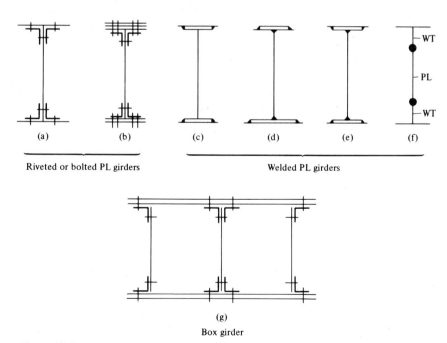

| (a) | (b) | (c) | (d) | (e) | (f) |

Riveted or bolted PL girders   Welded PL girders

(g)
Box girder

Figure 19-5

Buffalo Bayou Bridge, Houston Tex.—a 270-ft span. (Courtesy of the Lincoln Electric Company.)

Most steel highway bridges built today for spans less than about 80 ft are steel-beam bridges. For longer spans the plate girder begins to compete very well economically. Where loads are extremely large as for railroad bridges plate girders are competitive for spans as low as 45 or 50 ft.

The upper economical limits of plate girder spans depend on several factors, including whether the bridge is simple or continuous, whether a highway or railroad bridge is involved, the largest section which can be shipped in one piece, etc. Generally speaking plate girders are very economical for railroad bridges for spans from 50 to 130 ft (15 to 40 m) and for highway bridges for spans from 80 to 150 ft (24 to 46 m). However, they are often very competitive for much longer spans, particularly when continuous. In fact they are actually common for 200-ft (61-m) spans and have been used for many spans in excess of 400 ft (122 m). The main span of the continuous Bonn-Beuel plate girder bridge over the Rhine River in Germany is 643 ft.

Plate girders are not only used for bridges. They are also fairly common in various types of buildings where they are called upon to support heavy concentrated loads. Frequently a large ballroom or dining room with no interfering columns is desired on a lower floor of a multistory building. Such a situation is shown in Fig. 19-6. The plate girder shown must support some tremendous column loads for many stories above. The usual building plate girder of this type is simple to analyze because it probably does not have moving loads, although some building girders may be called upon to support traveling cranes.

The usual practical alternative to plate girders in the spans for which they are economical is the truss. In general plate girders have the following advantages with particular comparison made to trusses.

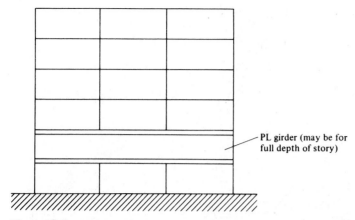

PL girder (may be for full depth of story)

Figure 19-6

1. The pound price for fabrication is lower than for trusses but higher than for rolled beam sections.
2. Erection is cheaper and faster than for trusses.
3. Due to their compactness, vibration and impact are not serious problems.
4. Plate girders require smaller vertical clearances than trusses.
5. The plate girder has fewer critical points for stresses than do trusses.
6. A bad connection here or there is not as serious as in a truss where such a situation could spell disaster.
7. There is less danger of injury to plate girders in an accident as compared with trusses. Should a truck run into a bridge plate girder it would probably just bend it a little, but a similar accident with a bridge truss member could cause a broken member and perhaps failure.
8. A plate girder is more easily painted than a truss.

On the other hand, plate girders are usually heavier than trusses for the same spans and loads, and they have a further disadvantage in the large number of connections required between webs and flanges.

19-4 PLATE GIRDER PROPORTIONS

Depth

The depths of plate girders vary from about $\frac{1}{6}$ to $\frac{1}{15}$ of their spans with average values of $\frac{1}{10}$ to $\frac{1}{12}$ depending on the particular conditions of the job. One condition that may limit the proportions of the girder is the largest size that can be fabricated in the shop and shipped to the job. There may be a transportation problem such as clearance requirements that limit maximum depths to 10 or 12 ft along the shipping route.

Indiana Harbor Works, East Chicago, Ind. (Courtesy of IR Construction Products Co.)

The shallower girders will probably be used when the loads are light, and the deeper ones when very large concentrated loads need to be supported as from the columns in a tall building. If there are no depth restrictions for a particular girder it will probably pay the designer to make rough designs and corresponding cost estimates to arrive at a depth decision.

19-5 PROPORTIONS OF WEBS

Buckling Considerations

The depths of plate girders have been previously discussed in Section 19-4 and were said to have average values varying from $\frac{1}{10}$ to $\frac{1}{12}$ of spans. After the total depth is assumed the web depth can be estimated to be from 2 to 4 in. less than the total depth and selected to the nearest inch. A plate-girder web must have sufficient thickness to prevent *vertical buckling of the compression flange.* As the compression flange of a simply supported beam deflects or curves downward, it pushes against the web subjecting it to vertical compression. The amount of this downward force can be estimated to equal the total bending stress in the compression flange times the sine of the angle made by the curved flange with the horizontal.

The total load that can be applied to the web in this manner before the web buckles can be estimated by the Euler formula. Expressing this another way, the Euler formula can be used to determine a limiting height-to-thickness ratio to prevent buckling. Based on such a derivation (including an assumption for residual stress variation) the maximum ratios for webs given in the last section of this chapter were developed.

Web Shear

The LRFD Specification for plate girders permits their design on the basis of postbuckling strength. Designs on this basis give better economy and provide a more realistic idea of the actual strength of a girder. Should a girder be loaded until initial buckling occurs, it will not collapse because of a phenomenon known as *tension field action.*

After initial buckling a plate girder acts much like a truss. The web acts as a truss with tension diagonals and is able to resist additional shear. A diagonal strip of the web acts similarly to the diagonal of a parallel chorded truss (see Fig. 19-7). The stiffeners keep the flanges from coming together and the flanges keep the stiffeners from coming together. The intermediate stiffeners, which before initial buckling were assumed to resist no load, will after buckling resist compressive loads (or will serve as the compression verticals of a truss) due to diagonal tension. The result is that a plate-girder web can probably resist loads equal to two or three times those present at initial buckling before complete collapse will occur.

Until the web buckles initially, deflections are relatively small but after initial buckling the girder's stiffness decreases considerably and deflections may increase to several times the values estimated by the usual deflection theory.

The estimated total or ultimate shear that a panel (a part of the girder between a pair of stiffeners) can withstand equals the shear initially causing web buckling plus the shear which can be resisted by tension field action. The amount of tension field action is dependent on the proportions of the panels.

The design shear strength of a plate girder is equal to $\phi_v V_n$ with $\phi_v = 0.90$ and V_n equal to the appropriate value determined from the various formulas given in LRFD Appendix G as follows:

If $h/t_w \leq 187 \sqrt{k/F_{yw}}$

$$V_n = 0.6 \, A_w F_{yw}$$

(LRFD Formula A-G3-1)

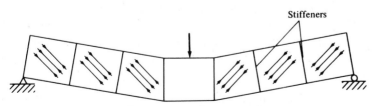

Stiffeners

Figure 19-7 Tension field action in plate-girder web.

If $h/t_w > 187 \sqrt{k/F_{yw}}$

$$V_n = 0.6 \, A_w F_{yw} \left[C_v + \frac{1 - C_v}{1.15 \sqrt{1 + (a/h)^2}} \right] \qquad \text{(LRFD Formula A-G3-2)}$$

For the end panels of nonhybrid girders, for all the panels of hybrid girders, and for web-tapered girders and when $a/h > 3.0$ or $>[260 \, (h/t_w)]^2$ tension field action is not permitted and

$$V_n = 0.6 \, A_w F_{yw} C_v \qquad \text{(LRFD Formula A-G3-3)}$$

In the preceding formulas

$$k = \text{a web buckling coefficient} = 5 + \frac{5}{(a/h)^2} \qquad \text{(LRFD Formula A-G3-4)}$$

which is to be taken $= 5.0$ if $a/h > 3.0$ or $>[260 \, (h/t_w)]^2$.

C_v is the ratio of the "critical" web stress to the shear yield stress of the web (F_{yw}) as follows:

If $187 \sqrt{k/F_{yw}} \leq h/t_w \leq 234 \sqrt{k/F_{yw}}$

$$C_v = \frac{187 \sqrt{k/F_{yw}}}{h/t_w} \qquad \text{(LRFD Formula A-G3-5)}$$

If $h/t_w > 234 \sqrt{k/F_{yw}}$

$$C_v = \frac{44,000 \, k}{(h/t_w)^2 \, F_{yw}} \qquad \text{(LRFD Formula A-G3-6)}$$

Transverse Stiffeners

As will be described in detail in Section 19-8 it is often necessary to stiffen the high thin webs of plate girders to prevent them from buckling. The LRFD Specification (A-G4) requires their use if h/t_w of the web is equal to or greater than $418/\sqrt{F_{yw}}$ or if the computed shear V_u is greater than $0.6 \, \phi A_w F_{yw} C_v$. Typical transverse stiffeners are shown in Fig. 19-9.

When transverse stiffeners are spaced with clear distances a and when that value is not greater than $1\frac{1}{2} h$ (where h is the clear distance between flanges less fillet radii for rolled shapes; and for built-up sections is the distance between adjacent lines of fasteners or the clear distance between flanges when welds are used) and

For $a/h \leq 1.5$

$$\text{Max.} \frac{h}{t_w} = \frac{2000}{\sqrt{F_{yf}}} \qquad \text{(LRFD Formula A-G1-1)}$$

For $a/h > 1.5$

$$\text{Max.} \frac{h}{t_w} = \frac{14{,}000}{\sqrt{F_{yf}\,(F_{yf} + 16.5)}} \qquad \text{(LRFD Formula A-G1-2)}$$

The compression flange of plate girders can theoretically fail not only by local buckling or lateral-torsional buckling but also by vertical buckling of the compression flange into the web. The last formula given above (A-G1-2) limits the h/t_w value to prevent this type of failure.

For this discussion it is assumed that a/h is greater than 1.5. If we use A36 steel and if we are going to design a plate girder its web may not have a thickness greater than $970/\sqrt{36} = h_c/162$ nor may it be less than $14{,}000/\sqrt{36(36 + 16.5)} = h_c/322$.

For a trial girder section using A36 steel a ratio of web depth to thickness somewhere between 162 and 322 is probably selected.

From a corrosion standpoint the usual practice is to use some absolute minimum web thickness. For bridge girders, $\frac{3}{8}$ in. is a common minimum, while $\frac{1}{4}$ or $\frac{5}{16}$ in. are probably the minimum values for the more sheltered building girders.

19-6 DESIGN OF PLATE GIRDERS WITH SLENDER WEBS BUT WITH FULL LATERAL BRACING FOR THEIR COMPACT COMPRESSION FLANGES

In Example 19-3 a plate girder with full lateral bracing for its compression flange is selected. As the girder proportions are not compact a plastic analysis is not permissible. Transverse stiffeners are considered in detail later in this chapter and thus their design is not included in this example or the next.

If we have a plate girder with a slender web (that is, with $h_c/t_w > 970/\sqrt{F_{yf}}$) its flexural strength is the lesser value computed from the limit states of tension-flange yielding and buckling as determined by the formulas to follow:

$$M_u = \phi_b M_n \qquad \text{with} \qquad \phi_b = 0.9$$

For tension-flange yielding

$$M_n = S_{xt} R_{PG} R_e F_{yt} \qquad \text{(LRFD Formula A-G2-1)}$$

For compression flange buckling

$$M_n = S_{xc} R_{PG} R_e F_{cr} \qquad \text{(LRFD Formula A-G2-2)}$$

in which

$$R_{PG} = 1 - 0.0005\, a_r \left(\frac{h_c}{t_w} - \frac{970}{\sqrt{F_{cr}}}\right) \le 1.0 \qquad \text{(LRFD Formula A-G2.3)}$$

R_e is a hybrid girder factor given in Appendix G2 of the Manual. It is to be taken as 1.0 for nonhybrid girders.

a_r is the ratio of the web area to the compression flange area.

Swinging a plate girder into place on Goat Island Bridge across the Niagara River, Niagara Falls, N.Y. (Courtesy of Bethlehem Steel Corporation.)

F_{cr} is the critical compression flange stress as determined in Chapter E of the Specification. It equals F_{yf} if the compression flange is fully braced laterally, that is, if $\lambda \leq \lambda_p$.

S_{xc} and S_{xt} are the section moduli values referred to the compression and tension flanges respectively.

EXAMPLE 19-3

Design a welded plate girder of A36 steel for a 70-ft span to support a uniform service dead load of 2 k/ft and a uniform service live load of 3.5 k/ft. The compression flange will be laterally braced for its entire length. The girder is to be simply supported at its ends and is not framed into columns. Do not consider stiffener design.

Solution

Maximum shear and moment

$$\text{Assume plate girder weight} = 250 \text{ lb/ft}$$

$$w_u = (1.2)(2.25) + (1.6)(3.5) = 8.3 \text{ k/ft}$$

$$M_u = \frac{(8.3)(70)^2}{8} = 5084 \text{ ft-k}$$

$$V_u = (35)(8.3) = 290.5 \text{ k}$$

Preliminary web design

Assume girder depth $= (\frac{1}{10})(12 \times 70) = 84$ in.

Assume web depth $= 84 - 2 = 82$ in.

Minimum web t_w if we assume $a/h > 1.5$

$$= \frac{82}{322} = 0.255 \text{ in.}$$

Maximum t_w for noncompact web (to be classed as plate girder) $= 82/162 = 0.506$ in. or 0.50 in.

Select minimum $t_w = \frac{5}{16}$ in. (rather than 0.25 in. because of corrosion minimum)

$$\frac{h_c}{t_w} = \frac{82}{5/16} = 262.4$$

Preliminary flange design

Using following formula to approximate flange area

$$A_f = \frac{M_u}{F_y h} \approx \frac{(12)(5084)}{(36)(82)} = 20.67 \text{ in.}^2$$

Try $1\frac{1}{8} \times 20$ plate ($A_f = 22.5$ in.^{2}). Is it compact?

$$\frac{b_f}{2t_f} = \frac{20}{(2)(1\frac{1}{8})} = 8.89 < \frac{65}{\sqrt{36}} = 10.8 \qquad \text{YES}$$

Check trial girder section

$$I = (\tfrac{1}{12})(\tfrac{5}{16})(82)^3 + (2)(1\tfrac{1}{8} \times 20)(41.56)^2 = 92,084 \text{ in.}^4$$

$$S_{xt} = \frac{92,084}{42.125} = 2186 \text{ in.}^3$$

As $\dfrac{h_c}{t_w} = 262.4 > \dfrac{970}{\sqrt{36}} = 162$ Appendix G2 applies

$$a_r = \frac{(\tfrac{5}{16})(82)}{(1\tfrac{1}{8})(20)} = 1.139$$

$$R_{PG} = 1 - (0.0005)(1.139)\left(262.4 - \frac{970}{\sqrt{36}}\right) = 0.943$$

Design flexural strength for tension-flange yield

$$M_n = S_{xt} R_{PG} R_e F_{yt}$$

$$= (2186)(0.943)(1.0)(36) = 74,210 \text{ in.-k} = 6184 \text{ ft-k}$$

$$M_u = \phi_b M_n = (0.9)(6184) = 5566 \text{ ft-k} > 5084 \text{ ft-k} \qquad \text{OK}$$

Use $\frac{5}{16} \times 82$ web with one $1\frac{1}{8} \times 20$ plate for each flange

Note: A check of shear values for this girder will show that transverse stiffeners are required.

19-7 DESIGN OF PLATE GIRDERS WITH NONCOMPACT FLANGES AND WITHOUT FULL LATERAL BRACING FOR COMPRESSION FLANGES

As indicated in Section 19-6 the critical flange stress F_{cr} for plate girders with slender webs ($h_c/t_w > 970/\sqrt{F_{yf}}$), compact flanges, and with full lateral bracing supplied for the compression flange equals F_{yf}. Should these latter two conditions not be provided it is necessary to consider the limit states of lateral-torsional buckling and local flange buckling as described below and use *the lower value obtained.*

If $\lambda_p < \lambda \le \lambda_r$

$$F_{cr} = C_b F_{yf}\left[1 - \frac{1}{2}\left(\frac{\lambda - \lambda_p}{\lambda_r - \lambda_p}\right)\right] \le F_{yf} \qquad \text{(LRFD Formula A-G2-5)}$$

If $\lambda > \lambda_r$

$$F_{cr} = \frac{C_{PG}}{\lambda^2} \qquad \text{(LRFD Formula A-G2-6)}$$

For the limit state of lateral torsion at buckling the following values are to be used in Formulas A-G2-5 and A-G2-6:

$$\lambda = \frac{L_b}{r_T} \qquad \text{(LRFD Formula A-G2-7)}$$

$$\lambda_p = \frac{300}{\sqrt{F_{yf}}} \qquad \text{(LRFD Formula A-G2-8)}$$

$$\lambda_r = \frac{756}{\sqrt{F_{yf}}} \qquad \text{(LRFD Formula A-G2-9)}$$

C_{PG} = plate girder coefficient

$\quad\;\; = 28,600\, C_b \qquad \text{(LRFD Formula A-G2-10)}$

$$C_b = 1.75 + 1.05\left(\frac{M_1}{M_2}\right) + 0.3\left(\frac{M_1}{M_2}\right)^2 \le 2.3$$

r_T = radius of gyration of compression flange plus one-sixth of web area about the axis of the web.

For the limit state of flange local buckling the following values are to be used:

$$\lambda = \frac{b_f}{2t_f} \qquad \text{(LRFD Formula A-G2-11)}$$

$$\lambda_p = \frac{65}{\sqrt{F_{yf}}} \qquad \text{(LRFD Formula A-G2-12)}$$

$$\lambda_r = \frac{150}{\sqrt{F_{yf}}} \qquad \text{(LRFD Formula A-G2-13)}$$

$$C_{PG} = 11{,}200 \qquad \text{(LRFD Formula A-G2-14)}$$

$$C_b = 1.0$$

Example 19-4 illustrates the selection of the cross section for a plate girder for which full lateral bracing is not provided for the compression flange.

EXAMPLE 19-4

The cross section for a welded plate girder is to be selected for a 54-ft span with simple end supports to support a uniform service dead load of 2 k/ft and concentrated service live loads of 150 k each at its one-third points. The design is to be made with A36 steel assuming that lateral bracing is provided for the compression flange only at the ends and one-third points only. Do not design stiffeners.

Arc-welded plate girders, Louisiana State Fair Grounds, Shreveport, La. (Courtesy of the Lincoln Electric Company.)

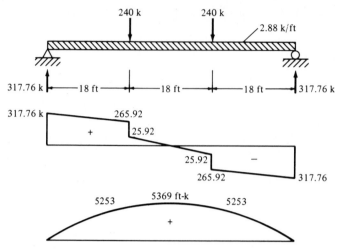

Figure 19-8

Solution

Maximum shear and moment (Fig. 19-8)

$$\text{Assume beam weight} = 400 \text{ lb/ft}$$

$$w_u = (1.2)(2.4) = 2.88 \text{ k/ft}$$

$$P_u = (1.6)(150) = 240 \text{ k}$$

Drawing shear and movement diagrams,

Preliminary web design

Assume girder depth $= (\frac{1}{10})(12 \times 54) = 64.8$ in.   Say 64 in.

Assume web depth $= 64 - 2 = 62$ in.

Minimum web t_w if we assume $\dfrac{a}{h} > 1.5 = \dfrac{62}{322} = 0.193$ in.

Maximum t_w for noncompact web (to be classed as plate girder) $= 62/162 = 0.383$ in.

Select $t_w = \frac{3}{8}$ in. noting a corrosion minimum of about $\frac{5}{16}$ in.

$$\frac{h_c}{t_w} = \frac{62}{3/8} = 165.3$$

Preliminary flange design

Using following formula to approximate flange area

$$A_f = \frac{M_u}{F_{yh}} \approx \frac{(12)(5369)}{(36)(62)} = 28.87 \text{ in.}^2$$

Try $1\frac{1}{4} \times 24$ plate ($A_f = 30$ in.^{2}). Is it compact?

$$\frac{b_f}{2t_f} = \frac{24}{(2)(1\frac{1}{4})} = 9.6 < \frac{65}{\sqrt{36}} = 10.8 \qquad \text{YES}$$

Check trial girder section

Computing section modulus

$$I = (\tfrac{1}{12})(3/8)(62)^3 + (2)(1\tfrac{1}{4})(24)(31.625)^2 = 67{,}456 \text{ in.}^4$$

$$S = \frac{67{,}456}{32.25} = 2091.7 \text{ in.}^3$$

$$\frac{h_c}{t_w} = \frac{62}{3/8} = 165.3 > \frac{970}{\sqrt{36}} = 162$$

$$\therefore \text{ Appendix G2 applies}$$

Properties of compression flange plus one-third of compression part of web

$$A_f + \tfrac{1}{6} A_w = (1.25)(24) + (\tfrac{1}{6})(\tfrac{3}{8} \times 62) = 33.88 \text{ in.}^2$$

$$I_y = (\tfrac{1}{12})(1.25)(24)^3 = 1440 \text{ in.}^4$$

$$r_T = \sqrt{\frac{1440}{33.88}} = 6.52 \text{ in.}$$

Checking limitations of Appendix G. If $a/h \le 1.5$,

$$\frac{h}{t_w} \text{ maximum} = \frac{2000}{\sqrt{36}} = 333 > 165.3 \qquad \text{OK}$$

Checking bending design strength of center 18-ft section.
$C_b = 1.0$ since the moment within a significant part of this unbraced 18-ft segment is greater than are the moments at its ends.

For the limit state of lateral-torsional buckling

$$\lambda = \frac{L_b}{r_T} = \frac{(12)(18)}{6.52} = 33.13$$

λ_p = limiting slenderness parameter for compact elements

$$= \frac{300}{\sqrt{F_{yf}}} = \frac{300}{\sqrt{36}} = 50 > 32.83$$

$$\therefore F_{cr} = F_{yf} = 36.0 \text{ ksi}$$

For the limit state of flange local buckling

$$\lambda = \frac{b_f}{2t_f} = \frac{24}{(2)(1.25)} = 9.6$$

$$\lambda_p = \frac{65}{\sqrt{F_{yf}}} = \frac{65}{\sqrt{36}} = 10.8 > 9.6$$

$$\therefore F_{cr} = F_{yf} = 36.0 \text{ ksi}$$

Design flexural strength

$$R_{PG} = 1 - (0.0005)\left(\frac{1.25 \times 24}{3/8 \times 62}\right)(165.3 - 162) = 0.998$$

$$M_n = (2091.7)(0.998)(1.0)(36) = 75,151 \text{ in.-k} = 6263 \text{ ft-k}$$

$$M_u = \phi_b M_n = (0.9)(6263) = 5637 \text{ ft-k} > 5369 \text{ ft-k} \qquad \text{OK}$$

Check bending strength of end 18-ft section

$$C_b = 1.75 + 1.05\frac{M_1}{M_2} + 0.3\left(\frac{M_1}{M_2}\right)^2 \leq 2.3$$

$$= 1.75 + 1.05\left(-\frac{0}{5253}\right) + 0.3\left(-\frac{0}{5253}\right)^2 = 1.75$$

For the limit state of lateral-torsional buckling

$$\lambda = \frac{L_b}{r_T} = \frac{(12)(18)}{6.52} = 33.12$$

$$\lambda_p = \frac{300}{\sqrt{F_{yf}}} = \frac{300}{\sqrt{36}} = 50 > 33.12$$

$$\therefore F_{cr} = 36.0 \text{ ksi}$$

For the limit state of flange local buckling. As it was for the center 18-ft section $F_{cr} = F_{yf} = 36.0$ ksi and $R_{PG} = 0.998$.

$$\phi_b M_n = 5636 \text{ ft-k} > 5369 \text{ ft-k} \qquad \text{OK}$$

Use $\frac{3}{8} \times 62$ web and one $1\frac{1}{4} \times 24$ plate each flange

19-8 DESIGN OF STIFFENERS

As previously indicated it is usually necessary to stiffen the high thin webs of plate girders to keep them from buckling. If the girders are bolted or riveted the stiffeners will probably consist of a pair of angles bolted or riveted to the girder webs. If the girders are welded (the normal situation) the stiffeners will probably consist of a pair of plates welded to the girder webs. Figure 19-9 illustrates these types of stiffeners.

Stiffeners are divided into two groups: *bearing stiffeners* which transfer heavy reactions or concentrated loads to the full depth of the web, and *intermediate or nonbearing stiffeners* which are placed at various intervals along the web to prevent buckling due to diagonal compression. An additional purpose of bearing stiffeners is the transfer of heavy loads to plate girder webs without putting all the loads on the flange connections.

Bearing Stiffeners

Bearing stiffeners are placed in pairs on the webs of plate girders at unframed girder ends and where required for concentrated loads. They should fit tightly

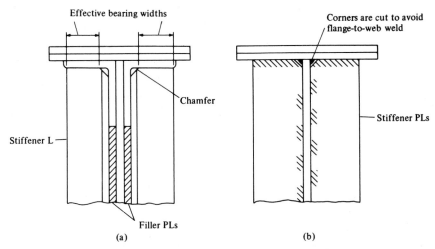

Effective bearing widths

Corners are cut to avoid flange-to-web weld

Chamfer

Stiffener L

Stiffener PLs

Filler PLs

(a)   (b)

Figure 19-9 (a) Angle bearing stiffeners. (b) Welded girder stiffeners.

against the flanges being loaded and should extend out toward the edges of the flange plates or angles as far as possible. If the load normal to the girder flange is tensile the stiffeners must be welded to the loaded flange. If the load is compressive it is necessary to obtain a snug fit, that is, a good bearing between the flange and the stiffeners. To accomplish this goal the stiffeners may be welded to the flange or the outstanding legs of the stiffeners may be milled.

A bearing stiffener is a special type of column which is difficult to accurately analyze because it must support the load in conjunction with the web. The amount of support provided by these two elements is difficult to estimate. The LRFD Specification (K1.8) states that the factored load or reaction may not exceed the design strength of a column consisting of the stiffener effective area plus a portion of the web equal to $12\,t_w$ at girder ends and $25\,t_w$ at interior concentrated loads. Only the part of the stiffeners outside of the fillets of the flange angles or the parts outside of the flange to web welds (see Fig. 19-9) are to be considered effective to support the bearing loads. The effective length of these "bearing stiffener columns" is assumed by the LRFD Specification (K1.8) as being equal to $0.75h$.

At an unframed girder end an end bearing stiffener is required if the factored reaction R_u is larger than ϕR_n. If an interior load or reaction is larger than the same ϕR_n an interior bearing stiffener is needed.

$$\phi = 1.0 \quad \text{and} \quad R_n = (5k + N)F_{yw}t_w \qquad \text{(LRFD Formula K1-2)}$$

Intermediate Stiffeners

Intermediate or nonbearing stiffeners which are also called stability or transverse intermediate stiffeners are required by the LRFD if $h/t_w > 418/\sqrt{F_{yw}}$ or if the factored shear V_u is $>0.6\phi A_w F_{yw} C_v$ where C_v is computed with the following expression using a value of $k = 5.0$.

$$C_v = \frac{187\ \sqrt{k/F_{yw}}}{h/t_w}$$ (LRFD Formula A-G3-5)

The LRFD Specification also imposes some additional arbitrary limits on panel aspect ratios (a/h) for plate girders even where shear stresses are small. The purpose of these limitations is to facilitate the handling of girders during fabrication and erection. The Specification (G3) states that tension field action is not permitted for end panels in nonhybrid plate girders, for all panels in hybrid and web tapered plate girders, and when $a/h > 3.0$ or $>[260\ (h/t_w)]^2$ and that

$$V_n = 0.6\ A_w F_{yw} C_v$$ (LRFD Formula A-G3-3)

The LRFD Specification (Appendix G4) states that the moment of inertia of a transverse intermediate stiffener about an axis at the center of the girder web if a pair of stiffeners is used or about the face in contact with the web when single stiffeners are used may not be less than

$$I_{st}\ \text{minimum} = at_w^3 j$$

where $$j = \frac{2.5}{(a/h)^2} - 2 \geq 0.5$$ (LRFD Formula A-G4-1)

Furthermore the stiffener area A_{st} may not be less than the value given by the following formula:

$$A_{st}\ \text{minimum} = \frac{F_{yw}}{F_{yst}}\left[0.15\ Dht_w(1 - C_v)\frac{V_u}{\phi_v V_n} - 18\ t_w^2\right] \geq 0$$

     (LRFD Formula A-G4-2)

In the preceding formula D is to be taken equal to 1.0 for stiffeners in pairs, 1.8 for single-angle stiffeners, and 2.4 for single-plate stiffeners. The terms C_v and V_n are defined in Appendix G3 while V_u is the factored shear at the section in question.

The values given for I_{st} minimum, j, and A_{st} minimum are the same as those required by the AASHTO in its specification for load factor design for steel bridges.[1]

Longitudinal Stiffeners

Longitudinal stiffeners, though not as effective as transverse ones, are frequently used for bridge plate girders because many designers feel they are more attractive. As such stiffeners are seldom used for plate girders in buildings they are not covered in this textbook.

[1]*Standard Specifications for Highway Bridges,* 13th ed., 1983 (Washington, D.C.: American Association of State Highway and Transportation Officials), p. 163.

Highway bypass bridge, Stroudsburg, Pa. (Courtesy of Bethlehem Steel Corporation.)

19-9 FLEXURE-SHEAR INTERACTION

When a plate girder whose web depends on tension field action is subjected to relatively large shear and bending moment at the same location the girder cannot provide its full capacity either in shear or in moment. As a result an empirical interaction equation is used to check the adequacy of the girder. This equation, which follows, agrees quite well with test results.

If stiffeners are required and

$$\frac{0.6\,V_n}{M_n} \le \frac{V_u}{M_u} \le \frac{V_n}{0.75\,M_n}$$

The following interaction equation is to be used

$$\frac{M_u}{M_n} + 0.625\,\frac{V_u}{V_n} \le 1375\phi \qquad \text{(LRFD Formula A-G5-1)}$$

In the preceding equation M_n is the nominal flexural strength of the girder as determined by Appendix G2, ϕ is 0.90, and V_n is the nominal shear strength of the girder as determined by Appendix G3. The value of M_u may not exceed ϕM_n nor may V_u exceed ϕV_n with $\phi = 0.90$ in both cases.

Bearing and intermediate stiffeners are designed in Example 19-5 for the plate girder of Example 19-4. In addition the girder is checked for flexure-shear interaction at several locations.

EXAMPLE 19-5

Design bearing and intermediate stiffeners for the plate girder of Example 19-4. Also check flexure-shear interaction at points 6, 12, and 18 ft from the girder ends.

Solution

End Bearing Stiffeners

(a)   Are end bearing stiffeners required?

Assume point bearing of the end reaction (that is, $N = 0$) and assume a $\frac{5}{16}$-in. web to flange fillet weld.

Check local web yielding

k = distance from outer edge of flange to web end of fillet weld

$$= 1\tfrac{1}{4} + \tfrac{5}{16} = 1.56 \text{ in.}$$

$$R_n = (5k + N)F_{yw}t_w$$

$$\phi R_n = (1.0)(5 \times 1.56 + 0)(36)(\tfrac{3}{8}) = 105.3 \text{ k}$$

$$< 317.76 \text{ k} \therefore \text{ End bearing stiffeners are required}$$

If local web yielding check had been satisfactory (that is, $\phi R_n > 317.76$ k) it would have then been necessary to also check the criteria set forth in Sections K1.4 and K1.5 of the LRFD Specification before we could definitely say that end bearing stiffeners were not required.

(b)   Design of end bearing stiffeners

Try two plate stiffeners $\frac{3}{4} \times 11$ as shown

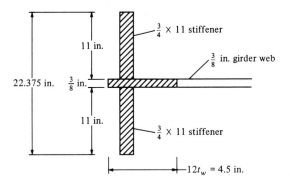

Check width-thickness ratio (Table B5.1 Specification)

$$\frac{11}{3/4} = 14.67 < \frac{95}{\sqrt{36}} = 15.83 \qquad \text{OK}$$

Check column strength of stiffener (area is shown crosshatched in preceding sketch)

$$I = (\tfrac{1}{12})(\tfrac{3}{4})(22.375)^3 = 700 \text{ in.}^4$$

(neglecting the almost infinitesimal contribution of the web)

$$A \text{ of column} = (2)(\tfrac{3}{4})(11) + (4.5)(\tfrac{3}{8}) = 18.19 \text{ in.}^2$$

$$r = \sqrt{\frac{700}{18.19}} = 6.20 \text{ in.}$$

$$KL = (0.75)(62) = 46.5 \text{ in.}$$

$$\frac{KL}{r} = \frac{46.5}{6.20} = 7.50$$

$\phi F_{cr} = 30.51$ ksi from Table 3-36 of Part 6 of Manual

or can be determined using Formulas E2-2 or E2-3 as applicable.

$$\phi P_n = (30.51)(18.19) = 555.0 \text{ k} > 317.76 \text{ k} \qquad \text{OK}$$

Check bearing criterion

$$\phi R_n = (0.75)(2.0 \, F_y A_{pd})$$

A_{pd} = projected bearing area of stiffeners subtracting $\tfrac{1}{2}$ in. from the 11-in. width of each stiffener for the cutouts for the web to flange welds as shown in Fig. 19-9(b).

$$A_{pd} = (2)(11 - 0.5)(\tfrac{3}{4}) = 15.75 \text{ in.}$$

$$\phi R_n = (0.75)(2.0)(36)(15.75) = 850 \text{ k} > 317.76 \text{ k} \qquad \text{OK}$$

Use two plates $\tfrac{3}{4} \times 11 \times 5$ ft $1\tfrac{3}{4}$ in. for bearing stiffeners (a depth of $61\tfrac{3}{4}$ in. is used instead of 62 in. for fitting purposes)

(c)   Design of interior bearing stiffeners

The design procedure for interior bearing stiffeners is the same as for end bearing stiffeners except a 25 t_w length of the web may be counted as part of the "column" as compared with the 12 t_w value used for the end stiffeners. Here the same size stiffeners are used at the concentrated loads as at the end reactions.

Intermediate Stiffeners

(a)   Are intermediate stiffeners required?
Check shear strength in unstiffened end panel.

$$\frac{h}{t_w} = \frac{62}{3/8} = 165.3 > \frac{418}{\sqrt{36}} = 70$$

$$\frac{a}{n} = \frac{(12)(18)}{62} = 3.48$$

$\therefore k = 5.0$ as per LRFD Specification F2.2

$$\frac{V_u}{A_w} = \frac{317.76}{(3/8)(62)} = 13.67 \text{ ksi}$$

$$\left(\frac{260}{h/t_w}\right)^2 = \left(\frac{260}{165.3}\right)^2 = 2.47$$

$\therefore$ Tension field action is not permitted in the end panels of this girder when $a/h > 3.0$ or $> (260/h/t_w)^2$. Thus Formula F2.3 (or A-G3-3, which is equivalent) is used to calculate V_n.

Using Formulas A-G3-5, A-G3-6, and A-G3-3)

$$\frac{h}{t_w} = 165.3 > 234 \sqrt{\frac{5.0}{36}} = 87.2$$

$$V_n = \left(\frac{3}{8} \times 62\right) \frac{(26,400)(5.0)}{(165.3)^2} = 112.32 \text{ k}$$

$$\frac{\phi_v V_n}{A_w} = \frac{(0.9)(112.32)}{3/8 \times 62} = 4.35 \text{ ksi} < \frac{V_u}{A_w} = 13.67 \text{ ksi}$$

$\therefore$ Intermediate stiffeners are required in end panel

(b)   Determine spacing of end panel stiffeners

Letting $\phi_v V_n / A_w = 13.67$ ksi, we can solve for a/h with LRFD Formula F2-3 or A-G3-3 or we can use Table 10-36 in the Manual. Here the table is used.

With $\dfrac{\phi_v V_n}{A_w} = 13.67 \text{ ksi}$ and $\dfrac{h}{t_w} = 165.3$

$$\frac{a_1}{h} = 0.688 \text{ by double interpolation}$$

$$a_1 = (0.688)(62) = 42.66 \text{ in.}$$

Use 42 in. to first stiffener
Are additional stiffeners required?

V_u at first interior stiffener $= 317.76 - \left(\frac{42}{12}\right)(2.88) = 307.68$ k

$$\frac{V_u}{A_w} = \frac{307.68}{3/8 \times 62} = 13.23 \text{ ksi} > 4.35 \text{ ksi}$$

$\dfrac{a_2}{h}$ by double interpolation $= 0.706$

$$a_2 = (0.706)(62) = 43.77 \text{ in.}$$

$\therefore$ Spacing continues to increase

Use 1 stiffener at 42 in. and remainder at $\dfrac{(12)(48) - 42}{4} = 43.5$ in. o.c.

(c)   Check center 18-ft panel

$$\frac{h}{t_w} = 165.3$$

$$\frac{a}{h} = \frac{(12)(18)}{62} = 3.48 > \left[\frac{260}{h/t_w}\right]^2 = 2.47$$

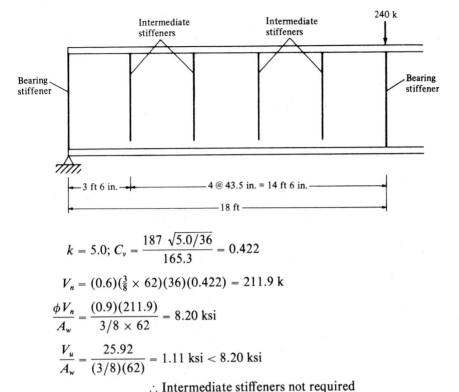

$$k = 5.0; \quad C_v = \frac{187 \sqrt{5.0/36}}{165.3} = 0.422$$

$$V_n = (0.6)(\tfrac{3}{8} \times 62)(36)(0.422) = 211.9 \text{ k}$$

$$\frac{\phi V_n}{A_w} = \frac{(0.9)(211.9)}{3/8 \times 62} = 8.20 \text{ ksi}$$

$$\frac{V_u}{A_w} = \frac{25.92}{(3/8)(62)} = 1.11 \text{ ksi} < 8.20 \text{ ksi}$$

$$\therefore \text{ Intermediate stiffeners not required}$$

Check flexure-shear interaction

Computing the values of V_u, M_u, and other values needed to substitute into interaction formula A-G5-1 at points 6, 12, and 18 ft from the end of the span.

Distance from end support (ft)	V_u (kips)	M_u (ft-k)	V_n (kips)	M_n (ft-k)	$\dfrac{0.6 V_n}{M_n}$	$\dfrac{V_u}{M_u}$	$\dfrac{V_n}{0.75 M_n}$
6	300.5	1854.7	353	6263	0.0338	0.162	0.0780
12	283.2	3605.8	353	6263	0.0338	0.079	0.0780
18	265.9	5253.1	353	6263	0.0338	0.051	0.0780

Since $0.6\,V_n/M_n < V_u/M_u > V_n/0.75\,M_n$ it is not necessary to check flexure-shear interaction.

PROBLEMS

19-1. Select a cover-plated W section limited to a maximum depth of 18.00 in. to support the service loads shown in the accompanying figure. Use A36 steel and assume the beam has full lateral bracing for its compression flange. (*Ans.* One solution W16×36 with 1 PL 1½×14 each flange)

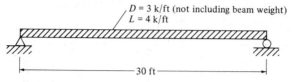

$D = 3$ k/ft (not including beam weight)
$L = 4$ k/ft

|← 30 ft →|

Problem 19.1

19-2. The architects specify that a beam no greater than 16.50 in. in depth be designed to support a dead service uniform load of 4 klf not including the beam weight and a live service uniform load of 6 klf for a 24-ft simple span. A572 steel with $F_y = 50$ ksi is to be used and full lateral bracing for the compression flange is to be provided.

19-3. Design a 48-in.-deep welded built-up wide-flange section with no intermediate stiffeners for a 50-ft simple span to support a service dead load of 1 klf (not including the beam weight) and a service live load of 1.8 klf. The section is to be framed between columns and is to have full lateral bracing for its compression flange. The design is to be made with A36 steel. (*Ans.* One solution $\frac{7}{16} \times 46$ in. web with one $\frac{5}{8} \times 10$ in. PL each flange)

19-4. Repeat Prob. 19-3 if the span is to be 60 ft and if A572 steel ($F_y = 50$ ksi) is to be used.

19-5. Design a welded plate girder with A36 steel for a 60-ft span to support a uniform service dead load of 2.0 klf (not including the girder weight) and a service live load of 4.5 klf. The compression flange will be laterally braced for its entire length. The girder is to be simply supported at its ends and is not framed into the columns. Do not design stiffeners. (*Ans.* $\frac{5}{16} \times 70$ web with one PL1 × 22 each flange)

19-6. Repeat Prob. 19-5 if A572 steel ($F_y = 50$ ksi) is used.

19-7. The cross section for a welded plate girder is to be selected for the span and service loads shown in the accompanying illustration. The design is to be made with A36 steel assuming that lateral bracing is provided for the compression flange at the ends and concentrated load points only. Do not design stiffeners. (*Ans.* $\frac{5}{16} \times 58$ web and one 2 × 25 cover plate each flange)

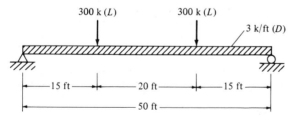

300 k (L)   300 k (L)   3 k/ft (D)

|← 15 ft →|← 20 ft →|← 15 ft →|
|← 50 ft →|

Problem 19.7

19-8. Repeat Prob. 19-7 using A572 steel ($F_y = 50$ ksi).

19-9. Design end bearing stiffeners and intermediate stiffeners for the plate girder selected in Prob. 19-7. Also check flexure-shear interaction at points 8, 15, and 20 ft from the girder ends.

19-10. Design end bearing stiffeners and intermediate stiffeners for the plate girder selected in Prob. 19-8. Also check flexure-shear interaction at points 8, 15, and 20 ft from the girder ends.

Chapter 20

Multistory Buildings

20-1 INTRODUCTION

Space is not taken in this introductory text to discuss the design of multistory buildings in detail. The material of this chapter is presented merely to give the student a general idea of the problems involved in the design of such buildings and not to present elaborate design examples.

Office buildings, hotels, apartment houses, and other buildings of many stories are quite common in the United States and the trend is toward an even larger number of tall buildings in the future. Available land for building in our heavily populated cities is becoming scarcer and scarcer and costs are becoming higher and higher. Tall buildings require a smaller amount of this expensive land to provide required floor space. Other factors contributing to the increased number of multistory buildings are new and better materials and construction techniques.

There are, on the other hand, several factors that may limit the heights to which buildings will be erected. These include the following.

1. Certain city building codes prescribe maximum heights to which buildings can be constructed.
2. Foundation conditions may not be satisfactory for supporting buildings of many stories.
3. Floor space may not be rentable above a certain height. Someone will always be available to rent the top floor or two of a 250-story building, but floor numbers 100 through 248 may not be so easy to rent.
4. There are several cost factors that tend to increase with taller buildings. Among these factors are elevators, plumbing, heating and air conditioning, glazing, exterior walls, wiring, etc.

Whether a multistory building is used for an office building, a hospital, a school, an apartment house, or whatever, the problems of design are generally the same. The construction is probably of the skeleton type in which the loads are transmitted to the foundation by a framework of steel beams and columns. The floor slabs, partitions, and exterior walls are all supported by the frame. This type of framing which can be erected to tremendous heights may also be referred to as beam-and-column construction. Bearing-wall construction is probably not used for buildings of more than a few stories although it has on occasion been used for buildings up to 20 or 25 stories. The columns of a skeleton frame are probably spaced 20, 25, or 30 ft on centers with beams and girders framing into them from both directions (see Fig. 15-1). On some of the floors, however, it may be necessary to have much larger open areas between columns for dining rooms, ballrooms, etc. For such cases very large beams (perhaps plate girders) may be needed to support column loads for many floors above.

It is usually necessary in multistory buildings to fireproof the members of the frame with concrete, gypsum, or some other material. The exterior walls are probably constructed with concrete or masonry units although an increasing number of modern buildings are being erected with large areas of glass in the exterior walls.

For these tall and heavy buildings the usual spread footings may not be sufficient to support the loads. If the bearing strength of the soil is high, steel grillage footings may be sufficient, but for poor soil conditions pile or pier foundations may be necessary.

For multistory buildings the beam-to-column systems are said to be superimposed on top of each other story by story or tier by tier. The column sections can be fabricated for one, two, or more floors with the two-story lengths probably

Broadview Apartments, Baltimore, Md. (Courtesy of the Lincoln Electric Company.)

being the most common. Theoretically, column sizes can be changed at each floor level but the costs of the splices involved usually more than cancel any savings in column weights. Columns of three or more stories in height are difficult to erect. The result is that the two-story heights work out very well most of the time.

20-2 COLUMN SPLICES

Column splices are usually placed 2 or 3 ft above floor levels to keep from interfering with the beam and column connections. Typical column splices are shown in Fig. 20-1. As shown in the figure, the column ends are usually milled so they can be placed firmly in contact with each other for purposes of load transfer.

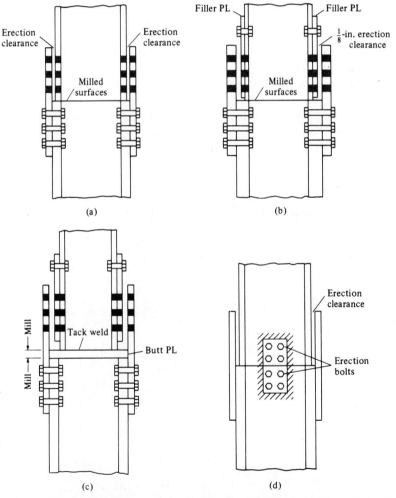

Figure 20-1 (a) Columns with same nominal depths. (b) Columns with small depth difference. (c) For columns of equal or different depths. (d) Splice made for all four sides of column.

When the contact surfaces are milled a large part of the axial compression (if not all) can be transferred through the contacting areas. It is obvious that splice plates are necessary even though full contact is made between the columns and only axial loads are involved. They are even more necessary when consideration is given to the shears and moments existing in practical columns subjected to off-center loads, lateral forces, moments, etc.

There is obviously a great deal of difference between tension splices and compression splices. In tension splices all load has to be transferred through the splice, whereas in splices for compression members a large part of the load can be transferred directly in bearing between the columns. The splice material is then needed to transfer only the remaining part of the load.

The amount of load to be carried by the splice plates is difficult to estimate. Should the column ends not be milled, the plates should be designed to carry 100 percent of the load. When the surfaces are milled and axial loads only are involved, the amount of load to be carried by the plates might be estimated to be from 25 to 50 percent of the total load. If bending is involved perhaps 50 to 75 percent of the total load may have to be carried by the splice material.

The bridge specifications spell out very carefully splice requirements for compression members but the LRFD does not. For example, the AREA says that compression member splices shall be designed to carry at least one-half of the total load. They further specify that compression members be spliced on all four sides.

Part (a) of Fig. 20-1 shows a splice that may be used for columns with the

Welded column splice for Colorado State Service Building, Denver, Colo. (Courtesy of the Lincoln Electric Company.)

same nominal depths. The student will notice in the LRFD Manual that wide-flange shapes of the same series (as W14) generally have the same inside distances between flanges although their total actual depths may vary as much as several inches (8.68 in. from a W14×730 to a W14×22). The type of splice shown in part (b) of the figure which has filler plates is used when the nominal depth of the top column is more than 2 in. less than that of the lower column.

Part (c) shows a type of splice which can be used for columns of equal or different nominal depths. For this type of splice the butt plate is shop-welded to the lower column and the clip angles used for field erection are shop-welded to the upper column. In the field the erection bolts shown are installed and the upper column is field-welded to the butt plate. The horizontal welds on this plate resist shears and moments in the columns.

Part (d) shows welded splices made on all four sides of the columns. The web splices are bolted in place in the field and field-welded to the column webs. The flange splices are shop-welded to the lower column and field-welded to the top column. The web plates may be referred to as *shear plates* and the flange plates as *moment plates*.

20-3 DISCUSSION OF LATERAL FORCES

For tall buildings, lateral forces must be considered as well as gravity forces. In past decades the usual practice was to consider wind forces in a building-frame design when the height was two or more times the least lateral dimension. The usual framing of buildings with a smaller height-width ratio has sufficient stiffness to resist forces of large magnitude without any special design provisions. Perhaps the common 2-to-1 ratio is a little high, however, for some modern buildings with their large areas of glass in the exterior walls.

High wind pressures on the sides of tall buildings produce overturning moments. These moments are probably resisted without difficulty by the axial strengths of the columns but the horizontal shears produced on each level may be of such magnitude as to require the use of special bracing or moment-resisting connections.

Unless they are fractured, the floors and walls provide a large part of the lateral stiffness of tall buildings. Although the amount of such resistance may be several times that provided by the lateral bracing, it is difficult to estimate and may not be reliable. Today so many modern buildings have light movable interior partitions, glass exterior walls, and lightweight floors that the steel frame should be assumed to provide all of the required lateral stiffness.

A building must not only be braced sufficiently laterally to prevent failure but also prevented from deflecting so much as to injure its various parts. Another item of importance is the provision of sufficient bracing to give the occupants a feeling of safety. They might not have this feeling in tall buildings which have a great deal of lateral movement in times of high winds. There have actually been tales of occupants of the upper floors of tall buildings complaining of seasickness on very windy days.

The horizontal deflection of a multistory building due to wind or seismic loading is called *drift*. It is represented by Δ in Fig. 20-2. Drift is measured by the *drift index*, Δ/h, where h is the height or distance to the ground.

The usual practice in the design of multistory steel buildings is to provide a structure with sufficient lateral stiffness to keep the drift index between approximately 0.0015 and 0.0030 radians for the worst storms that occur in a period of approximately 10 years. These so-called "10-year storms" might have wind in the range of about 90 mph depending on location and weather records. It is felt that when the index does not exceed these values, the users of the building will be reasonably comfortable.

In addition multistory buildings should be designed to withstand safely "50-year storms." For such cases however the drift index will be larger than the range mentioned with the result that the occupants will suffer some discomfort. The 1450-ft World Trade Center in New York City deflects or sways up to about 3 ft in "10-year storms" (drift index = 0.0021) while in hurricane winds it will sway up to about 7 ft (drift index = 0.0048).

Many areas of the world, including the western part of the United States, fall in earthquake territory, and in those areas it is necessary to consider seismic forces in design for tall or low buildings. During an earthquake there is an acceleration of the ground surface. This acceleration can be broken down into vertical and horizontal components. Usually the vertical component of the acceleration is assumed to be negligible but the horizontal component can be severe.

Most buildings can be designed with little extra expense to withstand the forces caused during an earthquake of fairly severe intensity. On the other hand, earthquakes during recent years have clearly shown that the average building which is not designed for earthquake forces can be destroyed by earthquakes that are not particularly severe. The usual practice is to design buildings for additional lateral loads (representing the estimate of the earthquake forces) which are equal to some percentage (5 to 10) of the weight of the building and its contents.

Many persons look upon the seismic loads to be used in design as being merely percentage increases of the wind loads. This is not altogether correct, however, as seismic loads are different in their action and are not proportional to the exposed area but to the building weight above the level in question.

The effect of lateral forces is to require the use of more steel even though load factors are reduced for wind and earthquake forces. This additional steel

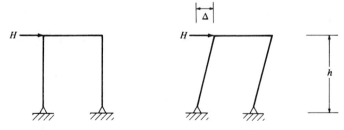

Figure 20-2

Chase Manhattan Bank Building, New York City. (Courtesy of Bethlehem Steel Corporation.)

will be used for bracing or moment-resisting connections as described in the section to follow.

20-4 TYPES OF LATERAL BRACING

A steel building frame with no lateral bracing is shown in Fig. 20-3(a). Should the beams and columns shown be connected together with the standard ("simple beam") connections, the frame would have little resistance to the lateral forces shown. Assuming the joints to act as frictionless pins, the frame would be laterally deflected as shown in part (b) of the figure.

To resist these lateral deflections, the simplest method from a theoretical standpoint is the insertion of full diagonal bracing as shown in part (c) of Fig. 20-3. From a practical standpoint, however, the student can easily see that in the average building full diagonal bracing would often be in the way of doors, windows, and other wall openings. Furthermore, many buildings have movable interior partitions and the presence of interior cross bracing would greatly reduce this flexibility. Usually diagonal bracing is convenient in solid walls in and around

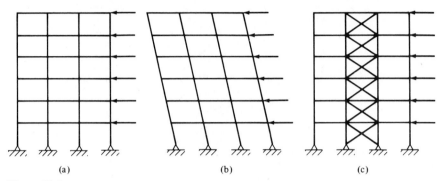

(a)   (b)   (c)

Figure 20-3

elevator shafts, stairwells, and other walls in which few or no openings are planned.

Another very common method of providing resistance to lateral forces involves the use of moment-resisting connections as illustrated in Fig. 20-4. This type of bracing, which is often called *bracket-type bracing*, may be used economically to provide lateral bracing for lower-rise buildings. As buildings become taller, however, bracket-type bracing is not very economical nor is it very satisfactory in limiting lateral deflections.

Some ways in which we can transmit lateral forces to the ground for buildings in the 20- to 60-story range are shown in Fig. 20-5. The X bracing system of part (a) works very well except that it might get in the way. Furthermore as compared with the systems shown in parts (b) and (c) of the same figure the floor beams have longer spans and will have to be larger.

In part (b) of Fig. 20-5 is shown the K bracing system. This system provides more freedom for the placing of openings than does the full X bracing. K braces are supported at midspan with the result that the floor beams will have smaller moments. The K bracing system (as does the knee bracing system to be discussed) uses less material than does X bracing. If we need more room than is available with the K bracing system we can go to the full-story knee bracing system shown in part (c) of Fig. 20-5.[1]

In the usual building the floor system (beams and slabs) is assumed to be rigid in the horizontal plane and the lateral loads are assumed to be concentrated at the floor levels. Floor slabs and girders acting together provide considerable resistance to lateral forces. Investigation of steel buildings that have withstood high wind forces have shown that the floor slabs distribute the lateral forces so that all of the columns on a particular floor have equal deflections. When rigid floors are present they spread the lateral shears to the columns or walls in the building. When lateral forces are particularly large, as in very tall buildings or where seismic forces are being considered, certain specially designed walls may

[1]E.H. Gaylord, Jr., and C.N. Gaylord, *Structural Engineering Handbook,* 2d ed., 1979 (New York: McGraw-Hill), pp. 19-77–19-111.

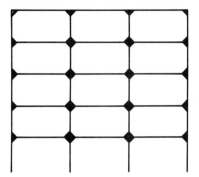

Figure 20-4   Bracket-type bracing.

be used to resist large parts of the lateral forces. These walls are called *shear walls*.

It is not necessary to brace every panel in a building. Usually the bracing can be placed in the outside walls with less interference than in the inside walls where movable partitions may be desired. Probably the bracing of the outside panels alone is not enough and some interior panels may need to be braced. It is assumed that the floors and beams are sufficiently rigid to transfer the lateral forces to the braced panels. Three possible arrangements of braced panels are shown in Fig. 20-6. A symmetrical arrangement is probably desirable to prevent uneven lateral deflection in the building and thus torsion.

As previously indicated bracing around elevator shafts is usually permissible while for other locations it will often interfere with windows, doors, movable partitions, glass exterior walls, open spaces, etc. Should bracing be used around an elevator shaft as shown in Fig. 20-7(a) and should calculations show that the drift index is too large it may be possible to use a "hat truss" on the top floor as shown in part (b) of the figure. Such a truss will substantially reduce drift. If a hat truss is not feasible because of interference with other items one or more "belt trusses" may be possible as shown in Fig. 20-7(c). A belt truss will appreciably reduce drift although not as much as a hat truss.

The bracing systems described so far are not efficient for buildings above approximately 60 stories. For these taller buildings there are very large lateral wind and perhaps seismic forces applied to buildings hundreds of feet above the ground. The designer needs to develop a system that will resist these loads without failure and in such a manner that lateral deflections are not so great as to

(a) X brace

(b) K brace

(c) Full-story
knee brace

Figure 20-5   (a) X brace. (b) K brace. (c) Full-story knee brace.

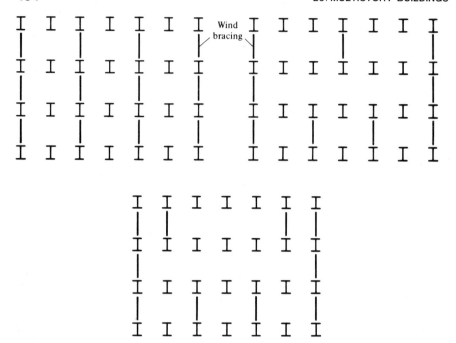

Figure 20-6

frighten the occupants. The bracing methods used for these buildings are usually based on a tubular frame concept.[2]

With the tubular system a vertical tubular cantilever frame is created much like the tube of Fig. 20-8. The "tube" consists of the building columns and girders in both the longitudinal and transverse directions of the building. The idea is to create a tube that will act like a continuous chimney or stack.

To build the tube the exterior columns are spaced close together from 3 or 4 ft up to 10 or 12 ft on center. They are connected with spandrel beams at the floor levels as shown in Fig. 20-9(a).

Further improvements of the tubular system can be made by cross bracing the frame with X bracing over many stories as illustrated in part (b) of Fig. 20-9. This latter system is very stiff and efficient and is very helpful in distributing gravity loads rather uniformly over the exterior columns.

Another variation on this system is the tube-within-a tube system. In this system interior columns and girders are used to create additional tubes. In tall buildings it is common to group elevator shafts, utility shafts, and stairs, and these systems can be used effectively to include shear walls and braced bents.[3]

---

[2]R.N. White, P. Gergely, and R.G. Sexsmith, 1974, *Structural Engineering,* vol. 3 (New York: Wiley), pp. 537–546.

[3]Op. cit.

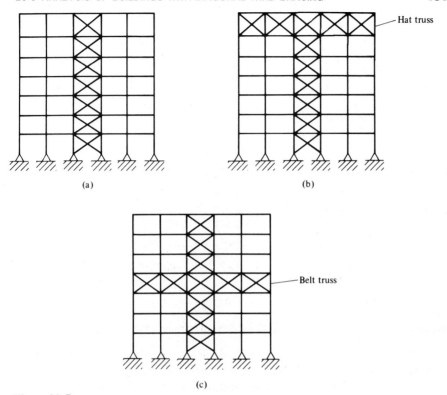

Hat truss

(a)   (b)

Belt truss

(c)

Figure 20-7

20-5 ANALYSIS OF BUILDINGS WITH DIAGONAL WIND BRACING FOR LATERAL FORCES

Full diagonal cross bracing has been described as being an economical type of wind bracing for tall buildings. Where this type of bracing cannot practically be used due to interference with windows, doors, or movable partitions, moment-resisting brackets are often used. Should, however, very tall narrow buildings

Figure 20-8   Solid-wall tube.

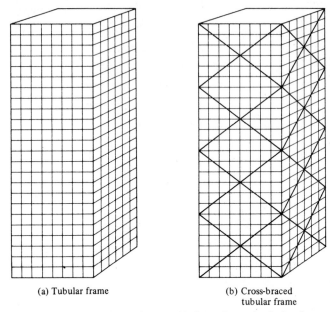

(a) Tubular frame   (b) Cross-braced
                                         tubular frame

Figure 20-9   (a) Tubular frame. (b) Cross-braced tubular frame.

(with height-to-least-width ratios of 5.0 or greater) be constructed, lateral deflections may become a problem with moment-resisting joints. The joints can satisfactorily resist the moments but the deflections may be excessive.

Should maximum wind deflections be kept under 0.002 times the building height there is little chance of injury to the building, says ASCE Subcommittee 31.[4] The deflection is to be computed neglecting any resistance supplied by floors and walls. To keep lateral deflections within this range it is necessary to use deep knee bracing, K bracing, or full diagonal cross bracing when the height-to-least-width ratio is about 5.0; and for greater values of the ratio, full diagonal cross bracing[5] or some other method such as the tubular frame is needed.

The student may often see bracing used in buildings in which he or she would think wind stresses were negligible. Such bracing stiffens up a building appreciably and serves the useful purpose of plumbing the steel frame during erection. Before bracing is installed the members of a steel building frame may be twisted in all sorts of directions. Connecting the diagonal bracing should pull them into their proper positions.

When diagonal cross bracing is used it is desirable to introduce initial tension into the diagonals. This prestressing will make the building frame tight and reduce its lateral deflection. Furthermore, the members can support compressive stress due to their pretensioning. Since the members can resist compression, the horizontal shear to be resisted will be assumed to split equally between the two diagonals. For buildings with several bay widths, equal shear distribution is usually assumed for each bay.

[4]"Wind Bracing in Steel Buildings," *Trans. ASCE* **105** (1940), pp. 1713–1739.
[5]L.E. Grinter, *Theory of Modern Steel Structures* (New York: Macmillan, 1962), p. 326.

Should the diagonals not be initially tightened rather stiff sections should be used so they will be able to resist appreciable compressive forces. The direct axial forces in the girders and columns can be found by joints from the shear forces assumed in the diagonals. Usually the girder axial forces so computed are too small to consider but the values for columns can be quite important. The taller the building becomes the more critical are the column axial forces caused by lateral forces.

For types of bracing other than cross bracing similar assumptions can be made for analysis. The student is referred to pages 333–339 of *Theory of Modern Steel Structures* by L. E. Grinter (New York: Macmillan, 1962) for discussion of this subject.

20-6 MOMENT-RESISTING JOINTS

For a large percentage of buildings under 8 to 10 stories in height the beams and girders are connected to each other and to the columns with simple end framed connections of the types previously described in Chapter 15. As buildings become taller it is absolutely necessary to use a definite wind-bracing system or moment-resisting joints. Moment-resisting joints may also be used in lower buildings where it is desired to take advantage of continuity and the consequent smaller beam sizes and depths and shallower floor construction. Moment-resisting brackets may also be necessary in some locations for loads that are applied eccentrically to columns.

Several types of moment-resisting connections which may be used as wind bracing are shown in Fig. 20-10. The design of connections of these types was also presented in Chapters 14 and 15. The average design company through the years will probably develop a file of moment-resisting connections from their previous designs. When they have a wind moment of such and such a value they merely refer to their file and select one of their former designs which will provide the required moment resistance.

In the following paragraphs a few comments are made about each of the types of connections shown in Fig. 20-10. The letter preceding each of these paragraphs corresponds to the letter in the figure for that connection.

(a)   The top-angle and seat-angle connection shown, sometimes called an *angle bracket,* provides the minimum acceptable type of wind connection. Even if web angles, shown dashed, are added, the moment resistance is not appreciably increased. This type of wind connection is not satisfactory for very tall buildings.

(b)   The split-beam bracket is one of the most common types of wind connections used in multistory buildings. When large girder reactions have to be transferred by this excellent type of connection the seat may have to be stiffened as showed by the dashed line.

(c)   This heavy bulky obsolete connection was formerly used to resist very large moments. Although it can reduce girder moments decidedly, its high fabrication cost probably cancels the saving in girder weight.

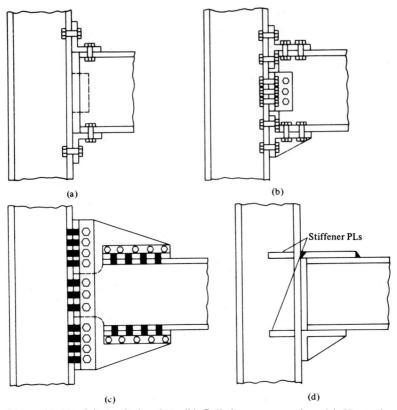

Figure 20-10 (a) Angle bracket. (b) Split-beam connection. (c) Heavy bracket. (d) Welded moment-resisting joint.

(d)   In this figure a very good welded type of connection for resisting wind moments is shown. The simple clean lines of this connection and much smaller amounts of connection steel used should be noted. Although the connections of parts (a) and (b) of Fig. 20-10 are shown to be riveted or bolted, they can also be satisfactorily welded.

20-7 ANALYSIS OF BUILDINGS WITH MOMENT-RESISTING JOINTS FOR LATERAL LOADS

Approximate methods have been very popular for many years for analyzing multistory buildings for lateral forces. Among the several reasons for this popularity have been the following.

1. Large building frames are statically indeterminate to a very high degree and their analysis by an "exact" method using a pocket calculator is a lengthy and difficult problem.

2. The resistance to lateral forces supplied by the walls and floors in a tall building is difficult to estimate accurately and the results of analysis by an "exact" method are therefore not very precise.
3. Before the members of a statically indeterminate structure can be designed their forces have to be determined, but these forces are dependent upon their sizes. Application of an approximate analysis should yield forces from which very good estimates of member sizes can be made for preliminary design.

The analysis of multistory buildings is a very difficult problem, and past practice has been to use one of the several approximate methods of analysis available. Today, however, with the availability of electronic computers it is feasible to make "exact" analyses in appreciably less time than required to make the approximate analyses (without the use of computers). The more accurate values obtained permit the use of smaller members. The results of computer usage are money saving in analysis time and in the use of smaller members.

In the pages that follow a brief review of the portal and cantilever approximate methods is presented. No consideration is given in either of these methods to the elastic properties of the members. These omissions can be very serious in unsymmetrical frames and in very tall buildings. To illustrate the seriousness of the matter, the changes in member sizes are considered in a very tall building. In such a building there will probably not be a great deal of change in beam sizes from the top floor to the bottom floor. For the same loadings and spans the changed sizes would be due to the large wind moments in the lower floors. The change, however, in column sizes from top to bottom would be tremendous. The result is that the relative sizes of columns and beams on the top floors are entirely different from their relative sizes on the lower floors. When this fact is not considered it causes large errors in the analysis.

If the height of a building is roughly five or more times its least lateral dimension, it is generally felt that a more precise method of analysis should be used than the portal or cantilever methods. There are several excellent approximate methods which make use of the elastic properties and which give values closely approaching the results of the "exact" methods. These include the Factor method, the Witmer method of K percentages, and the Spurr method. Should an exact method be desired, the slope deflection and moment distribution methods are available. If the slope deflection procedure is used, the designer will have the problem of solving a large number of simultaneous equations. The problem is not so serious, however, if a digital computer is used.

The Portal Method

The most common approximate method of analyzing tall building frames for lateral loads is the portal method. The chief advantage of this method, which is said to be satisfactory for most buildings up to 25 stories,[6] is its simplicity. A detailed

[6]"Wind Bracing in Steel Buildings," *Trans. ASCE* **105** (1940), p. 1723.

description of this method, including the basis for the assumptions made, can be found in most structural analysis textbooks. Only a brief listing of the steps involved in its application is given in this section. These steps are as follows.

1. The horizontal shears on each level are arbitrarily distributed between the columns. One commonly used procedure is to assume the shear divides between the columns in the ratio of one part to exterior columns and two parts to interior columns. Another common distribution (and the one used in the illustrative problem to follow) is to assume the shear V taken by each column is in proportion to the floor area it supports.

2. The moment M in each column is assumed to equal the column shear times half the column height (thus assuming a point of contraflexure at middepth).

3. The girder moments M are determined by joints by noting the sum of the girder moments at any joint equals the sum of the column moments at that joint. These calculations are easily made by starting at the upper left joint and working joint by joint across to the right; after which the next level is handled left to right, etc.

4. The shear V in each girder is assumed to equal its moment divided by half the girder length.

5. Finally, the column axial forces S are determined by summing up the beam shears and other column axial forces at each joint. These calculations are again handled conveniently by working from left to right and from the top floor down.

In Fig. 20-11 a building frame is shown which is to be analyzed by the portal method. The frames are assumed to be placed 20 ft 0 in. on centers and the exterior walls are assumed to be subjected to a wind pressure of 20 lb per vertical square foot. From these data the horizontal loads shown at each floor level are calculated.

This frame is analyzed in Fig. 20-12 by the portal method. The arrows shown on the figure give the direction of the girder shears and the column axial forces. The student can visualize the stress condition of the frame if he or she

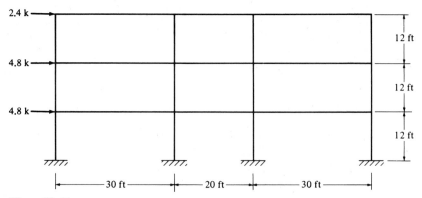

Figure 20-11

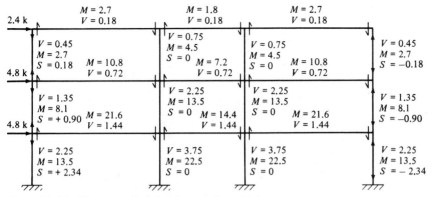

Figure 20-12 Frame analysis by the portal method.

assumes the frame is tending to be pushed over from left to right; by the wind, stretching the left exterior columns and compressing the right exterior columns.

The Cantilever Method

Another simple method of analyzing building frames for lateral forces is the cantilever method. This method is said to be a little more desirable for high narrow buildings than the portal method and may be used satisfactorily for buildings with heights not in excess of 25 to 35 stories.[7] It is not as popular as the portal method.

Instead of initially assuming the horizontal shears to be divided between the columns on each level in some proportion, the first assumption in the cantilever method pertains to the column axial forces. The axial force in a particular column is assumed to vary in direct proportion to its distance from the center of gravity of the group of columns on that level. With the wind blowing from left to right, the columns to the left of the center of gravity will be in tension while those to the right will be in compression. The steps involved in analyzing a building frame for lateral forces by the cantilever method are as follows.

1. To obtain the column axial forces moments are taken above an asumed plane of contraflexure through the middepth of the columns on each level. As an illustration of this procedure, moments are taken here to determine the axial forces in the columns on the top level of the building frame of Fig. 20-11. A free body is drawn in Fig. 20-13 of the part of the building above the assumed plane of contraflexure on the top level. The following equation is written.

$$\Sigma M_x = 0$$

$$(2.4)(6) + (30)(S) - (50)(S) - (80)(4S) = 0$$

$$S = 0.0424$$

[7]Op. cit.

The 20-story United Founders Life Tower in Oklahoma City, Okla. (Courtesy of the Lincoln Electric Company.)

2. The girder shears are determined by joints from the column axial forces.
3. The girder moments are determined by multiplying the girder shears by the half-girder lengths.
4. The column moments are found by joints from the girder moments.
5. The column shears are obtained by dividing the column moments by the half-column heights.

"Topping out" of Blue Cross-Blue Shield Building in Jacksonville, Florida. (Courtesy of Owen Steel Company, Inc.) The Christmas tree is an old North European custom used to ward off evil spirits. It is also used today to show that the steel frame was erected with no lost time accidents to personnel.

The analysis of the frame of Fig. 20-13 by the cantilever method is illustrated in Fig. 20-14. Arrows representing the directions of the column axial forces and the girder shears are again shown for the benefit of the student.

20-8 ANALYSIS OF BUILDINGS FOR GRAVITY LOADS

Simple Framing

If simple framing is used, the design of the girders is quite simple because the shears and moments in each girder can be determined by statics. The gravity loads applied to the columns are relatively easy to estimate but the column moments may be a little more difficult. If the girder reactions on each side of the

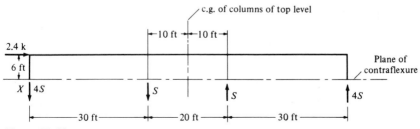

Figure 20-13

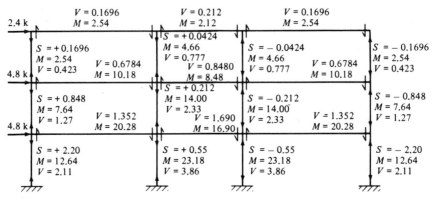

Figure 20-14   Frame analysis by the cantilever method.

interior column of Fig. 20-15 are equal, no moment will theoretically be produced in the column at that level. This situation is probably not realistic because it is highly possible for the live load to be applied on one side of the column and not on the other (or at least be unequal in magnitude). The results will be column moments. If the reactions are unequal the moment produced in the column will equal the difference between the reactions times the distances to the c.g. of the column.

When computing the moments in exterior columns there may often be moments due to spandrel beams opposing the moments caused by the floor loads on the inside of the column. Nevertheless, gravity loads will generally cause the exterior columns to have larger moments than the interior columns.

To estimate the moment applied to a column above a certain floor it is probably reasonable to assume that the unbalanced moment at that level splits evenly between the column above and below. In fact, such an assumption is often on the conservative side as the column below may be larger than the one above.

Rigid Framing

For buildings with moment-resisting joints it is a little more difficult to estimate the girder moments and make preliminary designs. If the ends of each girder are

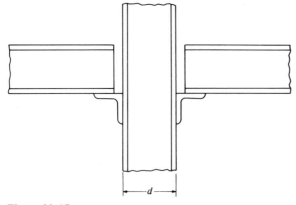

Figure 20-15

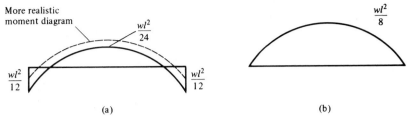

Figure 20-16   (a) Fixed-end beam. (b) Simple beam.

assumed to be completely fixed, the moments for uniform loads are as shown in Fig. 20-16(a). Conditions of complete fixity are probably not realized, with the result that the end moments are smaller than shown in part (a) of the figure. There is a corresponding increase in the positive center line moments approaching the simple moment $(wl^2/8)$ shown in part (b) of the figure. Probably a moment diagram somewhere in between the two extremes is more realistic. Such a moment diagram is represented by the dotted line of Fig. 20-16(a). A reasonable procedure is to assume a moment in the range of $wl^2/10$, where l is the clear span.

When the beams and girders of a building frame are rigidly connected to each other a continuous frame is the result. From a theoretical standpoint an accurate analysis of such a structure cannot be made unless the entire frame is handled as a unit.

Before this subject is pursued further it should be realized that in the upper floors of tall buildings the wind moments will be small and the framed and seated connections described in Chapter 15 will provide sufficient moment resistance. For this reason the beams and girders of the upper floors may very well be designed on the basis of simple beam moments, while those of the lower floors may be designed as continuous members with moment-resistant connections because of the larger wind moments.

For the design of beams and girders it is necessary to consider the load combinations of Section A4.1 of the LRFD Specification. For the upper floors the wind moments will not increase the girder sizes but on lower floors they will begin to cause increased sizes.

From a strictly theoretical viewpoint there are several live loading conditions which need to be considered to obtain maximum shears and moments at various points in a continuous structure. For the building frame shown in Fig. 20-17 it is desired to place live loads to cause maximum positive moment in span *AB*. A qualitative influence line for positive moment at the center line of this span is shown in part (a) of the figure. This influence line shows that to obtain maximum positive moment at the center line of span *AB* the live loads should be placed as shown in part (b) of the figure.

To obtain maximum negative moment at point *B* or maximum positive moment in span *BC*, other loading situations need to be considered. With the increased availability of electronic computers more of this detailed type of analysis is being done every day. The student, however, can see that unless the designer has access to a computer, he or she probably will not have sufficient time to go through all of these theoretical situations. Furthermore, it is doubtful if the accu-

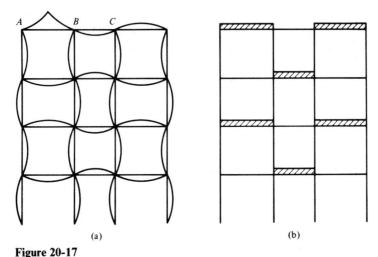

Figure 20-17

racy of our analysis methods would justify all of the work anyway. It is, however, often feasible to take out two stories of the building at a time as a free body and analyze that part by one of the "exact" methods such as the successive correction method of moment distribution.

Before such an "exact" analysis can be made it is necessary to make an estimate of the member sizes probably based on the results of analysis by one of the approximate methods.

20-9 DESIGN OF MEMBERS

Girders

The girders can be designed for the two load combinations mentioned in Section 20-8. For the top several floors the second condition will not control design, but further down in the building it will control the sizes for the reasons described in the next paragraph.

Each of the members of a building frame must be designed for all of the dead loads which it supports, but it may be possible to design some members for lesser loads than their full theoretical live-load values. For example, it seems unlikely in a building frame of several stories for the absolutely maximum-design live load to occur on every floor at the same time. The lower columns in a building frame are designed for all of the dead loads above but probably for a percentage of live load appreciably less than 100 percent. Some specifications require that the beams supporting floor slabs be designed for full dead and live loads but permit the main girders to be designed under certain conditions for reduced live loads. (This reduction is based on the thought that it seems rather improbable that a very large area of a floor would be loaded to its full live-load value at any one time.) The ANSI Code provides some very commonly used reduction expressions.[8]

Should simple framing be used the girders will be proportioned for simple beam moments plus the moments caused by the lateral loads. For continuous framing the girders will be proportioned for $wl^2/10$ (for uniform loads) plus the moments caused by the lateral loads. An interesting comparison of designs by the two methods is shown in pages 717–719 of Beedle et al., *Structural Steel Design* (New York: Ronald, 1964). It may be necessary to draw the moment diagram for the two cases (gravity loads and lateral loads) and add them together to obtain the maximum positive moment out in the span. The value so computed may very well control the girder size in some of the members.

A major part of the design of a multistory building is involved in setting up a column schedule which shows the loads to be supported by the various columns story by story. The gravity forces can be estimated very well while the shears, axial forces, and moments caused by lateral forces can be roughly approximated for the preliminary design from some approximate method (portal, cantilever, factor, or other).

If both axes of the columns are free to sway, the column effective lengths theoretically must be calculated for both axes as described in Chapter 5. If diagonal bracing is used in one direction, thus preventing sidesway, the K value will be less than 1.0 in that direction. When frames are braced in the narrow direction it is reasonable to use K = 1.0 for both axes.

After the sizes of the girders and columns are tentatively selected for a two-story height, an "exact" analysis can be performed and the members redesigned. The two stories taken out for analysis are often referred to as a *tier*. This process can be continued tier by tier down through the building. Many tall buildings have been designed and are performing satisfactorily in which this last step (the two-story "exact" analysis) was omitted.

---

[8]*American National Standard Minimum Design Loads for Buildings and Other Structures,* ANSI A58.1-1982 (New York: American Standards Institute, Inc.), page 11.

INDEX

AASHTO, 25, 35
Advantages of LRFD, 43–44
Advantages of structural steel, 1–3
AISI, 7
American Welding Society (AWS), 318
ANSI, 4, 5, 27, 28, 38
AREA, 35

Bars, 79–83
Bateman, E. H., 256–257
Batho, C., 256–257
Beam-columns, 236–252
Beams, 157–231
    bearing plates, 228–231
    bending coefficients, 190–191
    bending stresses, 158–159
    camber, 92, 211
    compact, 103
    continuous, 170–172, 203–205
    cover-plated, 457–459
    deflections, 209–212
    elastic buckling, 181, 194–197
    floor, 157
    girders, 157
    government anchors, 390
    holes, 185–187
    inelastic buckling, 180–181, 189–193
    joists, 157, 393–395
    lateral support, 188–197
    lintels, 157
    local flange bending, 214

    local web yielding, 214–215
    noncompact sections, 197
    plastic buckling, 178–180, 181–187
    ponding, 29–30, 212–213
    purlins, 79–83, 220–223
    scuppers, 213
    sections used as, 157–158
    shear, 205–209
    shear center, 223–228
    sidesway web buckling, 215–216
    spandrel, 157, 391
    stringers, 157
    unsymmetrical bending, 218–220
    web crippling, 215
    weight estimates, 183
Bearing plates
    beams, 228–231
    columns, 148–154
Bearing-wall construction, 389–390
Beedle, L. S., 17, 91, 95, 163, 507
Belford, D., 441
Bending and axial compression, 239–252
    design, 248–252
    drift index, 241, 490
    equivalent axial load method, 248–252
    first-order moments, 239–240
    magnification factors, 240–241
    modification factors, 241–242
    second-order moments, 239–240
Bending and axial force, 236–252
Bending and axial tension, 237–239

Bessemer, Sir H., 5
Block shear, 60–65
Bolts, 255–305, 308, 310
  advantages, 257
  bearing type connections, 261, 270–
    275
  common, 255
  direct tension indicator, 259–260
  eccentric shear, 284–296
  edge distances, 267, 268–269
  failure, 266–267
  fully tensioned, 258–261
  high strength, 256–305
  history, 256–257
  hole sizes, 262–263
  installation, 259–260
  mixed joints, 262
  prying action, 301–305
  shear and tension, 296–299
  slip-critical, 261, 276, 270
  snug-tight, 258–261
  spacing, 267
  tension loads, 299–301
Buckling, 98
  Euler, 95–99
Building codes, 25–26
Building connections, 360–385
  column web stiffeners, 382–385
  rigid (Type FR), 366, 378–382
  seated, 376–378
  semirigid (Type PR), 361, 363–365
  simple (Type PR), 361–363, 366–378
  single-plate, 375–376
  stiffened seat, 378
Built-up wide flange sections, 459–462
  defined, 459–460
Butler, L. J., 349
Butt joint, 265

Camber, 92, 211
Cantilever method, 501–503
Carbon steels, 14–16
Chesson, E., Jr., 57
Cochrane, V. H., 52
Coffin, C., 321
Columns, 89–154
  base plates, 148–154
  built-up members, 119–125
  column formulas, 95–98
  Euler formula, 95–99
  flexural-torsional buckling, 125–130
  formulas, 104–105
  intermediate, 104
  lacing, 130–133
  lateral bracing, 109–110, 117–118

  leaning on each other, 145–148
  long, 104
  LRFD design tables, 115–120
  sections used as, 92–95
  short, 104
  slenderness ratios, 106
  splices, 487–489
  stiffness reduction factors, 142–145
  tie plates, 113–115, 130–133
  web stiffeners, 382–385
Compact sections, 103
Composite beams, 407–436
  advantages, 408–410
  continuous sections, 430–431
  deflections, 424, 427
  design, 422–436
  economical spans, 422
  effective flange widths, 411
  encased sections, 431–436
  history of, 407–408
  lower bound moment of inertia, 424
  moment capacity, 417–422
  partially composite beams, 414
  shear connectors, 411–416
  shoring, 410, 423
  steel formed decks, 408
Composite columns, 440–454
  advantages, 440–442
  axial load and bending, 449–453
  disadvantages, 442–443
  flexural strength, 449
  lateral bracing, 443
  LRFD tables, 448–449
  specifications, 443–445
Compression members. See Columns
Cooper, P. B., 185
Coping, 367
Cornell, C. A., 43
Cost, 21
Cover-plated beams, 457–459
Crawford, S. F., 292

Davy, H., 320
Dead loads, 27
Deflections, 209–212, 424, 427
Disadvantages of structural steel, 3
Drift index, 241, 490
Ductility, 2, 11

Earthquake loads, 33
Eccentric loads, 284–296
  elastic methods, 284–290
  reduced eccentricity method, 285,
    290–291
  ultimate strength method, 285, 291–
    296

Edinger, J. A., 44
Effective lengths, 98–102, 137–142
Elasticity, 2
Elastic limit, 10
Elevators, 7
Ellingwood, B., 43
Equivalent axial load method, 248–252
Euler, L., 95
Euler formula, 95–99

Fatigue, 3, 83–85
Faying surface, 261
Fireproofing, 3, 405–406
First-order moments, 239–240
Fisher, J. W., 261
Floor construction types
  for buildings, 393–403
  composite, 397–398
  concrete-pan, 398–399
  flat-slab, 400–401
  one- and two-way slabs, 395–396
  open-web joists, 393–395
  precast concrete, 401–403
  steel decking, 400
Frames, 101
  braced, 101
  unbraced, 101

Gage, 52
Galambos, T. V., 37, 43
Gaylord, C. N., 57, 492
Gaylord, E. H., Jr., 57, 492
Gergely, P., 494
Government anchors, 390
Griffis, L. G., 441, 442, 443, 453
Griffiths, J. D., 366
Grinter, L. E., 496, 497
Gusset plates, 285

Higgins, T. R., 285
High-strength steels, 16–19
Historical data, 4–5
Hooke's Law, 10
Hughes, J., 324
Hybrid structures, 18, 181

Impact, 30
Ironworkers, 19
Instantaneous center of rotation, 349
Interaction equations, 219–220, 222–223, 237–252

Jenny, W. L., 5
Johnson, B. G., 138
Johnson, J. E., 228

Johnston, B., 365
Julian, O. G., 138

Kahn, F. R., 441
Kazinczy, G., 163
Kelly, W., 5
Kulak, G. L., 292, 295, 349
Kussman, R. L., 185

Lacing, 130–133
Lap joints, 264
Lawrence, L. S., 138
Limit states, 36
Load factors, 37–41
Loads, 26–35
Lothers, J. E., 365

MacGregor, J. G., 43
McGuire, W., 7, 284
Marino, F. J., 213
Modern structural steels, 13–17
Moment-resisting connections (Type FR), 378–382, 497–503
Mount, E. H., 365
Multistory buildings, 485–507
  analysis, 495–506
  belt truss, 493, 495
  design, 506–507
  drift index, 490
  hat truss, 493, 495
  lateral bracing, 491–495
  lateral forces, 489–503
  moment-resisting joints, 497–503
Munse, W. H., 57
Musschenbroek, P. V., 95

Net areas, 49–59
Noncompact section, 103

Open-web joists, 393–395

Pal, S., 349
Pitch, 52
Plastic analysis, 159–172
  collapse mechanisms, 164–165
  continuous beams, 170–172
  plastic hinges, 159–160
  plastic modulus, 160–163
  plastic moment, 159
  shape factor, 159
  theory of, 163–172
  virtual-work method, 165–172
  yield moment, 159
Plastic moment, 159

Plate girders, 459–460, 463–483
  advantages, 464–465
  defined, 459–460
  disadvantages, 465
  flexure-shear interaction, 479–483
  proportions, 465–469
  stiffeners, 468–469, 476–478
  web, 466–468
  web shear, 467–468
Ponding, 29–30, 212–213
Portal method, 499–501
Proportional elastic limit, 10
Proportional limit, 10
Prying action, 301–305
Purlins, 79–83, 220–223

Quenched and tempered steels, 17

Rains, W. A., 406
Ravindra, M. K., 37, 43
Reilly, C., 284
Reliability, 41–43
Residual stresses, 91–92
Resistance factors, 39–41
Richard, R. M., 376
Rivets, 305–310
  historical notes, 305–306
  specifications, 307–308
  types, 307
Rods, 79–83
Roof construction types, 403–406
Ryzin, G. V., 213

Safety, 21
Sag rods, 80–83
Salmon, C. G., 228
Scuppers, 30, 213
Second-order moments, 239–240
Section modulus, 159
Semirigid connections, 363–365
Setback, 370
Sexsmith, R. G., 494
Shape factor, 159
Shear, 205–209
Shear center, 125–126, 223–228
Shear lag, 56
Single-plate or shear tab connections,
    375–376
Skeleton construction, 390–391
Slender compression elements, 103
Slenderness ratios, 71, 90, 106, 113–115
Small-angle theory, 165
Snow loads, 28–29
Sparks, P. R., 32
Specifications, 25–26

Steel, defined, 4
Steel sections, 5–9
Stetina, H. J., 361
Stiffened elements, 102–103
Stiffness reduction factor, 142–145
Strain-hardening, 11
Stress-strain diagrams, 9–13
Structural Engineers Association of Cal-
    ifornia (SEAOC), 33
Structural Stability Research Council
    (SSRC), 139
Struik, J. H. A., 261
Sweet's Catalog File, 393, 394, 402

Tall L., 91
Tension members, 45–88
  analysis, 45–65
  bars, 46
  block shear, 60–65
  design, 70–85
  design strength, 48–49
  effective net areas, 56–58
  eyebars, 83
  fatigue loads, 83–85
  net areas, 49–59
  rods, 79–83
  selection of sections, 70–78
  shear lag, 56
  slenderness ratio, 71
  staggered holes, 52–56
  tie plates, 130–133
  upset rods, 79
Tie plates, 130–133
Timler, 349
Toughness, 2

Uncertainties in design, 40–41
Uniform Building Code, 33–35
Unstiffened elements, 102–103
Upset rods, 79

Van den Broek, J. A., 163

Web tear-out shear. See Block shear
Welding, 318–352
  advantages, 319–320
  classification, 325–327
  defined, 318
  electric arc, 320–322
  electrodes, 321–322
  end returns, 336
  fillers, 321
  fillet, 325–326, 332–352
  gas, 321

groove, 325–326, 328–331
inspection, 322–325
joints, 327
lands, 330
LRFD requirements, 333–337
partial-penetration, 331
plug, 325–326
position, 326–327
shear and bending, 351–352, 378–382
shear and torsion, 345–351
    elastic method, 346–348

shielded arc (SMAW), 321–322
    ultimate strength method, 348–351
slot, 325–326
submerged arc (SAW), 322
symbols, 327–328
types, 320–322
White, R. N., 494
Wind loads, 30–33

Yield point, 10
Yura, J. A., 108, 120, 145

ISBN 0-06-044346-4

90000